STRESS
AND
STRAIN
DATA
HANDBOOK

STRESS
AND
STRAIN
DATA
HANDBOOK

Teng H. Hsu

Gulf Publishing Company
Book Division
Houston, London, Paris, Tokyo

STRESS
AND
STRAIN
DATA
HANDBOOK

First Printing, August 1986
Second Printing, March 1987

Library of Congress Cataloging-in-Publication Data

Hsu, Teng H.
  Stress and strain data handbook.
  Includes index.
  1. Strains and stresses—Handbooks, manuals, etc.
I. Title.
TA648.3.H78   1986   624.1'76   86-4663
ISBN 0-87201-159-3

Contents

v

Preface

This book presents the general theories and principles of stress and strain for practical application. The primary intention is to provide a reference that meets the daily needs of the design engineer. The solutions and data required in engineering practice are often scattered through an extensive body of literature, and are not presented in a form that allows convenient application to the problem at hand. This handbook draws information from many sources into a useful and convenient single volume. Tedious derivations and detailed explanation of formulas are omitted. The data are presented in tables and graphs along with several examples to illustrate actual applications.

The scope of the book is indicated by the contents. It covers beams, frames, columns, beam-columns, plates, rings, torsion, shells, stability, and thermal stress. For each topic the general principles and theories are stated, followed by extensive tables and graphs for use in calculation of stress and strain. The data are arranged to provide a means of using general theories to solve practical engineering problems.

I express my gratitude to Mr. W. Yu and Dr. K. Pajouhi, who generously reviewed the manuscript; and to Mr. J. Garcia, who drew most of the graphs and illustrations. Finally, I would like to say that although every effort has been made to avoid errors, it is possible some could exist. I will be grateful for any suggestions you may have concerning needed corrections.

Teng H. Hsu

Notation

A   Cross-sectional area

A_r, A_u, A_x Coefficients of deflection

B_x, B_y Bending stress coefficients

B_i, B_j, B_u Coefficients of slope

B_r Bending stress coefficient for circular plates

C   Coefficient of constraint in elastic stability, numerical factor

C_c Torsional stiffness of circular section

C_m Reduction factor applied to bending term in interaction formula and dependent upon curvature caused by applied moments

C_o Torsional stiffness of non-circular section

C_w Warping constant of a section

D   Numerical factor, flexural rigidity of plate or shell

E   Modulus of elasticity in tension and compression, numerical factor

E_r Tangent modulus

F_a Axial stress permitted in the absence of bending moment

F_b Bending stress permitted in the absence of axial force

F_e Euler stress divided by factor of safety

F_y Specified minimum yield stress of steel

F_c Buckling stress of a column containing residual stresses

G   Modulus of elasticity in shear

H_o, H_h Horizontal thrust of a pinned and built-in arch

I   Moment of inertia of a section

I_e Moment of inertia of the elastic portion of a section

J   Polar moment of inertia, torsional constant

K   Effective length factor

K_m, K_A, K_B Moment coefficients of a circular ring

K_d Deflection coefficient of a circular ring

K_i, K_j, K_u Coefficients of end moments

K_1, K_2 Torsional factors

K_3 Stress concentration factor

K_x, K_y Deflection coefficient in x and y, directions respectively

K_ϕ, K_σ Force factors in shell

K_x, K_t Stress factor in shell

K_P, K_Q Force and shear factors

K_w Displacement factor

L Span length

M Moment

M_{ij}, M_{ji} End moments

M_x, M_y Bending moments per unit length of plate or shell sections

M_{xy} Twisting moment per unit length of plate or shell sections

M_r, M_t Radial and tangential moments

M_T Torsional moment

N_x, N_y, N_{xy} Normal and shearing forces per unit length of shell sections

N_σ, N_ϕ, $N_{\sigma\phi}$ Membrane forces per unit length of shell sections

P Applied load

P_{cr} Critical load

P_e Euler load

Q_x, Q_y Shearing force per unit length of plate or shell sections

Q Shearing force per unit length of cylindrical section

R_1, R_2 Reactions

R_i, R_o Inside and outside radii of a curved beam respectively

R Radius of curvature of the neutral axis

T Membrane tension

V Static shear on beam

V_c Volume of membrane hill of circular section

V_o Volume of membrane hill of non-circular sections

V_f Shear force introduced to a flange of I beam subjected to an end torsion

U Strain energy

a, b Distances from supports to loading point, distances

c Distance

e Eccentricity

f Stress

f_a, f_b Axial and bending stress, respectively

f_{bx}, f_{by} Bending stresses induced by M_x and M_y, respectively

f'_u A function of parameter u for plate deflection

g'_u A function of parameter u for plate bending

k Axial load factor for beam-column ($k^2 = P/EI$)

l Length, span

ln Natural logarithm

m A parameter which gives the number of half-waves into which a plate buckles

p, q Intensity of distributed load, pressure

r Radius of gyration, radius, radius of curvature of shell

r_n Radius of curvature of the middle surface of a circular plate

r_x, r_y Radius of gyration with respect to x and y axes, respectively

s_s, s_{xy} Shear stresses

t Thickness

u Axial load factor for beam-colums ($u = kl$), numerical factor

u, v, w Displacements in x, y, z directions

x, y, z Rectangular coordinates

Greek Symbols

α Angles, coefficient of thermal expansion, numerical factors

β Numerical factor or coefficient

γ Shearing unit strain, specific gravity

δ Deflection, displacement

δ_H, δ_V Horizontal and vertical displacements

ϵ Unit normal strain

θ Angle, polar coordinate, the angle of twist per unit length of section

ϵ_x, ϵ_y Unit normal strains in x and y directions

ν Poisson's ratio

ρ Radius of curvature

ρ_x, ρ_y Radii of curvature of the middle surface of a rectangular plate in the xz and yz planes

γ_{xz}, γ_{yz} Shear strains in xz and yz planes

σ Unit normal stress

σ_x, σ_y Unit normal stresses in x and y directions

σ_{cr} Critical stress

τ Unit shear stress

τ_{xy}, τ_{yz}, τ_{xz} Unit shear stresses on planes perpendicular to the x,y and z axes and parallel to the y,z and x axes

ϕ Angle, angular coordinate

χ Change of curvature in shell

χ_{xy} Twist of the middle surface of shell

ϕ Saint-Venant's torsion function

σ_{ϕ} Circumferential stress

ψ Numerical factor in calculation of reduction factor C_m for compression members braced against joint translation

Chapter 1

Flexure of Beams

BENDING OF BEAMS [1,2]

Forces or moments acting on a beam impart deflections perpendicular to the longitudinal axis of the beam and set up normal and shearing stresses on any cross section of the beam. It is convenient to imagine a beam being composed of an infinite number of fibers. The surface on the beam containing fibers that do not undergo any stress is called the *neutral surface*. The intersection of the neutral surface with any cross section of the beam perpendicular to its longitudinal axis is called the *neutral axis*. All fibers on one side of the neutral axis are in tension and those on the opposite side are in compression.

If all fibers in the beam are acting within the elastic range of the material, the following relations exist:

The bending stress at any point of a section is f_b.

$$f_b = \frac{My}{I} \tag{1-1}$$

where M is the bending moment at the section containing y, and y is the distance from the neutral axis to the point.

The shear stress at any point of a section is s_s.

$$s_s = \frac{V}{Ib} \int y \, da = \frac{VQ}{Ib} \tag{1-2}$$

where V = shear at the section
     da = area of that part of the section above the point
     y = distance from the neutral axis to the centroid of da
     b = width of the beam
     Q = first moment of the area da about the neutral axis

The radius of curvature of the elastic curve is R.

$$R = \frac{EI}{M} \tag{1-3}$$

The general differential equation of the elastic curve is:

$$M = EI \frac{d^2y}{dx^2} \tag{1-4}$$

The relations between the bending moment and the shear are:

$$V = \frac{dM}{dx} \tag{1-5a}$$

$$M = \int V \, dx \tag{1-5b}$$

The strain energy of flexure is U.

$$U = \int \frac{M^2}{2EI} \, dx \tag{1-6}$$

Example 1-1

A steel wire $^1/_{32}$ in. in diameter is coiled around a pulley 20 in. in diameter; calculate the maximum bending stress set up in the wire.

$E = 30 \times 10^3$ kips/in.^{2}

Radius of curvature $R = 20/2 = 10$ in.

Normal strain $\epsilon = y/R$

Maximum strain $\epsilon_{max} = 1/640 = 0.00156$ in.

Maximum stress $\sigma_{max} = E\epsilon_{max} = 30 \times 10^3 \times 0.00156$
$$= 46.88 \text{ kips/in.}^2$$

Example 1-2

If the greatest vertical shear of a rectangular beam is V, prove the maximum shear stress is 1.5 times the average shear stress.

From Equation 1-2,

$$s_s = \frac{VQ}{Ib} = \frac{12bh^2}{8b^2h^3} V = \frac{3}{2bh} \frac{V}{} = 1.5 \frac{V}{A}$$

V/A is the average shear stress of the section with vertical force V.

ELASTIC DEFLECTION OF BEAMS [3]

The deformation of a beam is expressed in terms of the deflection of the beam from its original unloaded position. The deflection is measured from the original neutral surface to the neutral surface of the deformed beam. The configuration of the deformed neutral surface is known as the *elastic curve* of the beam.

Design specifications frequently limit the deflections as well as the stresses. It is essential that the design engineer be able to calculate deflections, and numerous methods are available for the determination of beam deflections.

Double-Integration Method

For a given beam, if the load w_x at a point x can be expressed mathematically as a function of x, and if such load condition is known for the entire beam, then

$$V_x = \int w_x \, dx \tag{1-7a}$$

$$M_x = \int V_x \, dx \tag{1-7b}$$

$$\theta_x = \int \frac{M_x}{EI} \, dx \tag{1-7c}$$

$$y_x = \int \frac{\theta_x \, dx}{EI} = \int \int \frac{M_x}{EI} \, dx \tag{1-7d}$$

It is assumed that the beam is acting in the elastic range and that the deflections caused by shearing action are negligible compared to those caused by bending action.

Example 1-3

Determine the maximum deflection of a cantilever beam subject to a uniform load of w lb per unit length, as shown in Figure 1-1.

$$M_x = wx^2/2$$

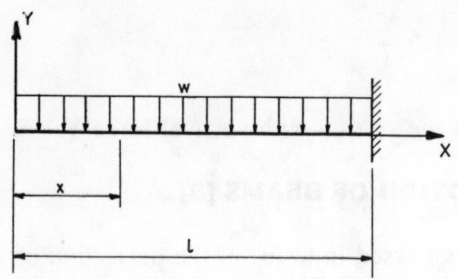

Figure 1-1. Cantilever beam for Example 1-3.

From Equations 1-7c and 1-7d,

$$EI\ \theta_x = \int M_x\ dx = wx^3/6 + C_1$$

$$EIy_x = \int \int M_x\ dx = \int (wx^3/6 + C_1)dx$$

$$EIy_x = wx^4/24 + C_1x + C_2$$

at $x = \ell$, $\theta_\ell = 0$ $C_1 = -w\ell^3/6$

at $x = \ell$, $y_\ell = 0$ $C_2 = w\ell^4/6 - w\ell^4/24 = w\ell^4/8$

at $x = 0$, $y = y_{max}$

$$y_{max} = C_2/EI = w\ell^4/(8EI)$$

Moment-Area Method

The first moment-area theorem states that the angle between the tangents at a and b of a deformed beam is equal to the area of the moment diagram between a and b, divided by EI:

$$\theta = \int_a^b \frac{M}{EI}\ dx \tag{1-8a}$$

The second moment-area theorem states that the vertical distance from point b of the deformed beam to the tangent at point a of the beam equals the moment with respect to b of the area of the bending moment diagram between a and b, divided by EI:

$$\delta = \int_a^b \frac{M}{EI}\ x\ dx \tag{1-8b}$$

Example 1-4

Determine the deflection of the free end of the cantilever beam in Example 1-3 using the moment-area method. The elastic curve and moment diagram are shown in Figure 1-2.

Area of the M/EI diagram $= w\ell^3/6$

Moment with respect to b $= w\ell^4/8$

$\delta = w\ell^4/8EI$

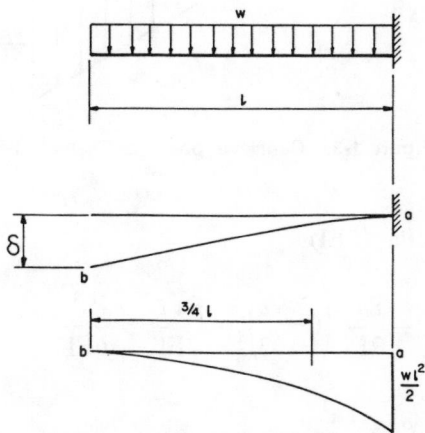

Figure 1-2. Elastic curve and moment diagram of a cantilever beam.

Conjugate Beam Method

In using this method, the moment diagram of the real beam is constructed, and a conjugate beam is then set up. It is loaded with the M/EI of the real beam. The vertical shear at any point of the conjugate beam equals the slope of the real beam at the same point. The bending moment at any point of the conjugate beam equals the deflection of the real beam at the same point. The boundary conditions of the conjugate beam should be selected so that the previous statements are satisfied.

Example 1-5

Determine the deflection and slope at the tip of a cantilever beam loaded by a concentrated force P, as shown in Figure 1-3.

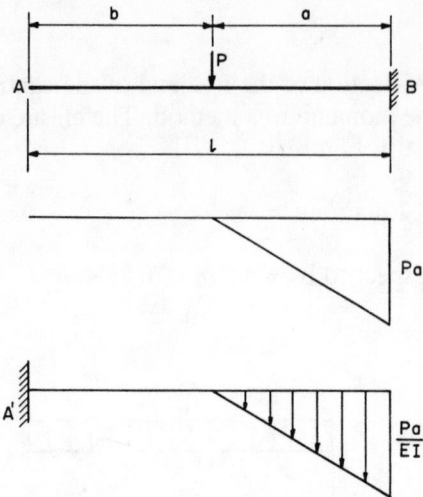

Figure 1-3. Cantilever beam for Example 1-5.

Shear at $A' = Pa^2/(2EI)$

Moment at $A' = \dfrac{Pa^2}{2EI}\left(\ell - \dfrac{a}{3}\right) = \dfrac{Pa^2\ell}{2EI} - \dfrac{Pa^3}{6EI}$

Slope at tip $\theta = \dfrac{Pa^2}{2EI}$

Deflection at tip $\delta = \dfrac{-Pa^2\ell}{2EI} + \dfrac{Pa^3}{6EI}$

Virtual Work Method

This method is used frequently for finding the deflection of a point on the beam. A unit virtual load is placed on the beam at the point where deflection is desired. Virtual moments caused by the unit load are determined along the beam. The internal energy of the beam after deflection is determined by integration. This is set equal to the external work done by the unit load.

Internal energy $U = \displaystyle\int \dfrac{M_x m_x}{EI}\, dx$

Work done $W = 1 \times y_x$

$$y_x = \int \frac{M_x m_x}{EI} \, dx \qquad (1-9)$$

where m = virtual bending moment at any point caused by 1
M = real bending moment at the same point

Example 1-6

Solve Example 1-4 using virtual work method.

A unit load is placed at the tip of a beam as shown in Figure 1-4.

$$\text{Internal energy} = \int \frac{M_x m_x}{EI} \, dx = \int \frac{wx^3}{2EI} \, dx = \frac{w\ell^4}{8EI}$$

Work done $= 1 \times y = y$

$$y = \frac{w\ell^4}{8EI}$$

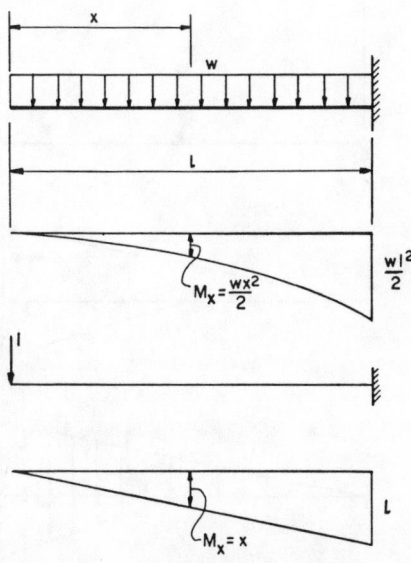

Figure 1-4. Cantilever beam for Example 1-6.

STATICALLY INDETERMINATE BEAMS [4]

For those beams where the number of unknown reactions exceeds the number of equilibrium equations available for the system, it is necessary to supplement the equilibrium equations with additional equations stemming from the deformation of the beam. In these cases, the beams are called *statically indeterminate*.

Figure 1-5 shows a beam with one degree indeterminacy. The problem can be solved by employing the moment-area method or conjugate beam method. The left-side reaction, R_i, is solved by the fact that the deflection at i must be zero.

$$\frac{R_i \ell^2}{2EI} \frac{2\ell}{3} - \frac{Pb^2}{2EI} \left(\ell - \frac{b}{3} \right) = 0$$

$$\frac{R_i \ell^3}{3} = \frac{Pb^2}{2} \left(\ell - \frac{b}{3} \right)$$

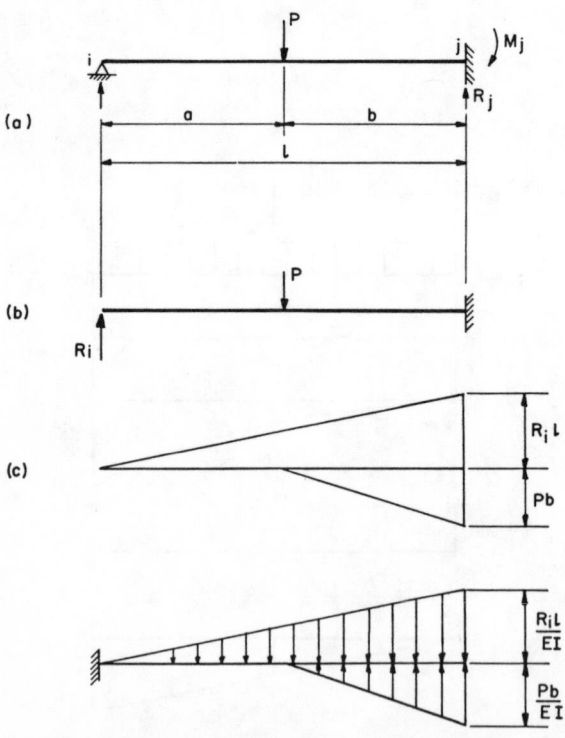

Figure 1-5. Beam with one degree indeterminacy.

$$R_i = \frac{3Pb^2}{2\ell^3}\left(\ell - \frac{b}{3}\right)$$

For a continuous beam, the problem of indeterminacy is usually solved by use of the three-moment theorem. The slope of any span of a continuous beam is the sum of the slope produced on a simply supported span by the given loading and the slope produced by end couples M_1 and M_2, as shown in Figure 1-6.

The slopes at both sides of joint 2 are set equal, giving the desired equation. If M_1 and M_3 are determinate as shown, the equation can be solved at once for M_2 and the reactions then found by statics. If the ends are fixed, the slopes at those points can be set equal to zero. This provides two additional equations, and M_1, M_2, and M_3 can be solved.

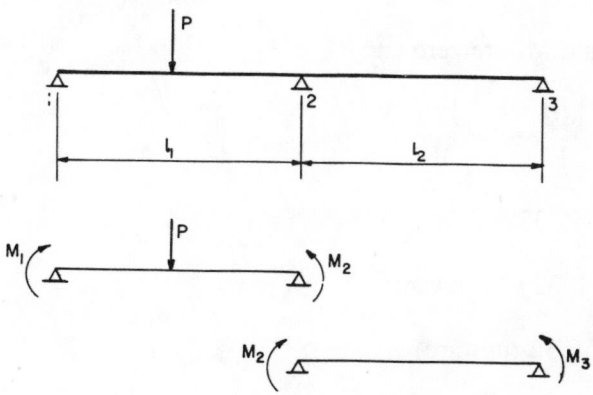

Figure 1-6. A continuous beam.

Example 1-7

The continuous beam shown in Figure 1-7 is loaded with a concentrated load P. Determine the reactions and moments.

The slope of the first span at joint 2 is

$$\theta_{21} = \frac{P\ell^2}{16EI} + \frac{1}{EI}\left(\frac{M_1\ell}{6} + \frac{M_2\ell}{3}\right)$$

The slope of the second span at joint 2 is

$$\theta_{23} = \frac{1}{EI}\left(-\frac{M_2\ell}{3} - \frac{M_3\ell}{6}\right)$$

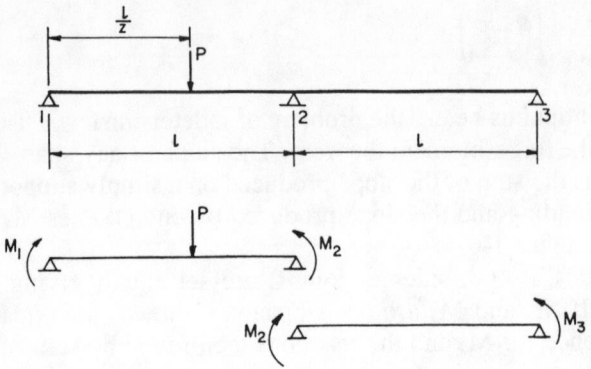

Figure 1-7. Continuous beam for Example 1-7.

Both M_1 and M_3 are zero and $\theta_{21} = \theta_{23}$

$$\frac{P\ell^2}{16EI} + \frac{M_2\ell}{3EI} = -\frac{M_2\ell}{3EI}$$

$$M_2 = -P\ell/32$$

$$R_3 = -P/32 \text{ (downward)}$$

$$R_1 = 15P/32 \text{ (upward)}$$

$$R_2 = 9P/16 \text{ (upward)}$$

One convenient method for solving statically indeterminate beams is moment distribution. This is sometimes referred to as the Hardy Cross method, in tribute to the originator of its basic concepts. If an external moment M is applied to the joint as shown in Figure 1-8, we can determine in what proportions the beams resist the moment M by means of distribution factors (DF). When M is applied at joint 2, it produces a rotation of θ. The equilibrium equation at joint 2 is

$$M_{21} + M_{23} = M \tag{1-10}$$

The stiffness factors of the beams are $4EK_{12}$ and $4EK_{23}$. We can state that

$$M_{21} = 4EK_{12}\,\theta$$

$$M_{23} = 4EK_{23}\,\theta$$

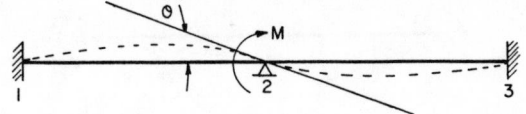

Figure 1-8. Moment distribution of a continuous beam.

Equation 1-10 can be written as

$$4E\,\theta\,(K_{12} + K_{23}) = M \text{ or}$$

$$M = 4E\,\theta\,\Sigma\,K$$

from which

$$\theta = \frac{M}{4E\Sigma K}$$

$$M_{21} = 4EK_{12}\,\frac{M}{4E\Sigma K} = \frac{K_{12}}{\Sigma K}\,M$$

$$M_{23} = 4EK_{23}\,\frac{M}{4E\Sigma K} = \frac{K_{23}}{\Sigma K}\,M$$

The ratio of $K_{12}/\Sigma K$ indicates the portion of the moment M that is resisted by beam 1-2 at joint 2. This ratio is referred to as the distribution factor of the beam. The distribution factor for the ith beam is defined as

$$DF_i = \frac{K_i}{\Sigma K} = \frac{I_i/\ell_i}{\Sigma I/\ell} \tag{1-11}$$

A beam is shown in Figure 1-9. Joint 2 is fixed, its stiffness factor is I/ℓ, and the moment at 2 is M_2; therefore,

$$M_1 = \frac{4EI}{\ell}\,\theta_1$$

$$M_2 = \frac{2EI}{\ell}\,\theta_1$$

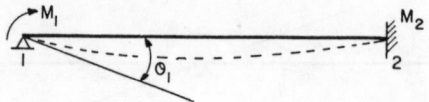

Figure 1-9.   Carry-over factor of a beam with fixed joint.

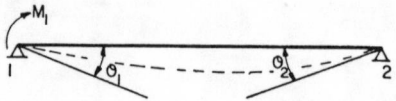

Figure 1-10.   Carry-over factor of a beam with pinned joint.

From these two equations we find that $M_2 = M_1/2$. This means one-half of the moment is generated at the fixed end. This is called the *carry-over moment*, and ½ is the *carry-over factor*.

If joint 2 is not fixed but pinned, as shown in Figure 1-10, then

$$M_2 = 2EI \, (\theta_1 + \theta_2)/\ell = 0$$

$$\theta_2 = -\theta_1/2$$

$$M_1 = 3EI \, \theta_1/\ell$$

The stiffness value is three-fourths of what it was for the fixed-end condition. The modified stiffness factor is $3K/4$.

Example 1-8

Calculate bending moments and reactions of the beam shown in Figure 1-11.

$$K_{12} = I/15$$

$$K_{23} = (I/10)(\tfrac{3}{4}) = 3I/40$$

$$K_{23}/K_{12} = 1.125$$

$$DF_{21} = 1/2.125 = 0.471$$

$$DF_{23} = 1.125/2.125 = 0.529$$

$$FM_{12} = 1 \times 15^2/12 = 18.75 \text{ kip-ft}$$

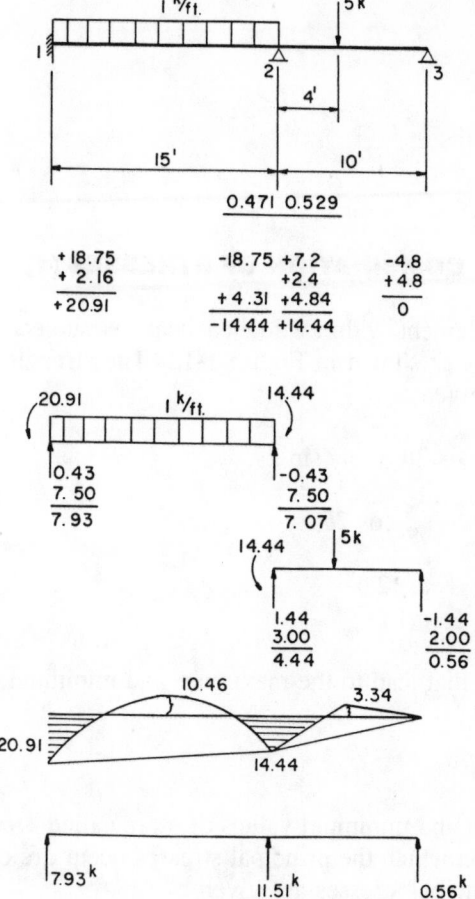

Figure 1-11. Moment distribution of Example 1-8.

$FM_{21} = -18.75$ kip-ft

$FM_{23} = 5 \times 4 \times 36/100 = 7.2$ kip-ft

$FM_{32} = 5 \times 6 \times 16/100 = -4.8$ kip-ft

From moment distribution we obtain

$M_1 = -20.91$ kip-ft

$M_2 = -14.44$ kip-ft

$M_3 = 0$

$R_1 = 7.93$ kips

$R_2 = 11.51$ kips

$R_3 = 0.56$ kips

COMBINATION OF STRESSES [1]

An isolated element within a loaded beam is subject to normal and shearing stresses as shown in Figure 1-12. The stresses on a plane inclined at θ are given as

$$\sigma = \sigma_1 - \sigma_2 \cos 2\theta + s_{xy} \sin 2\theta \tag{1-12a}$$

$$s = \sigma_2 \sin 2\theta + s_{xy} \cos 2\theta \tag{1-12b}$$

where $\sigma_1 = (\sigma_x + \sigma_y)/2$
$\sigma_2 = (\sigma_x - \sigma_y)/2$

The values of θ that lead to the maximum and minimum of σ are designated as θ_p, and

$$\tan \theta_p = -s_{xy}/\sigma_2 \tag{1-13}$$

The maximum and minimum values of σ are called *principal stresses*, and the planes on which the principal stresses occur are called *principal planes*. The principal stresses are given by

$$\sigma_{max} = \sigma_1 + \sqrt{\sigma_2^2 + s_{xy}^2} \tag{1-14a}$$

$$\sigma_{min} = \sigma_1 - \sqrt{\sigma_2^2 + s_{xy}^2} \tag{1-14b}$$

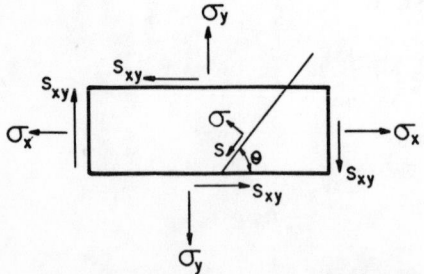

Figure 1-12. Stress combination.

Similarly, there are certain values of θ that lead to a maximum value of s for a given set of σ_x, σ_y, and s_{xy}. The angle is designated as θ_s, and

$$\tan 2\theta_s = \sigma_2/s_{xy} \tag{1-15}$$

The maximum value of s is given by

$$s_{max} = \sqrt{\sigma_2{}^2 + s_{xy}^2} \tag{1-16}$$

Note that the shearing stresses on principal planes are always zero, but the normal stress on the plane of maximum shearing stress is given by

$$\sigma' = \sigma_1$$

REFERENCES

1. Nash, W. A., *Strength of Materials*, 2nd Edition, McGraw-Hill Book Company, 1972.
2. Timoshenko, S. and MacCullough, G. H., *Elements of Strength of Materials*, 3rd Edition, D. Van Nostrand Company, Inc., 1957.
3. Roark, R. J., *Formulas for Stress and Strain*, 4th Edition, McGraw-Hill Book Company, 1965.
4. Norris, C. H. and Wilbur, J. B., *Elementary Structural Analysis*, 2nd Edition, McGraw-Hill Book Company, 1960.

Chapter 2

Beams with
Single Spans

BASIC CONCEPTS [1,3]

When a beam bends, fibers on the convex side lengthen and fibers on the concave side shorten. The neutral surface is normal to the loading plane, and the unit fiber stresses and strains are proportional to the distance from the neutral surface. In the case of a horizontal beam with vertical loads, the vertical deflection due to shear is negligible. Vertical deflection of a horizontal beam is largely due to bending.

The equations and data in this chapter are based on the following assumptions:

1. The beam is straight and the material is homogeneous.
2. The beam is long in proportion to its depth and not disproportionately wide.
3. All loads and reactions are perpendicular to the axis of the beam, and the neutral surface is normal to the plane of the loads.
4. Plane sections remain planar, and longitudinal displacements of the neutral surface are negligible.
5. The maximum bending stress does not exceed the proportional limit.

If these conditions are satisfied, the beam will bend in the plane of loading without twisting.

METHOD OF ANALYSIS

Analysis of a beam includes finding reactions (bending moments and shears) and deformations (deflections and rotations). A beam with single span, determinate or indeterminate, can be analyzed using any method discussed in "Elastic Deflection of Beams" in Chapter 1. A convenient method or methods can be selected when the type of loading and the boundary conditions of the beam are given. If a beam is supported so that the reactions can be obtained by applying the equations of static equilibrium, the beam is *statically determinate*. When the number of reactions

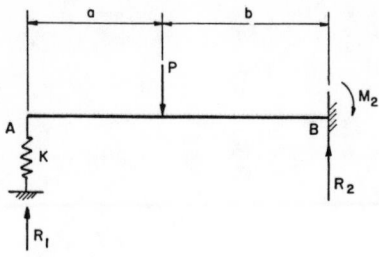

Figure 2-1.   Single-span beam with one degree indeterminacy.

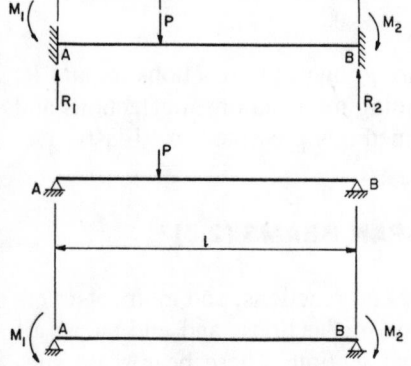

Figure 2-2.   Single-span beam with built-in ends.

exceeds the number of equilibrium equations available, the beam is *statically indeterminate*. For the beam shown in Figure 2-1, the unknown reactions are R_1, R_2, and M_2, and there are only two equilibrium equations ($\Sigma F_y = 0$, $\Sigma M = 0$); thus, determining the reactions requires additional equations. Two supplemental equations based on consideration of deformations of the beam can be derived. Based on the facts that deflection at A will be R_1/K and there is no slope at B, the reactions R_1 and M_2 can be obtained.

Figure 2-2 shows a built-in beam. In this case we have four unknown reactions and two available equations. Take the moments M_1 and M_2 at the supports for the statically indeterminate quantities. The solution can be obtained by combining the two simple beam cases, one loaded by P and the other loaded by M_1 and M_2, as shown in Figure 2-2. The rotating angle at A induced by P will be equal to the angle induced by M_1 and M_2, since the slope at A is zero. That is,

$$\frac{Pb(\ell^2 - b^2)}{6\ell EI} = \frac{M_1\ell}{3EI} + \frac{M_2\ell}{6EI} \tag{2-1}$$

The same condition applies at B; we obtain

$$\frac{Pab(2\ell - b)}{6\ell EI} = \frac{M_2\ell}{3EI} + \frac{M_1\ell}{6EI} \tag{2-2}$$

From Equations 2-1 and 2-2 we obtain

$$M_1 = \frac{Pab^2}{\ell^2}$$

$$M_2 = \frac{Pba^2}{\ell^2}$$

When the end moments M_1 and M_2 are known, the reactions R_1 and R_2 can be obtained by statics. After obtaining the reactions, deflections and rotations can be calculated by using methods presented in "Elastic Deflection of Beams" in Chapter 1.

DATA FOR SINGLE-SPAN BEAMS [2,3]

Formulas and dimensionless data for end reactions, end moments, vertical shears, bending moments, vertical deflections, and end slopes of single-span beams are provided in this section. These beams are supported and loaded in various ways. By superposition, the formulas and data can be applied to many types of loading and supports. The following notations are used in this chapter: M is positive when clockwise; V is positive when upward; y is positive when upward. Loads, applied couples, and constrained moments are positive when acting as shown.

w = unit load
W = total load
P = concentrated load
R_i, R_j = reactions at ends i and j, respectively
M_x = bending moment at x distance from end i
y = deflection at x distance from end i
θ_i, θ_j = slopes in radian at ends i and j
$C = a/\ell$
$D = x/\ell$
$E = c/\ell$
G_i, G_j = coefficients of reactions
J_x = coefficient of shear
K_x = coefficient of moment
A_x = coefficient of deflection
B_i, B_j = coefficient of slopes
K_i, K_j = coefficients of end moments

Example 2-1

The beam shown in Figure 2-3 is W12 × 26, the span is 16 ft, and w = 3 kips/ft. Calculate R_i, R_j, M_j, θ_i, and deflection at midspan.

From data provided in Case 11,

$R_i = w\ell/10 = 3 \times 16/10 = 4.8$ kips

$R_j = 2w\ell/5 = 2 \times 3 \times 16/5 = 19.2$ kips

At midspan

$D = 8/16 = 0.5$

$A_x = 0.2813, K_x = 0.583$

$$M_m = \frac{w\ell^2}{20} K_x = \frac{3 \times 16^2}{20} \times 0.583 = 22.387 \text{ kip-ft}$$

$$M_j = \frac{w\ell^2}{15} = \frac{3 \times 16^2}{5} = 51.2 \text{ kip-ft}$$

$$y_m = \frac{-w\ell^4}{120EI} A_x = \frac{-3 \times 16^4 \times 12^3 \times 0.2813}{120 \times 29 \times 10^3 \times 204} = -0.1346 \text{ in.}$$

$$\theta_i = \frac{-w\ell^3}{120EI} = \frac{-3 \times 16^3 \times 12^2}{120 \times 29 \times 10^3 \times 204} = -0.00249 \text{ rad}$$

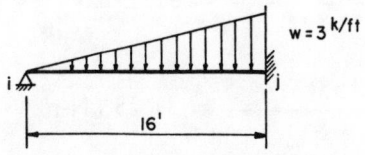

Figure 2-3. Single-span beam for Example 2-1.

Example 2-2

A beam is loaded as shown in Figure 2-4; calculate R_i, R_j, M_j, and the maximum moment between the ends.

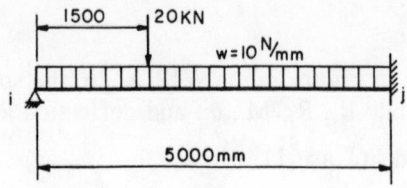

Figure 2-4.   Single-span beam for Example 2-2.

From Case 9,

$C = 1,500/5,000 = 0.3$

$G_i = 0.5635, G_j = 0.4365$

$K_j = 0.546$

$R_i = P\,G_i = 20 \times 0.5635 = 11.27$ kN

$R_j = P\,G_j = 20 \times 0.4365 = 8.73$ kN

$M_j = P\ell\,K_j/4 = 20 \times 5 \times 0.546/4 = 13.65$ kN-m

From Case 10,

$$R_i = \frac{3w\ell}{8} = \frac{3 \times 10 \times 5,000}{8} = 18.75 \text{ kN}$$

$$R_j = \frac{5w\ell}{8} = \frac{5 \times 10 \times 5,000}{8} = 31.25 \text{ kN}$$

$$M_j = \frac{w\ell^2}{8} = \frac{10 \times 5,000 \times 5}{8} = 31.25 \text{ kN-m}$$

$\Sigma\,R_i = 11.27 + 18.75 = 30.02$ kN

$\Sigma\,R_j = 8.73 + 31.25 = 39.98$ kN

$\Sigma\,M_j = 13.65 + 31.25 = 44.90$ kN-m

$M_{max} = 30.02 \times 1.5 - 10 \times 1.50 \times 1.5/2 = 33.78$ kN-m

Example 2-3

If the beam in Example 2-2 is a Japanese size $200 \times 150 \times 36.9$, calculate the slope at end i and the deflection at the concentrated load. From Case 9,

$$\theta_i = \frac{-P\ell^2}{40EI} B_i$$

$B_i = 1.47 \ (C = 0.3)$

$E = 29 \times 10^3 \ \text{ksi} = 199.827 \ \text{kN/mm}^2$

$$\theta_i = \frac{-20 \times 5,000 \times 5,000 \times 1.47}{40 \times 199.83 \times 3,330 \times 10^4} = -0.00276 \ \text{rad}$$

$$y = \frac{-P\ell^3 \ A_x}{120EI}$$

$A_x = 1.019 \ (C = 0.3, \ D = 0.3)$

$$y = \frac{-20 \times 5,000^3 \times 1.019}{120 \times 199.83 \times 3,330 \times 10^4} = -0.3191 \text{cm}$$

From Case 10,

$$\theta_i = \frac{-w\ell^3}{48EI}$$

$$\theta_i = \frac{-10 \times 5 \times 5,000 \times 5,000}{48 \times 199.83 \times 3,330 \times 10^4} = -0.0039 \ \text{rad}$$

$$y = \frac{-w\ell^4 \ A_x}{144EI} \ (A_x = 0.7056)$$

$$y = \frac{-10 \times 5 \times 5,000^3 \times 0.7056}{144 \times 199.83 \times 3,330 \times 10^4} = 0.4603 \ \text{cm}$$

$\Sigma \ \theta_i = -(0.00276 + 0.0039) = -0.00667 \ \text{rad}$

$\Sigma \ y = -(0.3191 + 0.4603) = -0.7794 \ \text{cm}$

REFERENCES

1. Timoshenko, S. and MacCullough, G. H., *Elements of Strength of Materials,* 3rd Edition, D. Van Nostrand Company, Inc., 1957.
2. Nash, W. A., *Strength of Materials,* 2nd Edition, McGraw-Hill Book Company, 1972.
3. Roark, R. J., *Formulas for Stress and Strain,* 4th Edition, McGraw-Hill Book Company, 1965.

Cases for Chapter 2

Case 1

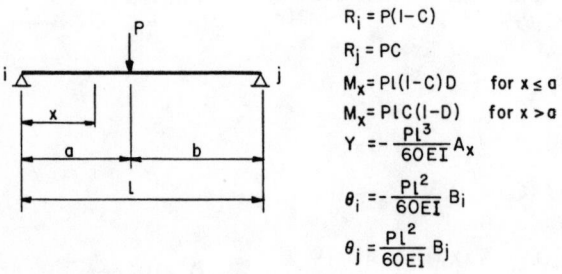

$$R_i = P(1-C)$$
$$R_j = PC$$
$$M_x = Pl(1-C)D \quad \text{for } x \le a$$
$$M_x = PlC(1-D) \quad \text{for } x > a$$
$$Y = -\frac{Pl^3}{60EI}A_x$$
$$\theta_i = -\frac{Pl^2}{60EI}B_i$$
$$\theta_j = \frac{Pl^2}{60EI}B_j$$

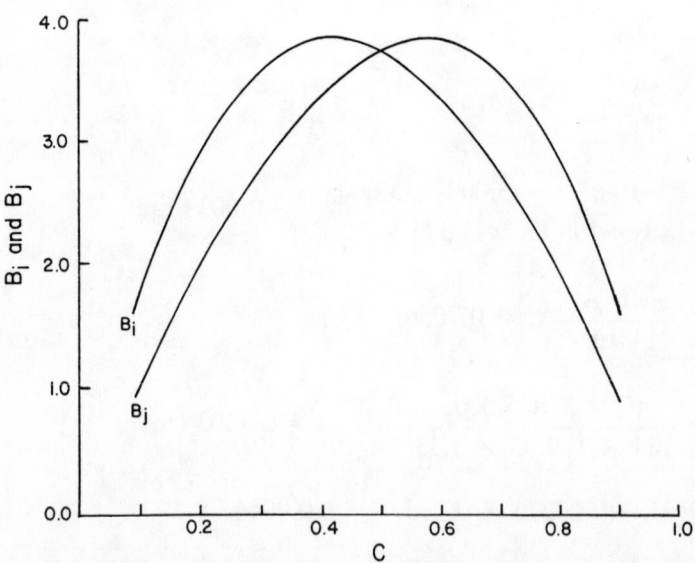

Case 1 continued

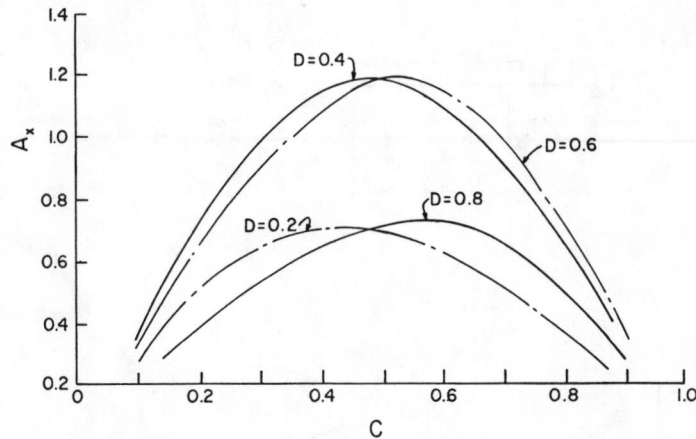

$$\mathbf{A_x}$$

D \ C	0.1	0.2	0.3	0.4	0.5	0.6	0.7	0.8	0.9
0.1	0.162	0.280	0.350	0.378	0.370	0.332	0.270	0.190	0.098
0.2	0.280	0.512	0.658	0.720	0.710	0.640	0.522	0.368	0.190
0.3	0.350	0.658	0.882	0.990	0.990	0.900	0.738	0.522	0.270
0.4	0.378	0.720	0.990	1.152	1.180	1.088	0.900	0.640	0.332
0.5	0.370	0.710	0.990	1.180	1.250	1.180	0.990	0.710	0.370
0.6	0.332	0.640	0.900	1.088	1.180	1.152	0.990	0.720	0.378
0.7	0.270	0.522	0.738	0.900	0.990	0.990	0.882	0.658	0.350
0.8	0.190	0.368	0.552	0.640	0.710	0.720	0.658	0.512	0.280
0.9	0.098	0.190	0.270	0.332	0.370	0.378	0.350	0.280	0.162

C	0.1	0.2	0.3	0.4	0.5	0.6	0.7	0.8	0.9
B_i	1.71	2.88	3.57	3.84	3.75	3.36	2.73	1.92	0.99
B_j	0.99	1.92	2.73	3.36	3.75	3.84	3.57	2.88	1.71

Case 2

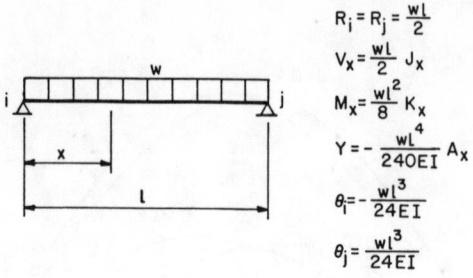

$$R_i = R_j = \frac{wl}{2}$$

$$V_x = \frac{wl}{2} J_x$$

$$M_x = \frac{wl^2}{8} K_x$$

$$Y = -\frac{wl^4}{240EI} A_x$$

$$\theta_i = -\frac{wl^3}{24EI}$$

$$\theta_j = \frac{wl^3}{24EI}$$

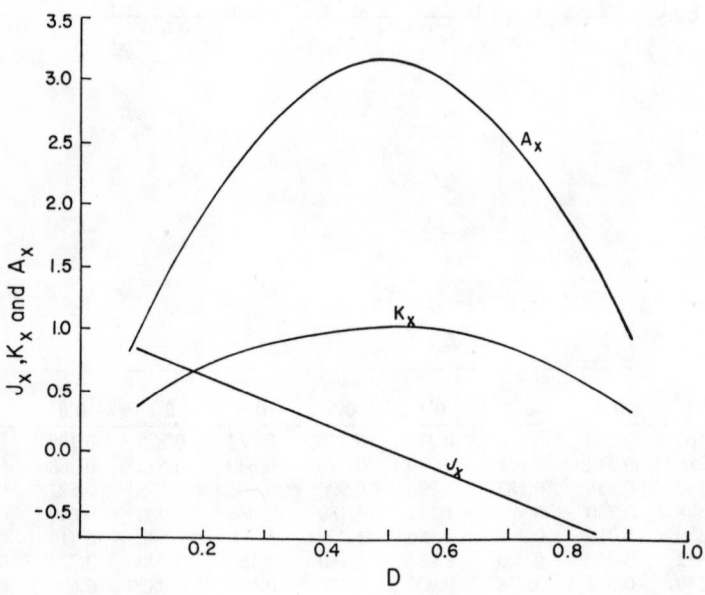

D	0.1	0.2	0.3	0.4	0.5	0.6	0.7	0.8	0.9
J_x	0.80	0.60	0.40	0.20	0.00	−0.20	−0.40	−0.60	−0.80
K_x	0.36	0.64	0.84	0.96	1.00	0.96	0.84	0.64	0.36
A_x	0.981	1.856	2.541	2.976	3.125	2.976	2.541	1.856	0.981

Case 3

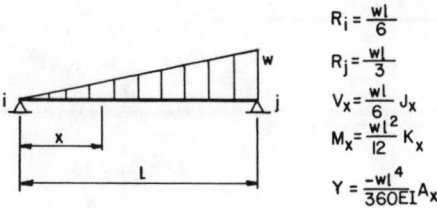

$$R_i = \frac{wl}{6}$$

$$R_j = \frac{wl}{3}$$

$$V_x = \frac{wl}{6} J_x$$

$$M_x = \frac{wl^2}{12} K_x$$

$$Y = \frac{-wl^4}{360EI} A_x$$

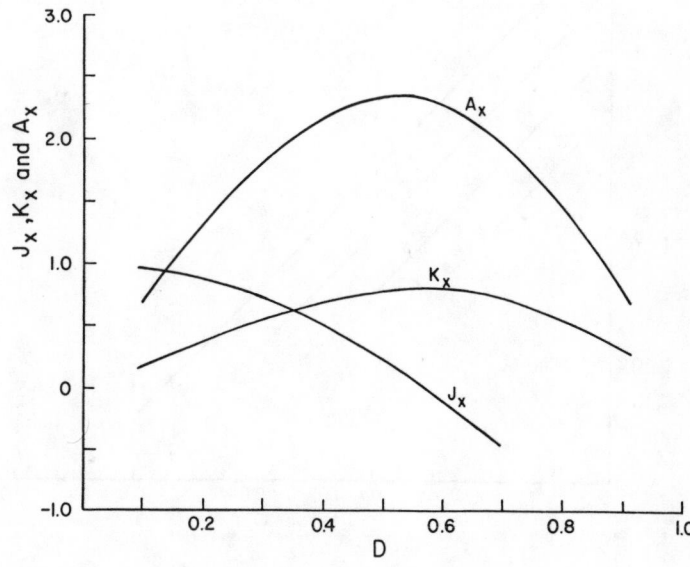

D	0.1	0.2	0.3	0.4	0.5	0.6	0.7	0.8	0.9
J_x	0.97	0.88	0.73	0.52	0.25	− 0.08	− 0.47	− 0.92	− 1.43
K_x	0.180	0.384	0.546	0.672	0.750	0.768	0.714	0.576	0.342
A_x	0.690	1.321	1.837	2.191	2.344	2.273	1.974	1.463	0.781

Case 4

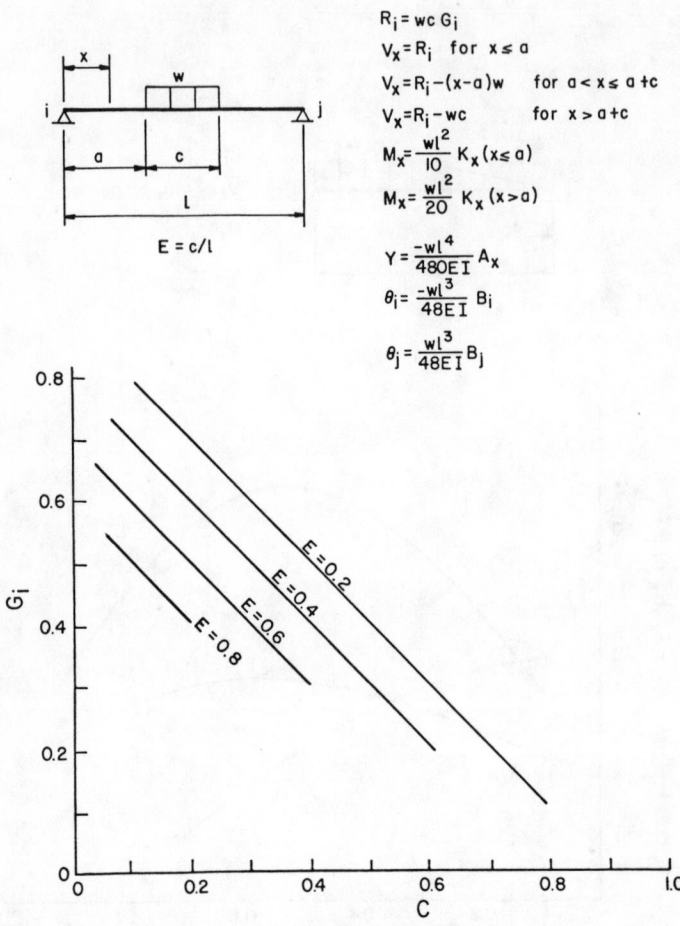

$$R_i = wc\, G_i$$
$$V_x = R_i \quad \text{for } x \le a$$
$$V_x = R_i - (x-a)w \quad \text{for } a < x \le a+c$$
$$V_x = R_i - wc \quad \text{for } x > a+c$$
$$M_x = \frac{wl^2}{10} K_x \,(x \le a)$$
$$M_x = \frac{wl^2}{20} K_x \,(x > a)$$

$$Y = \frac{-wl^4}{480EI} A_x$$
$$\theta_i = \frac{-wl^3}{48EI} B_i$$
$$\theta_j = \frac{wl^3}{48EI} B_j$$

$$E = c/l$$

Case 4 continued

$$K_x$$

E	C \ D	0.10	0.20	0.30	0.40	0.50	0.60	0.70	0.80	0.90
E = 0.10	0.10	0.085	0.240	0.210	0.180	0.150	0.120	0.090	0.060	0.030
	0.20	0.075	0.150	0.350	0.300	0.250	0.200	0.150	0.100	0.050
	0.30	0.065	0.130	0.195	0.420	0.350	0.280	0.210	0.140	0.070
	0.40	0.055	0.110	0.165	0.220	0.450	0.360	0.270	0.180	0.090
	0.50	0.045	0.090	0.135	0.140	0.175	0.440	0.330	0.220	0.110
	0.60	0.035	0.070	0.105	0.140	0.175	0.210	0.390	0.260	0.130
	0.70	0.025	0.050	0.075	0.100	0.125	0.150	0.175	0.300	0.150
	0.80	0.015	0.030	0.045	0.060	0.075	0.090	0.105	0.120	0.170
	0.90	0.005	0.010	0.015	0.020	0.025	0.030	0.035	0.040	0.045
E = 0.20	0.10	0.160	0.540	0.560	0.480	0.400	0.320	0.240	0.160	0.080
	0.20	0.140	0.280	0.740	0.720	0.600	0.480	0.360	0.240	0.120
	0.30	0.120	0.240	0.360	0.860	0.800	0.640	0.480	0.320	0.160
	0.40	0.100	0.200	0.300	0.400	0.900	0.800	0.600	0.400	0.200
	0.50	0.080	0.160	0.240	0.320	0.400	0.860	0.720	0.480	0.240
	0.60	0.060	0.120	0.180	0.240	0.300	0.360	0.740	0.560	0.280
	0.70	0.040	0.080	0.120	0.160	0.200	0.240	0.280	0.540	0.320
	0.80	0.020	0.040	0.060	0.080	0.100	0.120	0.140	0.160	0.260
E = 0.30	0.10	0.225	0.800	0.950	0.900	0.750	0.600	0.450	0.300	0.150
	0.20	0.195	0.390	1.070	1.160	1.050	0.840	0.630	0.420	0.210
	0.30	0.165	0.330	0.495	1.220	1.250	1.080	0.810	0.540	0.270
	0.40	0.135	0.270	0.405	0.540	1.250	1.220	0.990	0.660	0.330
	0.50	0.105	0.210	0.315	0.420	0.525	1.160	1.070	0.780	0.390
	0.60	0.075	0.150	0.225	0.300	0.375	0.450	0.950	0.800	0.450
	0.70	0.045	0.090	0.135	0.180	0.225	0.270	0.315	0.620	0.410
E = 0.40	0.10	0.280	1.020	1.280	1.340	1.200	0.960	0.720	0.480	0.240
	0.20	0.240	0.480	1.340	1.520	1.500	1.280	0.960	0.640	0.320
	0.30	0.200	0.400	0.600	1.500	1.600	1.500	1.200	0.800	0.400
	0.40	0.160	0.320	0.480	0.640	1.500	1.520	1.340	0.960	0.480
	0.50	0.120	0.240	0.360	0.480	0.600	1.340	1.280	1.020	0.560
	0.60	0.080	0.160	0.240	0.320	0.400	0.480	1.020	0.88	0.540
E = 0.50	0.10	0.325	1.200	1.550	1.700	1.650	1.400	1.050	0.700	0.350
	0.20	0.275	0.550	1.550	1.800	1.850	1.800	1.350	0.900	0.450
	0.30	0.225	0.450	0.675	1.700	1.850	1.800	1.550	1.100	0.550
	0.40	0.175	0.350	0.525	0.700	1.650	1.700	1.550	1.200	0.650
	0.50	0.125	0.250	0.375	0.500	0.625	1.400	1.350	1.100	0.650

Case 4 continued

$$A_x$$

E	C\D	0.10	0.20	0.30	0.40	0.50	0.60	0.70	0.80	0.90
E = 0.10	0.10	0.1801	0.3216	0.4074	0.4428	0.4350	0.3912	0.3186	0.2244	0.1158
	0.20	0.2550	0.4740	0.6230	0.6900	0.6850	0.6200	0.5070	0.3580	0.1850
	0.30	0.2938	0.5564	0.7566	0.8652	0.8750	0.8008	0.6594	0.4676	0.2422
	0.40	0.3014	0.5764	0.7986	0.9416	0.9810	0.9144	0.7614	0.5436	0.2826
	0.50	0.2826	0.5436	0.7614	0.9144	0.9810	0.9416	0.7986	0.5764	0.3014
	0.60	0.2422	0.4676	0.6594	0.8008	0.8750	0.8652	0.7566	0.5574	0.2948
	0.70	0.1850	0.3580	0.5070	0.6200	0.6850	0.6900	0.6230	0.4740	0.2550
	0.80	0.1182	0.2291	0.3257	0.4007	0.4468	0.4570	0.4239	0.3405	0.1802
	0.90	0.0394	0.0764	0.1086	0.1336	0.1490	0.1524	0.1414	0.1136	0.0666
E = 0.20	0.10	0.4352	0.7956	1.0304	1.1328	1.1200	1.0112	0.8256	0.5824	0.3008
	0.20	0.5488	1.0304	1.3796	1.5552	1.5600	1.4208	1.1664	0.8256	0.4272
	0.30	0.5952	1.1328	1.5552	1.8068	1.8560	1.7152	1.4208	1.0112	0.5248
	0.40	0.5840	1.1200	1.5600	1.8560	1.9620	1.8560	1.5600	1.1200	0.5840
	0.50	0.5248	1.0112	1.4208	1.7152	1.8560	1.8068	1.5552	1.1328	0.5952
	0.60	0.4272	0.8256	1.1664	1.4208	1.5600	1.5552	1.3796	1.0304	0.5488
	0.70	0.3008	0.5824	0.8256	1.0112	1.1200	1.1328	1.0304	0.7956	0.4352
	0.80	0.1552	0.3008	0.4272	0.5248	0.5840	0.5952	0.5488	0.4352	0.2468
E = 0.30	0.10	0.7290	1.3520	1.7870	1.9980	1.9950	1.8120	1.4850	1.0500	0.5430
	0.20	0.8502	1.6068	2.1782	2.4968	2.5410	2.3352	1.9278	1.3692	0.7098
	0.30	0.8778	1.6764	2.3166	2.7212	2.8370	2.6568	2.2194	1.5876	0.8262
	0.40	0.8262	1.5876	2.2194	2.6568	2.8370	2.7212	2.3166	1.6764	0.8778
	0.50	0.7098	1.3692	1.9278	2.3352	2.5410	2.4968	2.1782	1.6068	0.8502
	0.60	0.5430	1.0500	1.4850	1.8120	1.9950	1.9980	1.7870	1.3520	0.7290
	0.70	0.3402	0.6588	0.9348	1.1448	1.2690	1.2852	1.1718	0.9092	0.5018
E = 0.40	0.10	1.0304	1.9284	2.5856	2.9396	2.9760	2.7264	2.2464	1.5936	0.8256
	0.20	1.1328	2.1504	2.9396	3.4112	3.5220	3.2768	2.7264	1.9456	1.0112
	0.30	1.1200	2.1440	2.9760	3.5220	3.7120	3.5220	2.9760	2.1440	1.1200
	0.40	1.0112	1.9456	2.7264	3.2768	3.5220	3.4112	2.9396	2.1504	1.1328
	0.50	0.8256	1.5936	2.2464	2.7264	2.9760	2.9396	2.5856	1.9284	1.0304
	0.60	0.5824	1.1264	1.5936	1.9456	2.1440	2.1504	1.9256	1.4656	0.7956
E = 0.50	0.10	1.3130	2.4720	3.3470	3.8540	3.9570	3.6680	3.0450	2.1700	1.1270
	0.20	1.3750	2.6180	3.5990	4.2120	4.3970	4.1420	3.4830	2.5020	1.3050
	0.30	1.3050	2.5020	3.4830	4.1420	4.3970	4.2120	3.5990	2.6180	1.3750
	0.40	1.1270	2.1700	3.0450	3.6680	3.9570	3.8540	3.1850	1.9600	1.3130
	0.50	0.8650	1.6700	2.3550	2.8600	3.1250	3.0920	2.7270	2.042	1.0970

Case 4 continued

B$_i$

E \ C	0.10	0.20	0.30	0.40	0.50	0.60	0.70	0.80	0.90
0.10	0.1870	0.2610	0.2990	0.3058	0.2862	0.2450	0.1870	0.1170	0.0398
0.20	0.4480	0.5600	0.6048	0.5920	0.5312	0.4320	0.3040	0.1568	
0.30	0.7470	0.8658	0.8910	0.8370	0.7182	0.5490	0.3438		
0.40	1.0528	1.1520	1.1360	1.0240	0.8352	0.5888			
0.50	1.3390	1.3970	1.3230	1.1410	0.8750				
0.60	1.5840	1.5840	1.4400	1.1808					
0.70	1.7710	1.7010	1.4798						
0.80	1.8880	1.7408							
0.90	1.9278								

B$_j$

E \ C	0.10	0.20	0.30	0.40	0.50	0.60	0.70	0.80	0.90
0.10	0.1170	0.1870	0.2450	0.2862	0.3058	0.2990	0.2610	0.1870	0.0722
0.20	0.3040	0.4320	0.5312	0.5920	0.6048	0.5600	0.4480	0.2592	
0.30	0.5490	0.7182	0.8370	0.8910	0.8658	0.7470	0.5202		
0.40	0.8352	1.0240	1.1360	1.1520	1.0528	0.8192			
0.50	1.1410	1.3230	1.3970	1.3390	1.1250				
0.60	1.4400	1.5840	1.5840	1.4112					
0.70	1.7010	1.7710	1.6562						
0.80	1.8880	1.8432							
0.90	1.9602								

Case 5

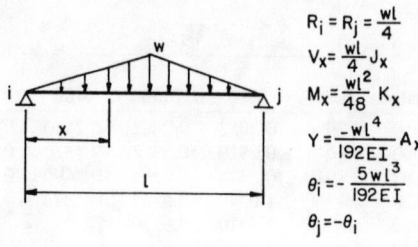

$$R_i = R_j = \frac{wl}{4}$$

$$V_x = \frac{wl}{4} J_x$$

$$M_x = \frac{wl^2}{48} K_x$$

$$Y = \frac{-wl^4}{192EI} A_x$$

$$\theta_i = -\frac{5wl^3}{192EI}$$

$$\theta_j = -\theta_i$$

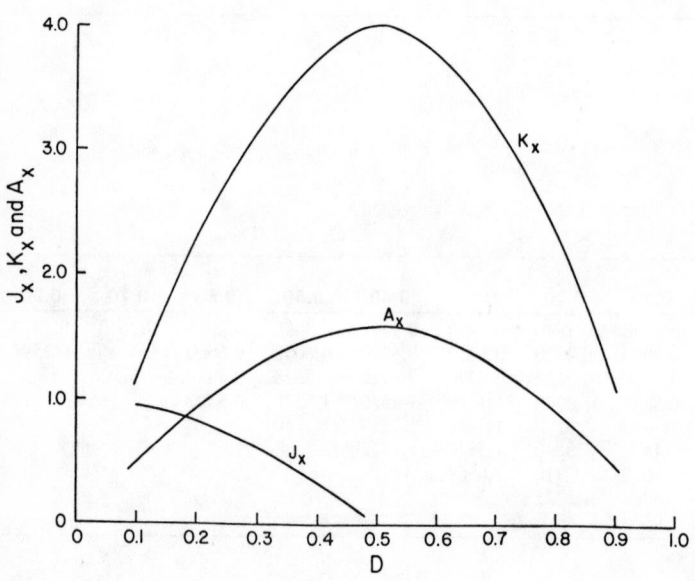

D	0.10	0.20	0.30	0.40	0.50	0.60	0.70	0.80	0.90
J_x	0.960	0.840	0.640	0.360	0.000	−0.360	−0.640	−0.840	−0.960
K_x	1.184	2.272	3.168	3.776	4.000	3.776	3.168	2.272	1.184
A_x	0.4920	0.9370	1.2918	1.5208	1.6000	1.5208	1.2918	0.9370	0.4920

Case 6

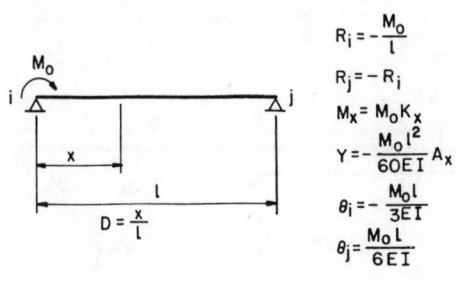

$$R_i = -\frac{M_0}{l}$$

$$R_j = -R_i$$

$$M_x = M_0 K_x$$

$$Y = -\frac{M_0 l^2}{60EI} A_x$$

$$\theta_i = -\frac{M_0 l}{3EI}$$

$$\theta_j = \frac{M_0 l}{6EI}$$

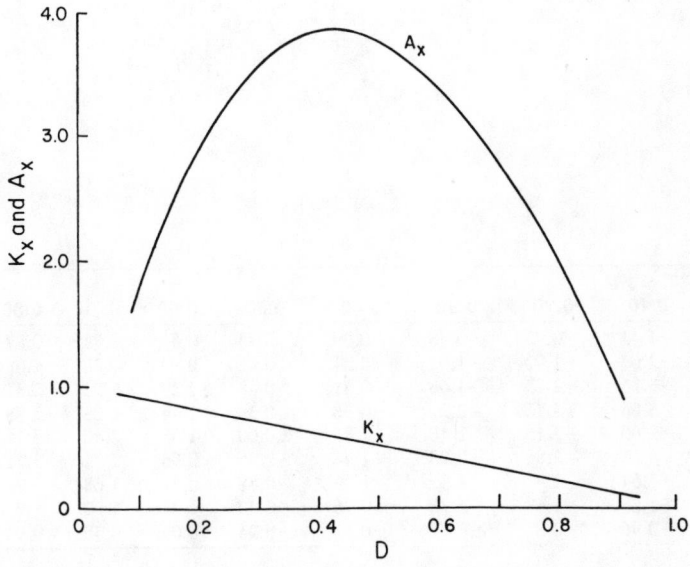

D	0.10	0.20	0.30	0.40	0.50	0.60	0.70	0.80	0.90
K_x	0.90	0.80	0.70	0.60	0.50	0.40	0.30	0.20	0.10
A_x	1.71	2.88	3.57	3.84	3.75	3.36	2.73	1.92	0.99

Case 7

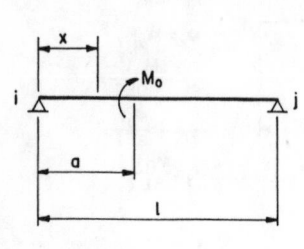

$$R_i = -\frac{M_o}{l}$$

$$R_j = -R_i$$

$$M_x = \frac{M_o}{l} x \quad \text{for } x \leq a$$

$$M_x = M_o - \frac{M_o}{l} x \quad \text{for } x > a$$

$$Y = \frac{M_o l^2}{60EI} A_x$$

$$\theta_i = -\frac{M_o l}{6EI} B_i$$

$$\theta_j = \frac{M_o l}{6EI} B_j$$

A_x

D \ C	0.10	0.20	0.30	0.40	0.50	0.60	0.70	0.80	0.90
0.10	−1.44	−0.93	−0.48	−0.09	0.24	0.51	0.72	0.87	0.96
0.20	−2.64	−1.92	−1.02	−0.24	0.42	0.96	1.38	1.68	1.86
0.30	−3.36	−2.73	−1.68	−0.51	0.48	1.29	1.92	2.37	2.64
0.40	−3.66	−3.12	−2.22	−0.96	0.36	1.44	2.28	2.88	3.24
0.50	−3.60	−3.15	−2.40	−1.35	0.00	1.35	2.40	3.15	3.60
0.60	−3.24	−2.88	−2.28	−1.44	−0.36	0.96	2.22	3.12	3.66
0.70	−2.64	−2.37	−1.92	−1.29	−0.48	0.51	1.68	2.73	3.36
0.80	−1.86	−1.68	−1.38	−0.96	−0.42	0.24	1.02	1.92	2.64
0.90	−0.96	−0.87	−0.72	−0.51	−0.24	0.09	0.48	0.93	1.44

C	0.10	0.20	0.30	0.40	0.50	0.60	0.70	0.80	0.90
B_i	1.43	0.92	0.47	0.08	−0.25	−0.52	−0.73	−0.88	−0.97
B_j	0.97	0.88	0.73	0.52	0.25	−0.08	−0.47	−0.92	−1.43

Case 8

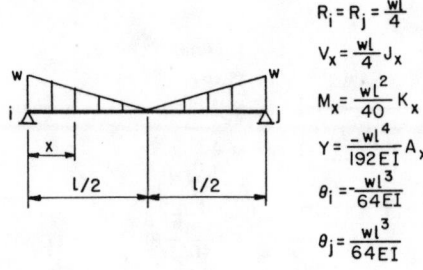

$$R_i = R_j = \frac{wl}{4}$$

$$V_x = \frac{wl}{4} J_x$$

$$M_x = \frac{wl^2}{40} K_x$$

$$Y = \frac{-wl^4}{192EI} A_x$$

$$\theta_i = -\frac{wl^3}{64EI}$$

$$\theta_j = \frac{wl^3}{64EI}$$

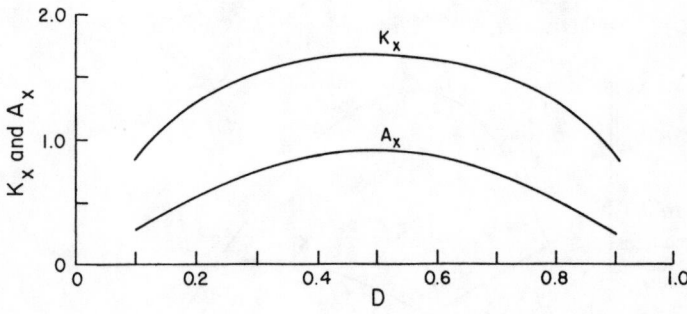

D	0.10	0.20	0.30	0.40	0.50	0.60	0.70	0.80	0.90
J_x	0.640	0.360	0.160	0.040	0	0.040	0.160	0.360	0.640
K_x	0.8130	1.3067	1.5600	1.6530	1.6667	1.6530	1.5600	1.3067	0.8130
A_x	0.2928	0.5478	0.7410	0.8600	0.9000	0.8600	0.7410	0.5478	0.2928

Case 9

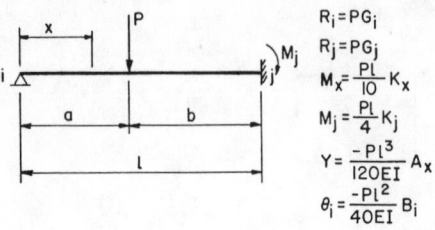

$$R_i = PG_i$$
$$R_j = PG_j$$
$$M_x = \frac{PL}{10} K_x$$
$$M_j = \frac{PL}{4} K_j$$
$$Y = \frac{-PL^3}{120EI} A_x$$
$$\theta_i = \frac{-PL^2}{40EI} B_i$$

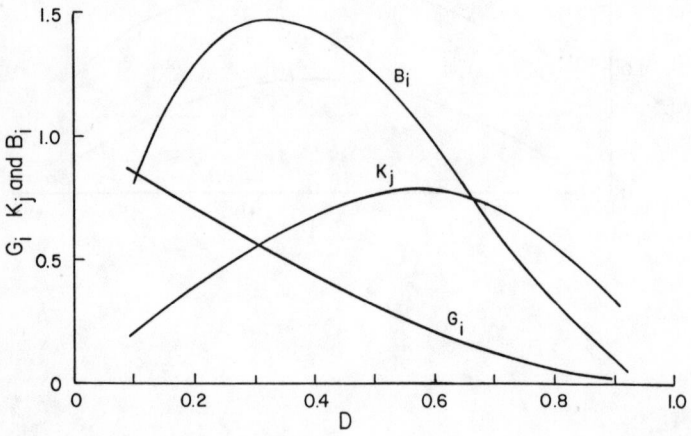

C	0.10	0.20	0.30	0.40	0.50	0.60	0.70	0.80	0.90
G_i	0.8505	0.7040	0.5635	0.4320	0.3125	0.2080	0.1215	0.0560	0.0145
G_j	0.1495	0.2960	0.4365	0.5680	0.6875	0.7920	0.8785	0.9440	0.9855
K_j	0.198	0.384	0.546	0.672	0.750	0.768	0.714	0.576	0.342
B_i	0.810	1.280	1.470	1.440	1.250	0.960	0.630	0.320	0.090

Case 9 continued

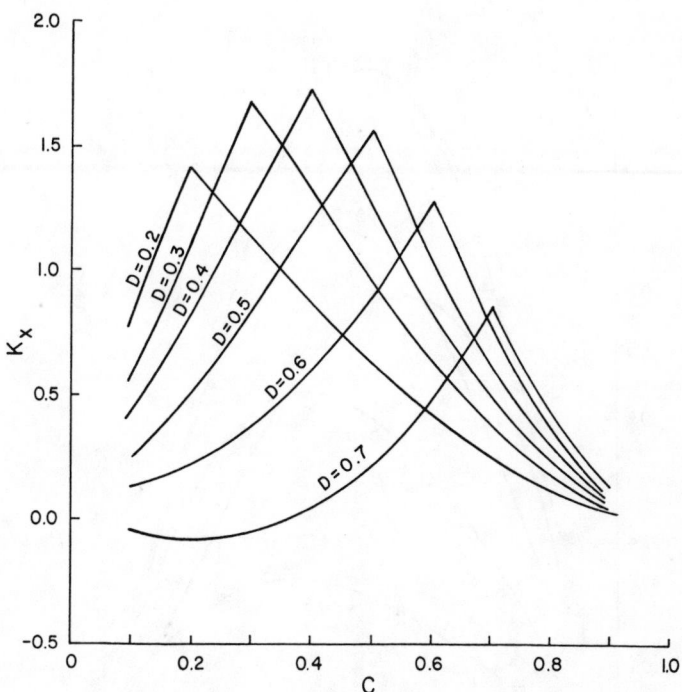

K_x

D \ C	0.10	0.20	0.30	0.40	0.50	0.60	0.70	0.80	0.90
0.10	0.851	0.704	0.564	0.432	0.313	0.208	0.122	0.056	0.015
0.20	0.801	1.408	1.127	0.864	0.625	0.416	0.243	0.112	0.029
0.30	0.552	1.112	1.691	1.296	0.938	0.624	0.365	0.168	0.044
0.40	0.402	0.816	1.254	1.728	1.250	0.832	0.486	0.224	0.058
0.50	0.253	0.520	0.818	1.160	1.563	1.040	0.608	0.280	0.073
0.60	0.103	0.224	0.381	0.592	0.875	1.248	0.729	0.336	0.087
0.70	−0.047	−0.072	−0.056	0.024	0.188	0.456	0.851	0.392	0.102
0.80	−0.196	−0.368	−0.492	−0.544	−0.500	−0.336	−0.028	0.448	0.116
0.90	−0.346	−0.664	−0.930	−1.112	−1.188	−1.128	−0.907	−0.496	0.131

Case 9 continued

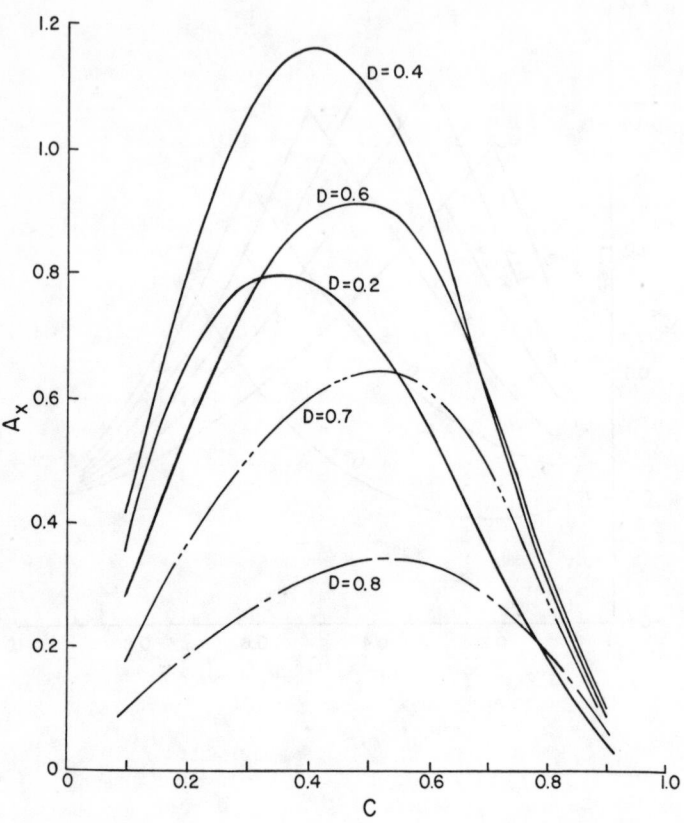

$$A_x$$

D\C	0.10	0.20	0.30	0.40	0.50	0.60	0.70	0.80	0.90
0.10	0.226	0.370	0.430	0.423	0.369	0.284	0.187	0.095	0.027
0.20	0.370	0.655	0.792	0.795	0.700	0.543	0.359	0.183	0.052
0.30	0.430	0.792	1.019	1.063	0.956	0.752	0.501	0.258	0.073
0.40	0.423	0.795	1.063	1.175	1.100	0.886	0.601	0.312	0.090
0.50	0.369	0.700	0.956	1.100	1.094	0.920	0.642	0.340	0.099
0.60	0.284	0.543	0.752	0.886	0.920	0.829	0.609	0.334	0.100
0.70	0.187	0.359	0.501	0.600	0.641	0.609	0.490	0.289	0.090
0.80	0.095	0.183	0.258	0.312	0.340	0.334	0.288	0.194	0.067
0.90	0.027	0.052	0.073	0.089	0.099	0.099	0.090	0.068	0.032

Case 10

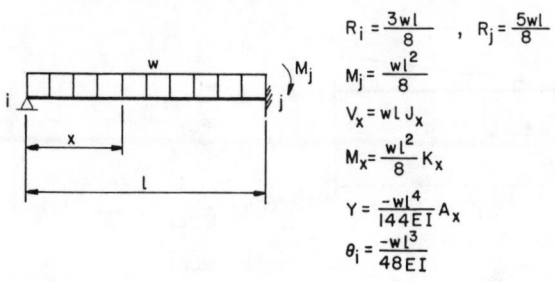

$$R_i = \frac{3wl}{8} \quad , \quad R_j = \frac{5wl}{8}$$

$$M_j = \frac{wl^2}{8}$$

$$V_x = wl\,J_x$$

$$M_x = \frac{wl^2}{8}K_x$$

$$Y = \frac{-wl^4}{144EI}A_x$$

$$\theta_i = \frac{-wl^3}{48EI}$$

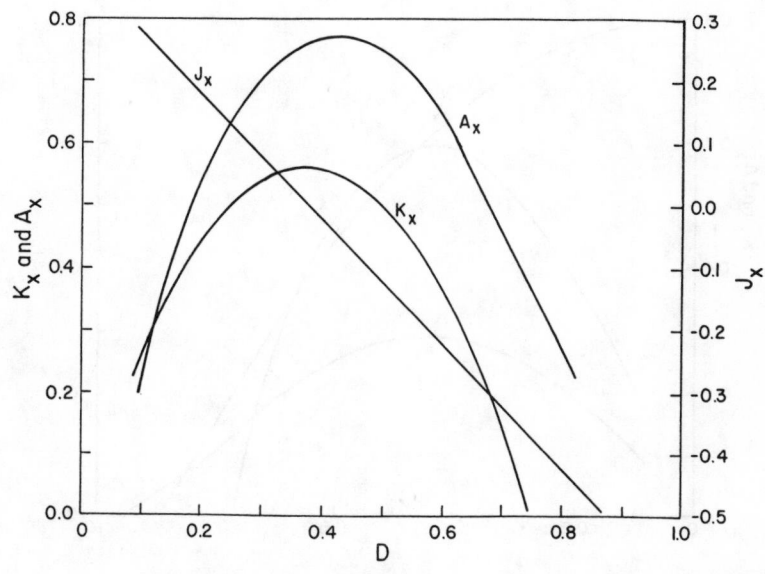

D	0.10	0.20	0.30	0.40	0.50	0.60	0.70	0.80	0.90
J_x	0.275	0.175	0.075	−0.025	−0.125	−0.225	−0.325	−0.425	−0.525
K_x	0.260	0.440	0.540	0.560	0.500	0.360	0.140	−0.160	−0.540
A_x	0.2106	0.5376	0.7056	0.7776	0.7500	0.6336	0.4536	0.2496	0.0756

Case 11

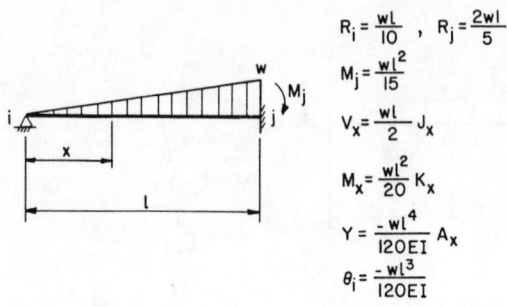

$$R_i = \frac{wL}{10} \quad , \quad R_j = \frac{2wl}{5}$$

$$M_j = \frac{wl^2}{15}$$

$$V_x = \frac{wl}{2} J_x$$

$$M_x = \frac{wl^2}{20} K_x$$

$$Y = \frac{-wl^4}{120EI} A_x$$

$$\theta_i = \frac{-wl^3}{120EI}$$

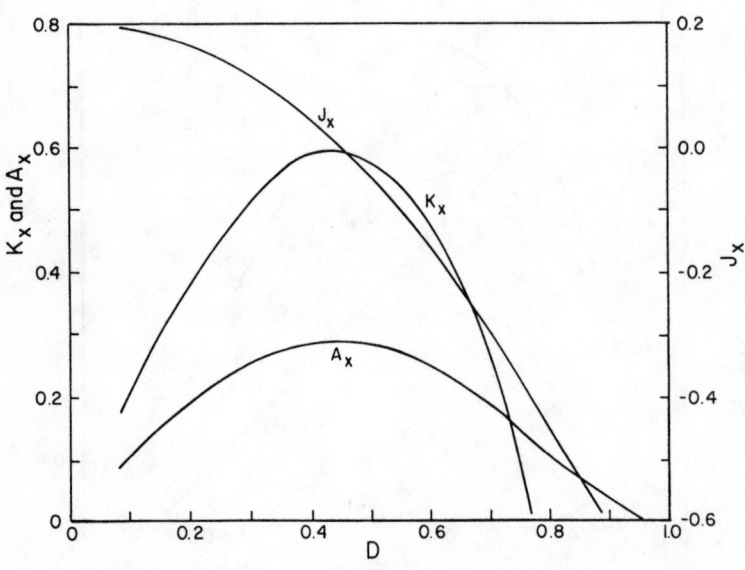

D	0.10	0.20	0.30	0.40	0.50	0.60	0.70	0.80	0.90
J_x	0.190	0.160	0.110	0.040	−0.050	−0.160	−0.290	−0.440	−0.610
K_x	0.197	0.373	0.510	0.587	0.583	0.480	0.257	−0.107	−0.630
A_x	0.0980	0.1843	0 2484	0.2822	0.2813	0.2458	0.1821	0.1037	0.0325

Case 12

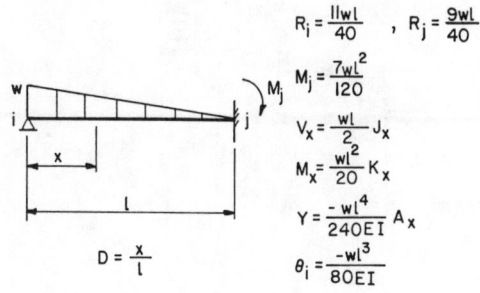

$$R_i = \frac{11wl}{40} \quad , \quad R_j = \frac{9wl}{40}$$

$$M_j = \frac{7wl^2}{120}$$

$$V_x = \frac{wl}{2} J_x$$

$$M_x = \frac{wl^2}{20} K_x$$

$$Y = \frac{-wl^4}{240EI} A_x$$

$$\theta_i = \frac{-wl^3}{80EI}$$

$$D = \frac{x}{l}$$

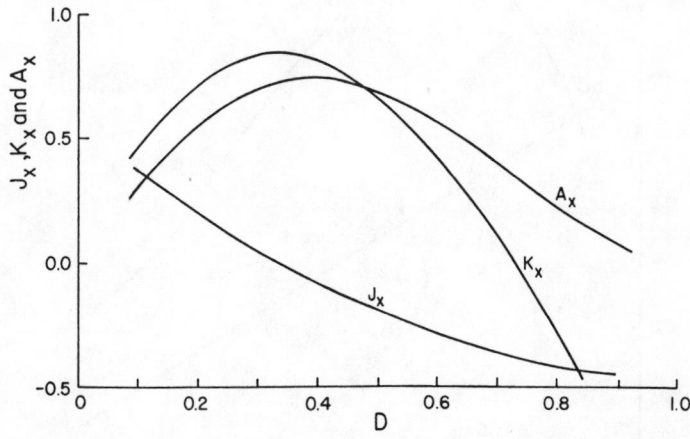

D	0.10	0.20	0.30	0.40	0.50	0.60	0.70	0.80	0.90
J_x	0.36	0.19	0.04	−0.09	−0.20	−0.29	−0.36	−0.41	−0.44
K_x	0.453	0.727	0.840	0.813	0.667	0.420	0.093	−0.293	−0.720
A_x	0.2900	0.5274	0.6791	0.7315	0.6875	0.5645	0.3919	0.2086	0.0610

Case 13

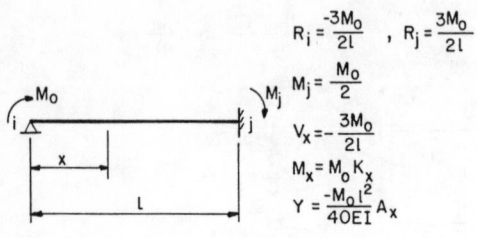

$$R_i = \frac{-3M_0}{2l} \quad , \quad R_j = \frac{3M_0}{2l}$$

$$M_j = \frac{M_0}{2}$$

$$V_x = -\frac{3M_0}{2l}$$

$$M_x = M_0 K_x$$

$$Y = \frac{-M_0 l^2}{40EI} A_x$$

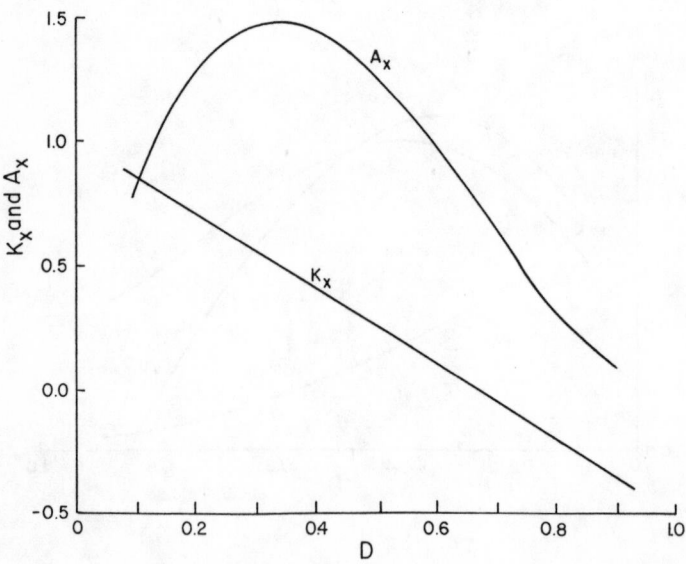

D	0.10	0.20	0.30	0.40	0.50	0.60	0.70	0.80	0.90
K_x	0.85	0.70	0.55	0.40	0.25	0.10	−0.05	−0.20	−0.35
A_x	0.810	1.280	1.470	1.440	1.250	0.960	0.630	0.320	0.090

Case 14

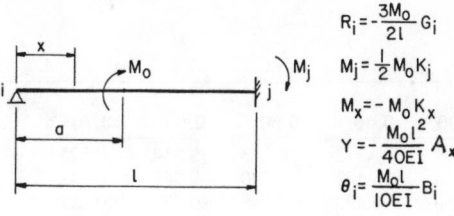

$$R_i = -\frac{3M_0}{2l} G_i$$

$$M_j = \frac{1}{2} M_0 K_j$$

$$M_x = -M_0 K_x$$

$$Y = -\frac{M_0 l^2}{40EI} A_x$$

$$\theta_i = \frac{M_0 l}{10EI} B_i$$

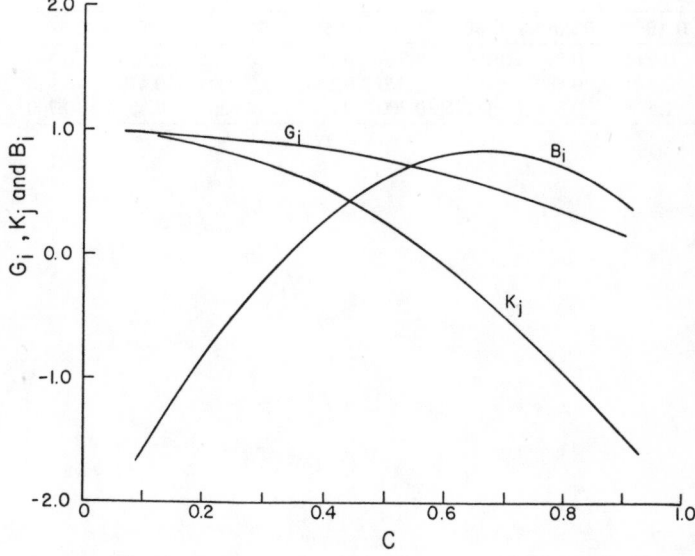

K_x

C D	0.10	0.20	0.30	0.40	0.50	0.60	0.70	0.80	0.90
0.10	0.1485	0.1440	0.1365	0.1260	0.1140	0.0960	0.0765	0.0540	0.0285
0.20	0.7030	0.2880	0.2730	0.2520	0.2280	0.1920	0.1530	0.1080	0.0570
0.30	0.5545	0.5680	0.4095	0.3780	0.3420	0.2880	0.2295	0.1620	0.0855
0.40	0.4060	0.4240	0.4540	0.5040	0.4560	0.3840	0.3060	0.2160	0.1140
0.50	0.2575	0.2800	0.3175	0.3700	0.5700	0.4800	0.3825	0.2700	0.1425
0.60	0.1090	0.1360	0.1810	0.2440	0.3160	0.5760	0.4590	0.3240	0.1710
0.70	0.0395	−0.0080	0.0445	0.1180	0.2020	0.3280	0.5355	0.3780	0.1995
0.80	−0.1880	−0.1520	−0.0920	−0.0080	0.0880	0.2320	0.3880	0.4320	0.2280
0.90	−0.3365	−0.2960	−0.2285	−0.1340	−0.2600	0.1360	0.3115	0.5140	0.2565

Case 14 continued

$$A_x$$

D \ C	0.10	0.20	0.30	0.40	0.50	0.60	0.70	0.80	0.90
0.10	0.640	0.330	0.079	−0.112	−0.243	−0.314	−0.325	−0.276	−0.168
0.20	1.139	0.717	0.213	−0.173	−0.440	−0.589	−0.619	−0.531	−0.325
0.30	1.357	1.019	0.456	−0.133	−0.548	−0.787	−0.852	−0.743	−0.459
0.40	1.354	1.094	0.662	0.058	−0.520	−0.870	−0.994	−0.890	−0.558
0.50	1.187	1.000	0.687	0.250	−0.313	−0.800	−1.013	−0.950	−0.613
0.60	0.918	0.794	0.586	0.294	−0.080	−0.538	−0.878	−0.902	−0.610
0.70	0.606	0.533	0.411	0.241	0.023	−0.245	−0.561	−0.725	−0.538
0.80	0.309	0.275	0.219	0.141	0.040	−0.083	−0.229	−0.397	−0.387
0.90	0.087	0.078	0.064	0.044	0.018	−0.014	−0.052	−0.096	−0.145

C	0.10	0.20	0.30	0.40	0.50	0.60	0.70	0.80	0.90
G_i	0.990	0.960	0.910	0.840	0.760	0.640	0.510	0.360	0.190
K_j	0.970	0.880	0.730	0.520	0.250	−0.080	−0.470	−0.920	−1.430
B_i	−1.575	−0.800	−0.175	0.300	0.625	0.800	0.825	0.700	0.425

Case 15

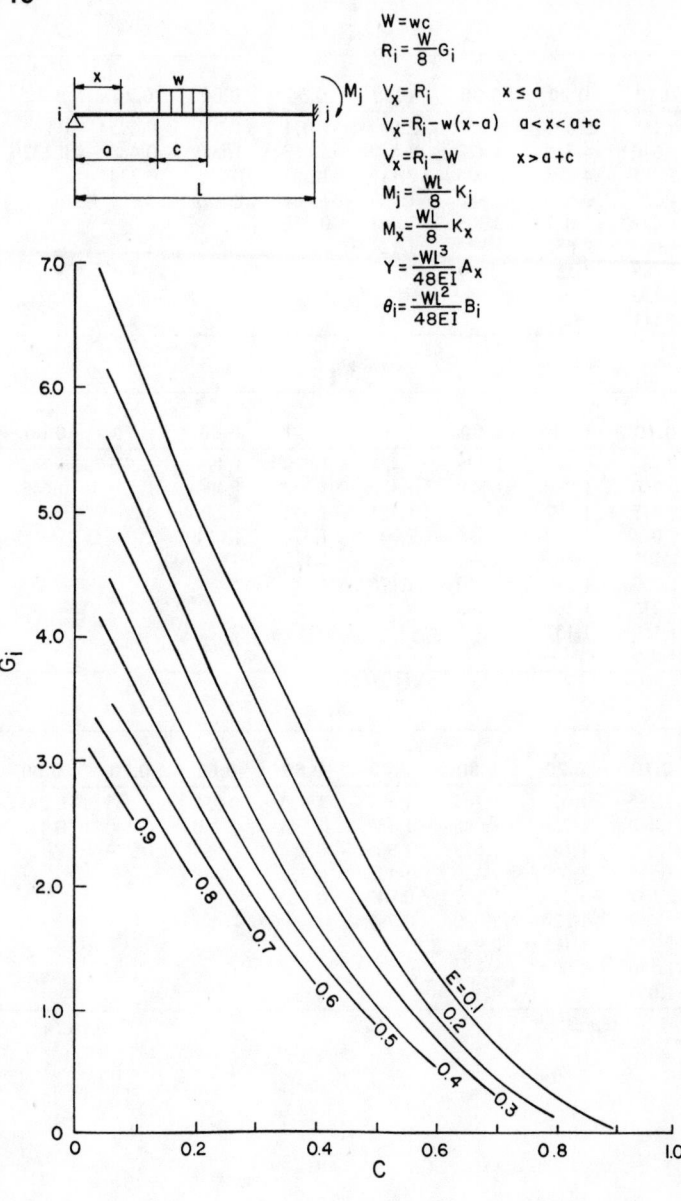

$$W = wc$$
$$R_i = \frac{W}{8} G_i$$

$$V_x = R_i \qquad\qquad x \le a$$
$$V_x = R_i - w(x-a) \qquad a < x < a + c$$
$$V_x = R_i - W \qquad\qquad x > a + c$$
$$M_j = \frac{WL}{8} K_j$$
$$M_x = \frac{WL}{8} K_x$$
$$Y = \frac{-WL^3}{48EI} A_x$$
$$\theta_i = \frac{-WL^2}{48EI} B_i$$

Case 15 continued

G_i

E\C	0.10	0.20	0.30	0.40	0.50	0.60	0.70	0.80	0.90
0.10	6.215	5.065	3.975	2.969	2.071	1.305	0.695	0.265	0.039
0.20	5.640	4.520	3.472	2.520	1.688	1.000	0.480	0.152	
0.30	5.085	4.003	3.005	2.115	1.357	0.755	0.333		
0.40	4.556	3.520	2.580	1.760	1.084	0.576			
0.50	4.059	3.037	2.203	1.461	0.875				
0.60	3.600	2.680	1.880	1.224					
0.70	3.185	2.335	1.617						
0.80	2.820	2.048							
0.90	2.511								

K_j

E\C	0.10	0.20	0.30	0.40	0.50	0.60	0.70	0.80	0.90
0.10	0.585	0.935	1.225	1.431	1.529	1.495	1.305	0.935	0.361
0.20	0.760	1.080	1.328	1.480	1.512	1.400	1.120	0.648	
0.30	0.915	1.197	1.395	1.485	1.443	1.245	0.867		
0.40	1.044	1.280	1.420	1.440	1.316	1.024			
0.50	1.141	1.363	1.397	1.339	1.125				
0.60	1.200	1.320	1.320	1.176					
0.70	1.215	1.265	1.183						
0.80	1.180	1.152							
0.90	1.089								

B_i

E\C	0.10	0.20	0.30	0.40	0.50	0.60	0.70	0.80	0.90
0.10	1.285	1.675	1.765	1.627	1.333	0.955	0.565	0.235	0.037
0.20	1.480	1.720	1.696	1.480	1.144	0.760	0.400	0.136	
0.30	1.575	1.689	1.575	1.305	0.951	0.585	0.279		
0.40	1.588	1.600	1.420	1.120	0.772	0.448			
0.50	1.537	1.351	1.249	0.943	0.625				
0.60	1.440	1.320	1.080	0.792					
0.70	1.315	1.165	0.931						
0.80	1.180	1.024							
0.90	1.053								

Case 15 continued

$$K_x$$

E	C \ D	0.10	0.20	0.30	0.40	0.50	0.60	0.70	0.80	0.90
E = 0.10	0.10	0.6215	0.8430	0.6645	0.4860	0.3075	0.1290	−0.0495	−0.2280	−0.4065
	0.20	0.5065	1.0130	1.1195	0.8260	0.5326	0.2390	−0.0545	−0.3480	−0.6415
	0.30	0.3975	0.7950	1.1925	1.1900	0.7875	0.3850	−0.0175	−0.4200	−0.8225
	0.40	0.2969	0.5938	0.8907	1.1876	1.0845	0.5814	0.0783	−0.4248	−0.9279
	0.50	0.2071	0.4142	0.6213	0.8284	1.0355	0.8426	0.2497	−0.3432	−0.9361
	0.60	0.1305	0.2610	0.3915	0.5220	0.6525	0.7830	0.5135	−0.1560	−0.8255
	0.70	0.0695	0.1390	0.2085	0.2780	0.3475	0.4170	0.4865	0.1560	−0.5745
	0.80	0.0265	0.0530	0.0795	0.1060	0.1325	0.1590	0.1855	0.2120	−0.1615
	0.90	0.0039	0.0078	0.0117	0.0156	0.0195	0.0234	0.0273	0.0312	0.0351
E = 0.20	0.10	0.5640	0.9280	0.8920	0.6560	0.4200	0.1840	−0.0520	−0.2880	−0.5240
	0.20	0.4520	0.9040	1.1560	1.0080	0.6600	0.3120	−0.0360	−0.3840	−0.7320
	0.30	0.3472	0.6944	1.0416	1.1888	0.9360	0.4832	0.0304	−0.4224	−0.8752
	0.40	0.2520	0.5040	0.7560	1.0080	1.0600	0.7120	0.164	−0.3840	−0.9320
	0.50	0.1688	0.3376	0.5064	0.6752	0.8440	0.8128	0.3816	−0.2496	−0.8808
	0.60	0.1000	0.2000	0.3000	0.4000	0.5000	0.6000	0.5000	0.0000	−0.7000
	0.70	0.0480	0.0960	0.1440	0.1920	0.2400	0.2880	0.3360	0.1840	−0.3680
	0.80	0.0152	0.0304	0.0456	0.0608	0.0760	0.0912	0.1064	0.1216	−0.0632
E = 0.30	0.10	0.5085	0.8837	0.9922	0.8340	0.5425	0.2510	−0.0405	−0.3320	−0.6235
	0.20	0.4003	0.8006	1.0676	1.0679	0.8015	0.4018	0.0021	−0.3976	−0.7973
	0.30	0.3005	0.6010	0.9015	1.0687	0.9692	0.6030	0.1035	−0.3960	−0.8955
	0.40	0.2115	0.4230	0.6345	0.8460	0.9242	0.7357	0.2805	−0.3080	−0.8965
	0.50	0.1357	0.2714	0.4071	0.5428	0.6785	0.6809	0.4166	−0.1144	−0.7787
	0.60	0.0755	0.1510	0.2265	0.3020	0.3775	0.4530	0.3952	−0.0707	−0.5205
	0.70	0.0333	0.0666	0.0999	0.1332	0.1665	0.1998	0.2331	0.1331	−0.2336
E = 0.40	0.10	0.4556	0.8112	0.9668	0.9224	0.6780	0.3336	−0.0108	−0.3552	−0.6996
	0.20	0.3520	0.7040	0.9560	1.0080	0.8600	0.5120	0.0640	−0.3840	−0.8320
	0.30	0.2580	0.5160	0.7740	0.9320	0.8900	0.6480	0.2060	−0.1360	−0.6780
	0.40	0.1760	0.3520	0.5280	0.7040	0.7800	0.6560	0.3320	−0.1920	−0.8160
	0.50	0.1084	0.2168	0.3252	0.4336	0.5420	0.5504	0.3588	0.0328	−0.6244
	0.60	0.0576	0.1152	0.1728	0.2304	0.2880	0.3456	0.3032	0.0608	−0.3816
E = 0.50	0.10	0.4059	0.7318	0.8977	0.9036	0.7495	0.4354	0.0413	−0.3528	−0.7469
	0.20	0.3037	0.6074	0.8311	0.8948	0.7985	0.5422	0.1259	−0.3704	−0.8667
	0.30	0.2203	0.4406	0.6609	0.8012	0.7815	0.6018	0.2621	−0.2376	−0.8173
	0.40	0.1461	0.2922	0.4383	0.5844	0.6505	0.5566	0.3027	−0.1112	−0.6851
	0.50	0.0875	0.1750	0.2625	0.3500	0.4375	0.4450	0.2925	−0.0200	−0.4925

Case 15 continued

$$A_x$$

E	C \ D	0.10	0.20	0.30	0.40	0.50	0.60	0.70	0.80	0.90
E = 0.10	0.10	0.1223	0.2093	0.2477	0.2463	0.2156	0.1666	0.1098	0.0559	0.0158
	0.20	0.1624	0.2945	0.3678	0.3759	0.3344	0.2610	0.1732	0.0887	0.0251
	0.30	0.1725	0.3212	0.4222	0.4536	0.4186	0.2944	0.2221	0.1148	0.0327
	0.40	0.1597	0.3017	0.4079	0.4608	0.4444	0.3679	0.2505	0.1347	0.0379
	0.50	0.1312	0.2500	0.3440	0.4007	0.4076	0.3545	0.2528	0.1361	0.0400
	0.60	0.0942	0.1806	0.2513	0.2985	0.3144	0.2911	0.2229	0.1259	0.0382
	0.70	0.0558	0.1075	0.1507	0.1815	0.1956	0.1889	0.1571	0.0982	0.0318
	0.80	0.0232	0.0449	0.0634	0.0771	0.0844	0.0838	0.0736	0.0523	0.0203
	0.90	0.0037	0.0071	0.0101	0.0123	0.0136	0.0138	0.0125	0.0096	0.0049
E = 0.20	0.10	0.1424	0.2519	0.3077	0.3111	0.2750	0.2138	0.1415	0.0723	0.0205
	0.20	0.1675	0.3079	0.3950	0.4147	0.3750	0.2957	0.1657	0.1018	0.0289
	0.30	0.1661	0.3114	0.4151	0.4562	0.4300	0.3477	0.2363	0.1231	0.0353
	0.40	0.1455	0.2759	0.3760	0.4307	0.4260	0.3597	0.2517	0.1338	0.0389
	0.50	0.1127	0.2153	0.2976	0.3496	0.3610	0.3228	0.2378	0.1310	0.0391
	0.60	0.0750	0.1440	0.2010	0.2400	0.2550	0.2400	0.1900	0.1120	0.0350
	0.70	0.0395	0.0762	0.1071	0.1293	0.1400	0.1363	0.1154	0.0753	0.0261
	0.80	0.0135	0.0260	0.0367	0.0447	0.0490	0.0488	0.0431	0.0310	0.0126
E = 0.30	0.10	0.1524	0.2750	0.3459	0.3586	0.3219	0.2527	0.1684	0.0865	0.0246
	0.20	0.1649	0.3058	0.3993	0.4301	0.3981	0.3188	0.2153	0.1117	0.0319
	0.30	0.1545	0.2910	0.3914	0.4384	0.4226	0.3500	0.2418	0.1275	0.0369
	0.40	0.1284	0.2441	0.3344	0.3867	0.3888	0.3368	0.2421	0.1311	0.0387
	0.50	0.0938	0.1794	0.2487	0.2936	0.3059	0.2782	0.2109	0.1200	0.0367
	0.60	0.0578	0.1110	0.1551	0.1857	0.1981	0.1879	0.1512	0.0921	0.0301
	0.70	0.0276	0.0531	0.0747	0.0903	0.0979	0.0955	0.0811	0.0534	0.0190
E = 0.40	0.10	0.1543	0.2817	0.3614	0.3841	0.3525	0.2807	0.1889	0.0978	0.0279
	0.20	0.1565	0.2918	0.3855	0.4227	0.4005	0.3277	0.2246	0.1178	0.0339
	0.30	0.1394	0.2634	0.3564	0.4034	0.3955	0.3352	0.2371	0.1271	0.0372
	0.40	0.1103	0.2099	0.2885	0.3354	0.3405	0.2998	0.2208	0.1229	0.0370
	0.50	0.0761	0.1457	0.2023	0.2394	0.2505	0.2296	0.1766	0.1031	0.0326
	0.60	0.0442	0.0850	0.1188	0.1423	0.1520	0.1444	0.1165	0.0715	0.0238
E = 0.50	0.10	0.1496	0.2753	0.3579	0.3874	0.3635	0.2955	0.2017	0.1054	0.0303
	0.20	0.1321	0.2459	0.3237	0.3524	0.3283	0.2570	0.1540	0.1239	0.0359
	0.30	0.1227	0.2322	0.3152	0.3590	0.3555	0.3060	0.2211	0.1213	0.0361
	0.40	0.0928	0.1769	0.2435	0.2837	0.2893	0.2566	0.1914	0.1088	0.0336
	0.50	0.0616	0.1180	0.1639	0.1940	0.2031	0.1864	0.1438	0.0844	0.0270

Case 16

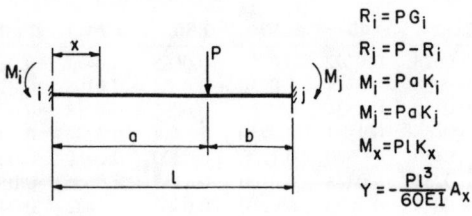

$$R_i = P\,G_i$$
$$R_j = P - R_i$$
$$M_i = P\,a\,K_i$$
$$M_j = P\,a\,K_j$$
$$M_x = P\,l\,K_x$$
$$Y = -\frac{P\,l^3}{60EI}\,A_x$$

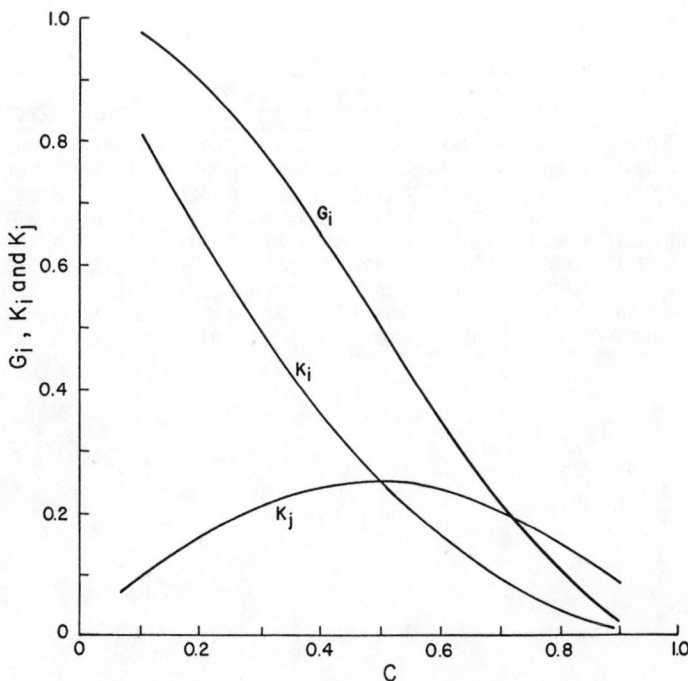

C	0.10	0.20	0.30	0.40	0.50	0.60	0.70	0.80	0.90
G_i	0.972	0.896	0.784	0.648	0.500	0.352	0.216	0.104	0.028
K_i	0.810	0.640	0.490	0.360	0.250	0.160	0.090	0.040	0.010
K_j	0.090	0.160	0.210	0.240	0.250	0.240	0.210	0.160	0.090

Case 16 continued

K_x

D \ C	0.10	0.20	0.30	0.40	0.50	0.60	0.70	0.80	0.90
0.10	0.0162	−0.0384	−0.0688	−0.0792	−0.0750	−0.0608	−0.0414	−0.0216	−0.0063
0.20	0.0134	0.0512	0.0098	−0.0144	−0.0250	−0.0256	−0.0198	−0.0112	−0.0034
0.30	0.0106	0.0408	0.0882	0.0504	0.0250	0.0096	0.0018	−0.0008	−0.0006
0.40	0.0078	0.0304	0.0666	0.1152	0.0750	0.0448	0.0234	0.0096	0.0022
0.50	0.0050	0.0200	0.0450	0.0800	0.1250	0.0800	0.0450	0.0200	0.0050
0.60	0.0022	0.0096	0.0234	0.0448	0.0750	0.1152	0.0666	0.0304	0.0078
0.70	0.0006	−0.0008	0.0018	0.0096	0.0250	0.0504	0.0882	0.0408	0.0106
0.80	−0.0034	−0.0112	−0.0198	−0.0256	−0.0250	−0.0144	0.0098	0.0512	0.0134
0.90	−0.0062	−0.0216	−0.0424	−0.0608	−0.0750	−0.0792	−0.0686	−0.0384	0.0162

A_x

D \ C	0.10	0.20	0.30	0.40	0.50	0.60	0.70	0.80	0.90
0.10	0.0146	0.0294	0.0363	0.0367	0.0325	0.0253	0.0167	0.0086	0.0024
0.20	0.0294	0.0819	0.1137	0.1209	0.1100	0.0870	0.0583	0.0301	0.0086
0.30	0.0363	0.1137	0.1852	0.2138	0.2025	0.1642	0.1118	0.0583	0.0167
0.40	0.0367	0.1210	0.2138	0.2765	0.2800	0.2355	0.1646	0.0870	0.0253
0.50	0.0325	0.0100	0.2025	0.2800	0.3125	0.2800	0.2025	0.1100	0.0325
0.60	0.0253	0.0870	0.1642	0.2355	0.2800	0.2765	0.2138	0.1210	0.0367
0.70	0.0167	0.0583	0.1118	0.1642	0.2025	0.2138	0.1852	0.1137	0.0363
0.80	0.0086	0.0301	0.058	0.0870	0.1100	0.1210	0.1137	0.0820	0.0294
0.90	0.0024	0.0086	0.0167	0.0253	0.0325	0.0367	0.0363	0.0294	0.0146

Case 17

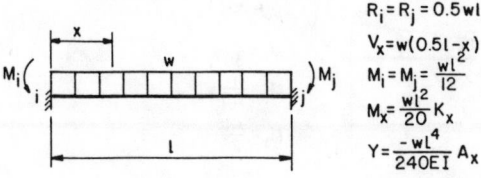

$R_i = R_j = 0.5wl$

$V_x = w(0.5l - x)$

$M_i = M_j = \dfrac{wl^2}{12}$

$M_x = \dfrac{wl^2}{20} K_x$

$Y = \dfrac{-wl^4}{240EI} A_x$

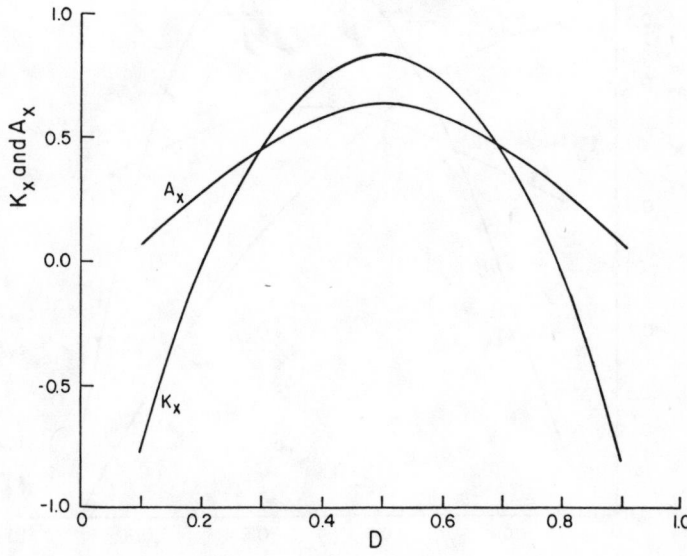

D	0.10	0.20	0.30	0.40	0.50	0.60	0.70	0.80	0.90
K_x	−0.767	−0.067	0.433	0.733	0.833	0.733	0.433	−0.067	−0.767
A_x	0.081	0.256	0.441	0.576	0.625	0.576	0.441	0.256	0.081

Case 18

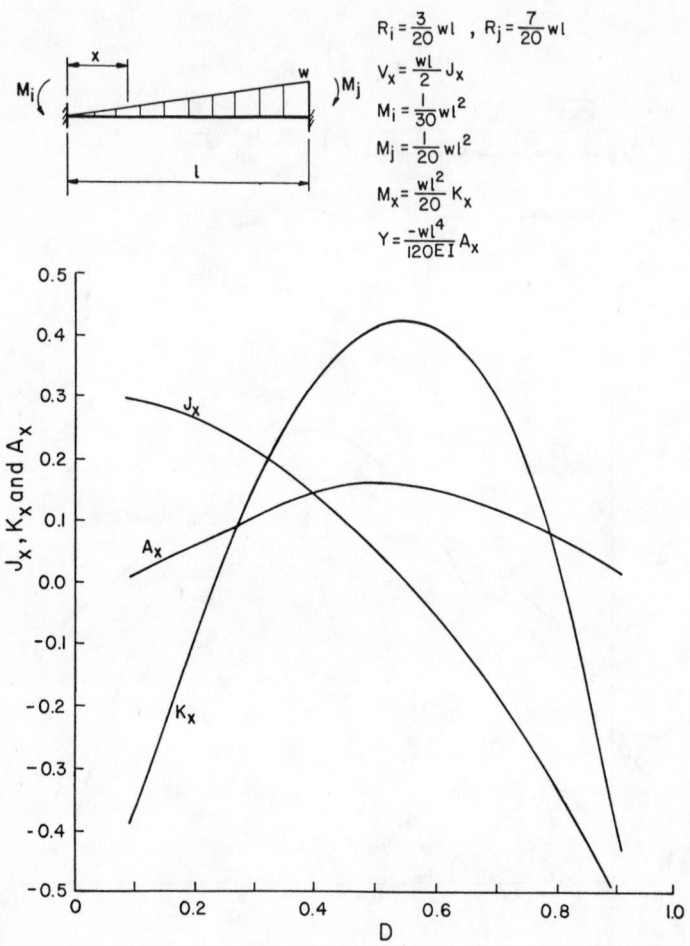

$$R_i = \frac{3}{20}wl \;,\; R_j = \frac{7}{20}wl$$

$$V_x = \frac{wl}{2}J_x$$

$$M_i = \frac{1}{30}wl^2$$

$$M_j = \frac{1}{20}wl^2$$

$$M_x = \frac{wl^2}{20}K_x$$

$$Y = \frac{-wl^4}{120EI}A_x$$

D	0.10	0.20	0.30	0.40	0.50	0.60	0.70	0.80	0.90
J_x	0.29	0.26	0.21	0.14	0.05	−0.06	−0.19	−0.34	−0.51
K_x	−0.370	−0.094	0.143	0.320	0.417	0.413	0.290	0.026	−0.397
A_x	0.017	0.056	0.100	0.138	0.156	0.150	0.119	0.072	0.024

Case 19

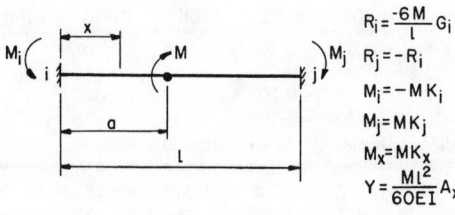

$$R_i = \frac{-6M}{l} G_i$$
$$R_j = -R_i$$
$$M_i = -M K_i$$
$$M_j = M K_j$$
$$M_x = M K_x$$
$$Y = \frac{M l^2}{60 E I} A_x$$

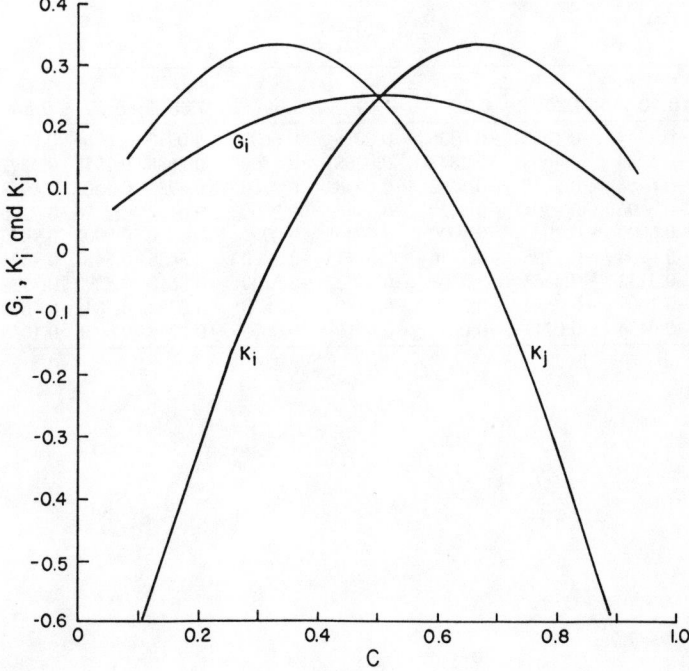

C	0.10	0.20	0.30	0.40	0.50	0.60	0.70	0.80	0.90
G_i	0.09	0.16	0.21	0.24	0.25	0.24	0.21	0.16	0.09
K_i	−0.63	−0.32	−0.07	0.12	0.25	0.32	0.33	0.28	0.17
K_j	0.17	0.28	0.33	0.32	0.25	0.12	−0.07	−0.32	−0.63

Case 19 continued

K_x

D \ C	0.10	0.20	0.30	0.40	0.50	0.60	0.70	0.80	0.90
0.10	−0.684	−0.416	−0.196	−0.024	0.100	0.176	0.204	0.184	0.116
0.20	0.262	−0.512	−0.322	−0.168	−0.050	0.032	0.078	0.088	0.062
0.30	0.208	0.392	−0.448	−0.312	−0.200	−0.112	−0.048	−0.008	0.008
0.40	0.154	0.296	0.426	−0.456	−0.350	−0.256	−0.174	−0.104	−0.046
0.50	0.100	0.200	0.300	0.400	−0.500	−0.400	−0.300	−0.200	−0.100
0.60	0.046	0.104	0.174	0.256	0.350	−0.544	−0.426	−0.296	−0.154
0.70	−0.008	0.008	0.048	0.112	0.200	0.312	−0.552	−0.392	−0.208
0.80	−0.062	−0.088	−0.078	−0.032	0.050	0.168	0.322	−0.488	−0.262
0.90	−0.116	−0.184	−0.204	−0.176	−0.100	0.024	0.196	0.416	−0.316

A_x

D \ C	0.10	0.20	0.30	0.40	0.50	0.60	0.70	0.80	0.90
0.10	−0.1944	−0.1056	−0.0336	0.0216	0.0600	0.0816	0.0864	0.0744	0.0456
0.20	−0.5280	−0.4608	−0.1848	0.0288	0.1800	0.2688	0.2952	0.3592	0.1608
0.30	−0.6468	−0.8232	−0.5292	−0.0648	0.2700	0.4752	0.5508	0.4968	0.3132
0.40	−0.6696	−0.9504	−0.8424	−0.3456	0.2400	0.6144	0.7776	0.7296	0.4704
0.50	−0.6000	−0.9000	−0.9000	−0.6000	0.00	0.6000	0.9000	0.9000	0.6000
0.60	−0.4704	−0.7296	−0.7776	−0.6144	−0.2400	0.3456	0.8424	0.9504	0.6696
0.70	−0.3132	−0.4968	−0.5508	−0.4752	−0.2700	0.0648	0.5292	0.8232	0.6468
0.80	−0.1608	−0.2592	−0.2952	−0.2688	−0.1800	−0.0288	0.1848	0.4608	0.4992
0.90	−0.0456	−0.0744	−0.0864	−0.0816	−0.0600	−0.0216	0.0336	0.1056	0.1944

Case 20

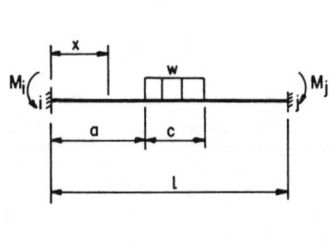

$$R_i = \frac{\overline{W}}{4} G_i \qquad \overline{W} = wc$$

$$V_x = R_i \qquad x < a$$

$$V_x = R_i - \overline{w}\frac{x-a}{c} \qquad a < x < a + c$$

$$V_x = R_i - \overline{w} \qquad x > a + c$$

$$M_i = \frac{\overline{W}l}{24} K_i$$

$$M_j = \frac{\overline{W}l}{24} K_j$$

$$M_x = \frac{\overline{W}l}{24} K_x$$

$$Y = \frac{-\overline{w}l^3}{480EI} A_x$$

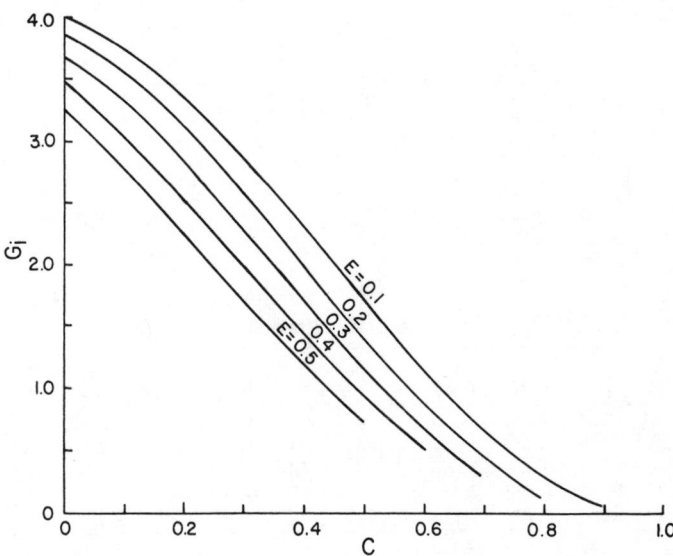

Case 20 continued

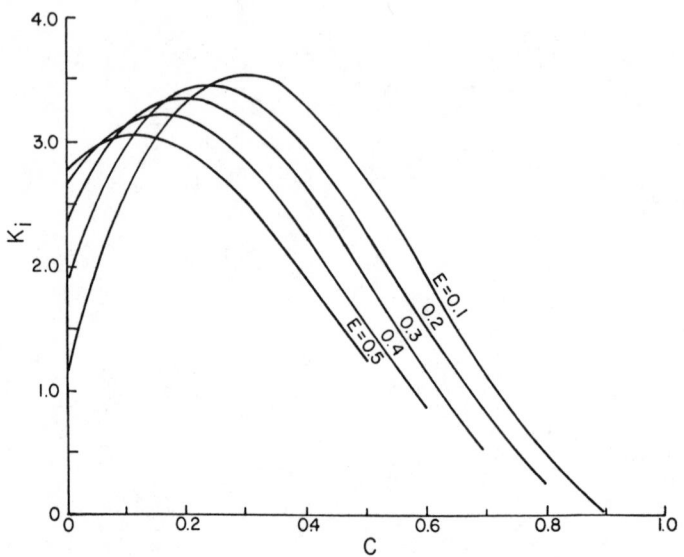

G_i, K_i, and K_j

	E \ C	0.0	0.10	0.20	0.30	0.40	0.50	0.60	0.70	0.80	0.90
	0.10	3.962	3.750	3.370	2.870	2.298	1.702	1.130	0.630	0.250	0.038
	0.20	3.856	3.560	3.120	2.584	2.000	1.416	0.880	0.440	0.144	
G_i	0.30	3.694	3.330	2.846	2.290	1.710	1.154	0.67	0.306		
	0.40	3.488	3.072	2.560	2.000	1.440	0.928	0.512			
	0.50	3.250	2.798	2.274	1.726	1.202	0.750				
	0.10	1.046	2.570	3.350	3.530	3.254	2.666	1.910	1.130	0.470	0.074
	0.20	1.808	2.960	3.440	3.392	2.960	2.288	1.520	0.800	0.272	
K_i	0.30	2.322	3.150	3.378	3.150	2.610	1.902	1.170	0.558		
	0.40	2.624	3.176	3.200	2.840	2.240	1.544	0.896			
	0.50	2.750	3.074	2.942	2.498	1.886	1.250				
	0.10	0.074	0.470	1.130	1.910	2.666	3.254	3.530	3.350	2.570	1.046
	0.20	0.272	0.800	1.520	2.288	2.960	3.392	3.440	2.960	1.808	
K_j	0.30	0.558	1.170	1.902	2.610	3.150	3.378	3.150	2.322		
	0.40	0.896	1.544	2.240	2.840	3.200	3.176	2.624			
	0.50	1.250	1.886	2.498	2.942	3.074	2.750				

Case 20 continued

$$K_x$$

E	C	0.10	0.20	0.30	0.40	0.50	0.60	0.70	0.80	0.90
E = 0.10	0.10	−0.3200	0.7300	0.5800	0.4300	0.2800	0.1300	−0.0200	−0.1700	−0.3200
	0.20	−1.328	0.6940	1.5160	1.1380	0.7600	0.3820	−0.0040	−0.3740	−0.7520
	0.30	−1.8080	−0.0860	1.6360	2.1580	1.4800	0.8020	0.1240	−0.5540	−1.2320
	0.40	−1.8752	−0.4964	0.8824	2.2612	2.4400	1.4188	0.3976	−0.6236	−1.6448
	0.50	−1.6448	−0.6236	0.3976	1.4188	2.4400	2.2612	0.8824	−0.4964	−1.8752
	0.60	−1.2320	−0.5540	0.1240	0.8020	1.4800	2.1580	1.6360	−0.0860	−1.8080
	0.70	−0.7520	−0.3740	0.0040	0.3820	0.7600	1.1380	1.5160	0.6940	−1.3280
	0.80	−0.3200	−0.1700	−0.0200	0.1300	0.2800	0.4300	0.5800	0.7300	−0.3200
	0.90	−0.0512	−0.0284	−0.0056	0.0172	0.0400	0.0628	0.0856	0.1084	0.1312
E = 0.20	0.10	−0.8240	0.7120	1.0480	0.7840	0.5200	0.2560	−0.0080	−0.2720	−0.5360
	0.20	−1.5680	0.3040	1.5760	1.6480	1.1200	0.5920	0.0640	−0.4640	−0.9920
	0.30	−1.8416	−0.2912	1.2592	2.2096	1.9600	1.1104	0.2608	−0.5888	−1.4384
	0.40	−1.7600	−0.5600	0.6400	1.8400	2.4400	1.8400	0.6400	−0.5600	−1.7600
	0.50	−1.4384	−0.5888	0.2608	1.1104	1.9600	2.2096	1.2592	−0.2912	−1.8416
	0.60	−0.9920	−0.4640	0.0640	0.5920	1.1200	1.6480	1.5760	0.3040	−1.5680
	0.70	−0.5360	−0.2720	−0.0080	0.2560	0.5200	0.7840	1.0480	0.7120	−0.8240
	0.80	−0.1856	−0.0992	−0.0128	0.0736	0.1600	0.2464	0.3328	0.4192	−0.0944
E = 0.30	0.10	−1.1520	0.4460	1.2440	1.2420	0.8400	0.4380	0.0360	−0.3660	−0.7680
	0.20	−1.6704	0.0372	1.3448	1.8524	1.5600	0.8676	0.1752	−0.5172	−1.2096
	0.30	−1.7760	−0.4020	0.9720	1.9460	2.1200	1.4940	0.4680	−0.5580	−1.5840
	0.40	−1.5840	−0.5580	0.4680	1.4940	2.1200	1.9460	0.9720	−0.4020	−1.7760
	0.50	−1.2096	−0.5172	0.1752	0.8676	1.5600	1.8524	1.3448	0.0372	−1.6704
	0.60	−0.7680	−0.3660	0.0360	0.4380	0.8400	1.2420	1.2440	0.4460	−1.1520
	0.70	−0.3744	−0.1908	−0.0072	0.1764	0.3600	0.5436	0.7273	0.5108	−0.5056
E = 0.40	0.10	−1.3328	0.2104	1.1536	1.4968	1.2400	0.6832	0.1264	−0.4304	−0.9872
	0.20	−1.6640	−0.1280	1.1080	1.7440	1.7800	1.2160	0.3520	−0.5120	−1.3760
	0.30	−1.6400	−0.4400	0.7600	1.6600	1.9600	1.6600	0.7600	−0.4400	−1.6400
	0.40	−1.3760	−0.5120	0.3520	1.2160	1.7800	1.7440	1.1080	−0.1280	−1.6640
	0.50	−0.9872	−0.4304	0.1264	0.6832	1.2400	1.4968	1.1536	0.2104	−1.3328
	0.60	−0.5888	−0.2816	0.0256	0.3328	0.6400	0.9472	0.9544	0.3616	−0.8312
E = 0.50	0.10	−1.3952	0.0436	1.0024	1.4812	1.4800	0.9988	0.2776	−0.4436	−1.1648
	0.20	−1.5776	0.2132	0.9112	1.5556	1.7200	1.4044	0.6088	−0.4268	−1.4624
	0.30	−1.4624	−0.4268	0.6088	1.4044	1.7200	1.5556	0.9112	−0.2132	−1.5776
	0.40	−1.1648	−0.4436	0.2776	0.9988	1.4800	1.4812	1.0024	0.0436	−1.3952
	0.50	−0.8000	−0.3500	0.1000	0.5500	1.0000	1.2100	0.9400	0.1900	−1.0400

Case 20 continued

$$A_x$$

E	C \ D	0.10	0.20	0.30	0.40	0.50	0.60	0.70	0.80	0.90
E = 0.10	0.10	0.1820	0.4480	0.5880	0.6120	0.5500	0.4320	0.2880	0.1480	0.0420
	0.20	0.2676	0.8008	1.2152	1.3464	1.2500	1.0016	0.6768	0.3512	0.1004
	0.30	0.2956	0.9528	1.6272	1.9944	1.9500	1.6096	1.1088	0.5832	0.1684
	0.40	0.2794	0.9339	1.6877	2.2650	2.4100	2.0870	1.4803	0.7941	0.2326
	0.50	0.2326	0.7941	1.4803	2.0870	2.4100	2.2650	1.6877	0.9339	0.2794
	0.60	0.1684	0.5832	1.1088	1.6096	1.9500	1.9944	1.6272	0.9528	0.2956
	0.70	0.1004	0.3512	0.6768	1.0016	1.2500	1.3464	1.2152	0.8008	0.2676
	0.80	0.0420	0.1480	0.2880	0.4320	0.5500	0.6120	0.5880	0.4480	0.1820
	0.90	0.0066	0.0235	0.0461	0.0698	0.0900	0.1022	0.1019	0.0845	0.0454
E = 0.20	0.10	0.2248	0.6244	0.9016	0.9792	0.9000	0.7168	0.4824	0.2496	0.0712
	0.20	0.2816	0.8768	1.4212	1.6704	1.6000	1.3056	0.8928	0.4672	0.1344
	0.30	0.2875	0.9434	1.6574	2.1297	2.1800	1.8483	1.2946	0.6886	0.2005
	0.40	0.2560	0.8640	1.5840	2.1760	2.4100	2.1760	1.5840	0.8640	0.2560
	0.50	0.2005	0.6886	1.2946	1.8483	2.1800	2.1297	1.6574	0.9434	0.2875
	0.60	0.1344	0.4672	0.8928	1.3056	1.6000	1.6704	1.4212	0.8768	0.2816
	0.70	0.0712	0.2496	0.4824	0.7168	0.9000	0.9792	0.9016	0.6244	0.2248
	0.80	0.0243	0.0858	0.1670	0.2509	0.3200	0.3571	0.3450	0.2662	0.1137
E = 0.30	0.10	0.2484	0.7339	1.1435	1.3176	1.2500	1.0144	0.6912	0.3608	0.1036
	0.20	0.2809	0.8958	1.5100	1.8686	1.8700	1.5661	1.0886	0.5762	0.1671
	0.30	0.2692	0.8936	1.5984	2.1154	2.2567	1.9872	1.4256	0.7704	0.2268
	0.40	0.2268	0.7704	1.4256	1.9872	2.2567	2.1155	1.5984	0.8936	0.2692
	0.50	0.1671	0.5762	1.0886	1.5661	1.8700	1.8686	1.5100	0.8958	0.2809
	0.60	0.1036	0.3608	0.6912	1.0144	1.2500	1.3176	1.1435	0.7339	0.2484
	0.70	0.0497	0.1742	0.3370	0.5011	0.6300	0.6869	0.6350	0.4444	0.1650
E = 0.40	0.10	0.2562	0.7839	1.2795	1.5544	1.5400	1.2826	0.8885	0.4691	0.1358
	0.20	0.2688	0.8704	1.5026	1.9232	2.0050	1.7408	1.2384	0.6656	0.1952
	0.30	0.2440	0.8160	1.4760	1.9890	2.1800	1.9890	1.4760	0.8160	0.2440
	0.40	0.1952	0.6656	1.2384	1.7408	2.0050	1.9232	1.5026	0.8704	0.2688
	0.50	0.1358	0.4691	0.8885	1.2826	1.5400	1.5544	1.2795	0.7839	0.2562
	0.60	0.0794	0.2765	0.5299	0.7782	0.9600	1.0138	0.8831	0.5715	0.1976
E = 0.50	0.10	0.2514	0.7859	1.3197	1.6610	1.7140	1.4790	1.0483	0.5621	0.1646
	0.20	0.2487	0.8130	1.4238	1.8605	1.9940	1.7915	1.3162	0.7230	0.2153
	0.30	0.2153	0.7230	1.3162	1.7915	1.9940	1.8605	1.4238	0.8130	0.2487
	0.40	0.1646	0.5621	1.0483	1.4790	1.7140	1.6610	1.3197	0.7859	0.2514
	0.50	0.1100	0.3800	0.7200	1.0400	1.2500	1.2640	1.0440	0.6440	0.2140

Case 21

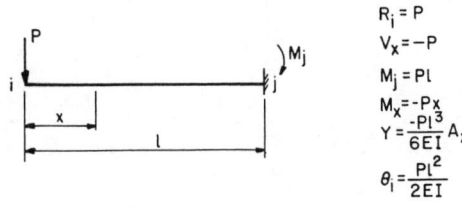

$$R_i = P$$
$$V_x = -P$$
$$M_j = Pl$$
$$M_x = -Px$$
$$Y = \frac{-Pl^3}{6EI} A_x$$
$$\theta_i = \frac{Pl^2}{2EI}$$

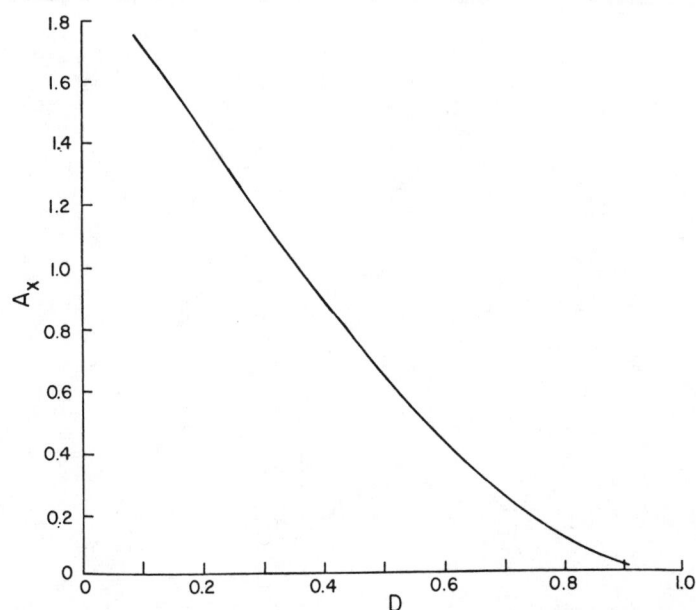

D	0.10	0.20	0.30	0.40	0.50	0.60	0.70	0.80	0.90
A_x	1.701	1.408	1.127	0.864	0.625	0.416	0.243	0.112	0.029

Case 22

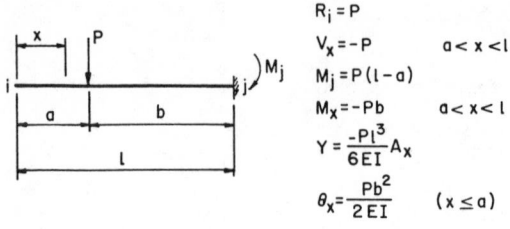

$$R_i = P$$
$$V_x = -P \qquad a < x < l$$
$$M_j = P(l-a)$$
$$M_x = -Pb \qquad a < x < l$$
$$Y = \frac{-Pl^3}{6EI} A_x$$
$$\theta_x = \frac{Pb^2}{2EI} \qquad (x \leq a)$$

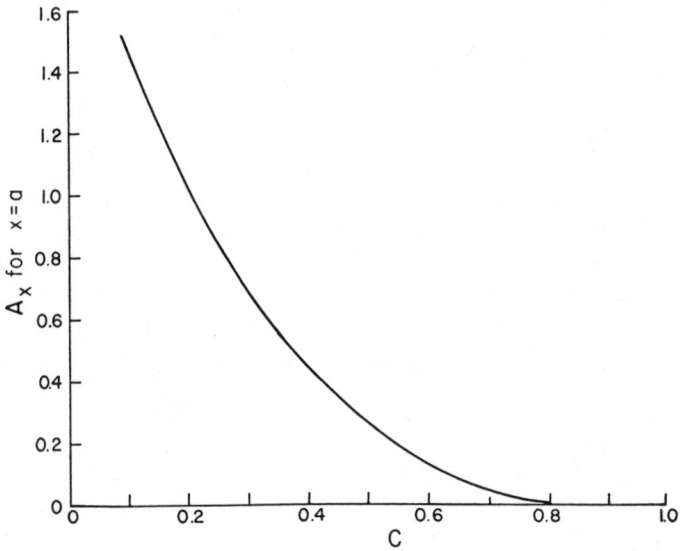

A_x

D \ C	0.10	0.20	0.30	0.40	0.50	0.60	0.70	0.80	0.90
0.10	1.458	1.216	0.980	0.756	0.550	0.368	0.216	0.100	0.026
0.20	1.216	1.024	0.833	0.648	0.475	0.320	0.189	0.088	0.023
0.30	0.980	0.833	0.686	0.540	0.400	0.272	0.162	0.076	0.020
0.40	0.756	0.648	0.540	0.432	0.325	0.224	0.135	0.064	0.017
0.50	0.550	0.475	0.400	0.325	0.250	0.176	0.108	0.052	0.014
0.60	0.368	0.320	0.272	0.224	0.176	0.128	0.081	0.040	0.011
0.70	0.216	0.189	0.162	0.135	0.108	0.081	0.054	0.028	0.008
0.80	0.100	0.088	0.076	0.064	0.052	0.040	0.028	0.016	0.005
0.90	0.026	0.023	0.020	0.017	0.014	0.011	0.008	0.005	0.002

Case 23

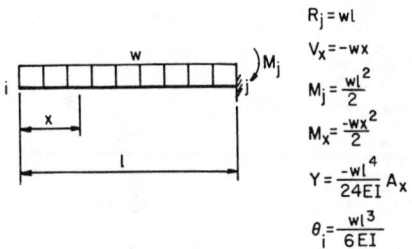

$$R_j = wl$$
$$V_x = -wx$$
$$M_j = \frac{wl^2}{2}$$
$$M_x = \frac{-wx^2}{2}$$
$$Y = \frac{-wl^4}{24EI} A_x$$
$$\theta_i = \frac{wl^3}{6EI}$$

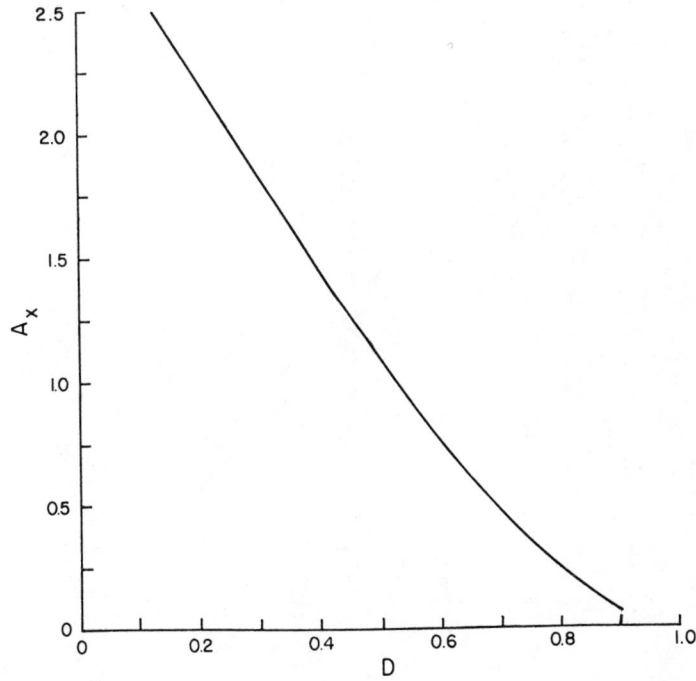

D	0.10	0.20	0.30	0.40	0.50	0.60	0.70	0.80	0.90
A_x	2.6001	2.2016	1.8081	1.4256	1.0625	0.7296	0.4401	0.2096	0.0561

Case 24

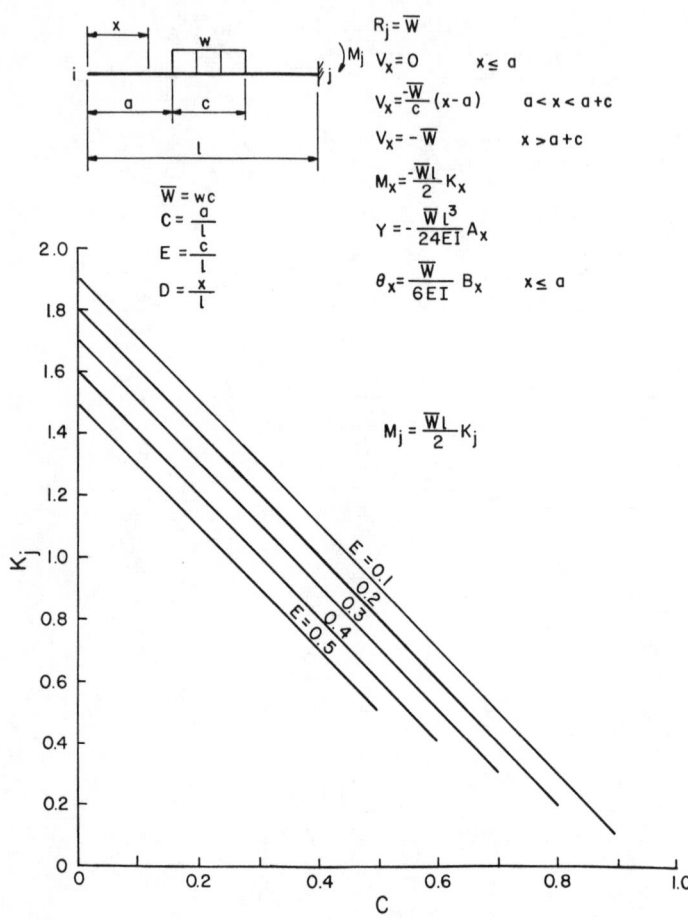

$R_j = \overline{W}$

$V_x = 0 \qquad x \leq a$

$V_x = \dfrac{\overline{W}}{c}(x-a) \qquad a < x < a+c$

$V_x = -\overline{W} \qquad x > a+c$

$M_x = -\dfrac{\overline{W}l}{2} K_x$

$Y = -\dfrac{\overline{W}l^3}{24EI} A_x$

$\theta_x = \dfrac{\overline{W}}{6EI} B_x \qquad x \leq a$

$\overline{W} = wc$

$C = \dfrac{a}{l}$

$E = \dfrac{c}{l}$

$D = \dfrac{x}{l}$

$M_j = \dfrac{\overline{W}l}{2} K_j$

Case 24 continued

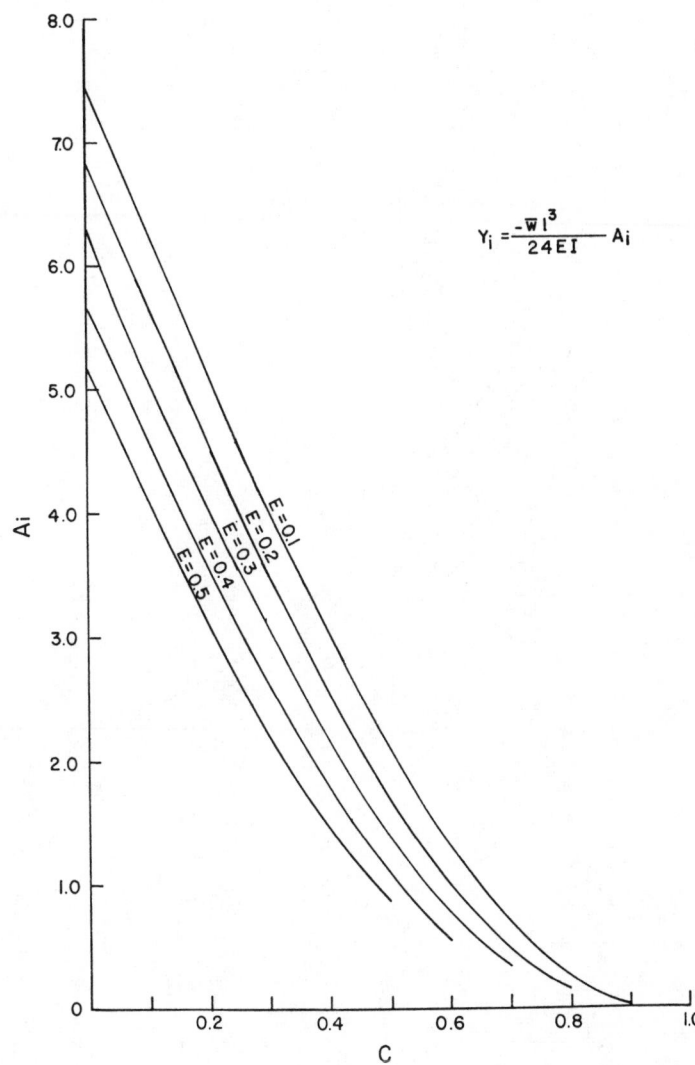

$$Y_i = \frac{-\overline{w} \, l^3}{24EI} A_i$$

Case 24 continued

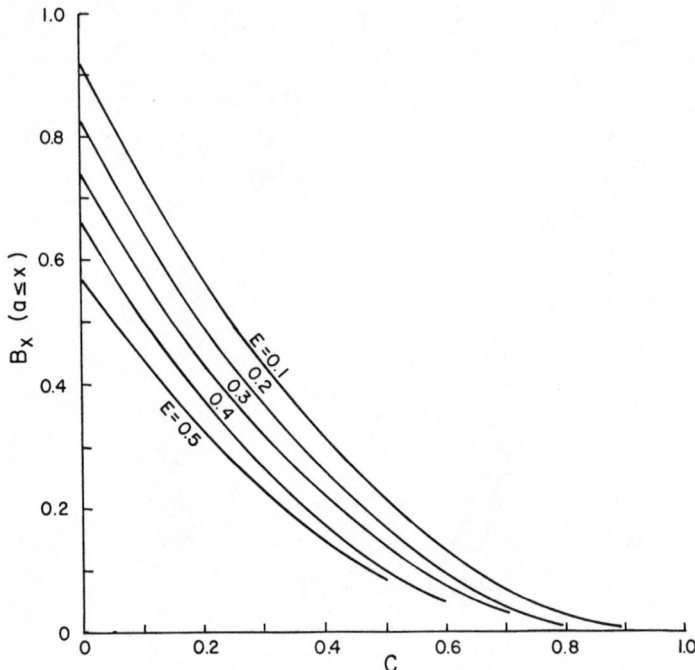

Case 24 continued

$$K_x$$

E	C	0.10	0.20	0.30	0.40	0.50	0.60	0.70	0.80	0.90	1.00
E = 0.10	0.0	0.10	0.30	0.50	0.70	0.90	1.10	1.30	1.50	1.70	1.90
	0.10		0.10	0.30	0.50	0.70	0.90	1.10	1.30	1.50	1.70
	0.20			0.10	0.30	0.50	0.70	0.90	1.10	1.30	1.50
	0.30				0.10	0.30	0.50	0.70	0.90	1.10	1.30
	0.40					0.10	0.30	0.50	0.70	0.90	1.10
	0.50						0.10	0.30	0.50	0.70	0.90
	0.60							0.10	0.30	0.50	0.70
	0.70								0.10	0.30	0.50
	0.80									0.10	0.30
	0.90										0.10
E = 0.20	0.0	0.05	0.20	0.40	0.60	0.80	1.00	1.20	1.40	1.60	1.80
	0.10		0.05	0.20	0.40	0.60	0.80	1.00	1.20	1.40	1.60
	0.20			0.05	0.20	0.40	0.60	0.80	1.00	1.20	1.40
	0.30				0.05	0.20	0.40	0.60	0.80	1.00	1.20
	0.40					0.05	0.20	0.40	0.60	0.80	1.00
	0.50						0.05	0.20	0.40	0.60	0.80
	0.60							0.05	0.20	0.40	0.60
	0.70								0.05	0.20	0.40
	0.80									0.05	0.20
E = 0.30	0.0	0.033	0.133	0.300	0.500	0.700	0.900	1.100	1.300	1.500	1.700
	0.10		0.033	0.133	0.300	0.500	0.700	0.900	1.100	1.300	1.500
	0.20			0.033	0.133	0.300	0.500	0.700	0.900	1.100	1.300
	0.30				0.033	0.133	0.300	0.500	0.700	0.900	1.100
	0.40					0.033	0.133	0.300	0.500	0.700	0.900
	0.50						0.033	0.133	0.300	0.500	0.700
	0.60							0.033	0.133	0.300	0.500
	0.70								0.033	0.133	0.300
E = 0.40	0.0	0.025	0.100	0.225	0.400	0.600	0.800	1.000	1.200	1.400	1.600
	0.10		0.025	0.100	0.225	0.400	0.600	0.800	1.000	1.200	1.400
	0.20			0.025	0.100	0.225	0.400	0.600	0.800	1.000	1.200
	0.30				0.025	0.100	0.225	0.400	0.600	0.800	1.000
	0.40					0.025	0.100	0.225	0.400	0.600	0.800
	0.50						0.025	0.100	0.225	0.400	0.600
	0.60							0.025	0.100	0.225	0.400
E = 0.50	0.0	0.02	0.08	0.18	0.32	0.50	0.70	0.90	1.10	1.30	1.50
	0.10		0.02	0.08	0.18	0.32	0.50	0.70	0.90	1.10	1.30
	0.20			0.02	0.08	0.18	0.32	0.50	0.70	0.90	1.10
	0.30				0.02	0.08	0.18	0.32	0.50	0.70	0.90
	0.40					0.02	0.08	0.18	0.32	0.50	0.70
	0.50						0.02	0.08	0.18	0.32	0.50

Case 24 continued

A_x

E	C \ D	0.0	0.10	0.20	0.30	0.40	0.50	0.60	0.70	0.80	0.90
E = 0.10	0.0	7.401	6.318	5.248	4.214	3.240	2.350	1.568	0.918	0.424	0.110
	0.10	6.215	5.347	4.480	3.626	2.808	2.050	1.376	0.810	0.376	0.098
	0.20	5.065	4.389	3.713	3.038	2.376	1.750	1.184	0.702	0.328	0.086
	0.30	3.975	3.467	2.959	2.451	1.944	1.450	0.992	0.594	0.280	0.074
	0.40	2.939	2.605	2.241	1.877	1.513	1.150	0.800	0.486	0.232	0.062
	0.50	2.071	1.827	1.583	1.339	1.095	0.851	0.608	0.378	0.184	0.050
	0.60	1.305	1.157	1.009	0.861	0.713	0.565	0.417	0.270	0.136	0.038
	0.70	0.695	0.619	0.543	0.467	0.391	0.315	0.239	0.163	0.088	0.026
	0.80	0.265	0.237	0.209	0.181	0.153	0.125	0.097	0.069	0.041	0.014
	0.90	0.039	0.035	0.031	0.027	0.023	0.019	0.015	0.011	0.007	0.003
E = 0.20	0.0	6.808	5.833	4.864	3.920	3.024	2.200	1.472	0.864	0.400	0.104
	0.10	5.640	4.868	4.097	3.332	2.592	1.900	1.280	0.756	0.352	0.092
	0.20	4.520	3.928	3.336	2.745	2.160	1.600	1.088	0.648	0.304	0.080
	0.30	3.472	3.036	2.600	2.164	1.729	1.300	0.896	0.540	0.256	0.068
	0.40	2.520	2.216	1.912	1.608	1.304	1.001	0.704	0.432	0.208	0.056
	0.50	1.688	1.492	1.296	1.100	0.904	0.708	0.513	0.324	0.160	0.044
	0.60	1.000	0.888	0.776	0.664	0.552	0.440	0.328	0.217	0.112	0.032
	0.70	0.480	0.428	0.376	0.324	0.272	0.220	0.168	0.116	0.065	0.020
	0.80	0.152	0.136	0.120	0.104	0.088	0.072	0.056	0.040	0.024	0.009
E = 0.30	0.0	6.227	5.351	4.480	3.626	2.808	2.050	1.376	0.810	0.376	0.098
	0.10	5.085	4.401	3.717	3.038	2.376	1.750	1.184	0.702	0.328	0.086
	0.20	4.003	3.487	2.971	2.455	1.944	1.450	0.992	0.594	0.280	0.074
	0.30	3.005	2.633	2.261	1.889	1.517	1.150	0.800	0.486	0.232	0.062
	0.40	2.115	1.863	1.611	1.359	1.107	0.855	0.608	0.378	0.184	0.050
	0.50	1.357	1.201	1.045	0.889	0.733	0.577	0.421	0.270	0.136	0.038
	0.60	0.755	0.671	0.587	0.503	0.419	0.335	0.251	0.167	0.088	0.026
	0.70	0.333	0.297	0.261	0.225	0.189	0.153	0.117	0.081	0.045	0.014
E = 0.40	0.0	5.664	4.871	4.100	3.332	2.592	1.900	1.280	0.756	0.352	0.092
	0.10	4.556	3.952	3.348	2.748	2.160	1.600	1.088	0.648	0.304	0.080
	0.20	3.520	3.072	2.624	2.176	1.732	1.300	0.896	0.540	0.256	0.068
	0.30	2.580	2.264	1.948	1.632	1.316	1.004	0.704	0.432	0.208	0.056
	0.40	1.760	1.552	1.344	1.136	0.928	0.720	0.516	0.324	0.160	0.044
	0.50	1.084	0.960	0.836	0.712	0.588	0.464	0.340	0.220	0.112	0.032
	0.60	0.576	0.512	0.448	0.388	0.320	0.256	0.192	0.128	0.068	0.020
E = 0.50	0.0	5.125	4.425	3.728	3.041	2.376	1.750	1.184	0.702	0.328	0.086
	0.10	4.056	3.527	2.995	2.466	1.947	1.450	0.992	0.594	0.280	0.074
	0.20	3.077	2.689	2.301	1.913	1.528	1.153	0.800	0.486	0.232	0.062
	0.30	2.203	1.935	1.667	1.399	1.131	0.866	0.611	0.378	0.184	0.050
	0.40	1.461	1.289	1.087	0.945	0.773	0.601	0.432	0.273	0.136	0.038
	0.50	0.875	0.775	0.675	0.575	0.475	0.375	0.275	0.178	0.091	0.026

$B_x \; (a \leqslant x)$

E \ C	0.0	0.1	0.2	0.3	0.4	0.5	0.6	0.7	0.8	0.9
0.1	0.9033	0.7233	0.5633	0.4233	0.3033	0.2033	0.1233	0.0633	0.0233	0.0033
0.2	0.8133	0.6933	0.4933	0.3633	0.2533	0.1633	0.0933	0.0433	0.0133	
0.3	0.7300	0.5700	0.4300	0.3100	0.2100	0.1300	0.0700	0.030		
0.4	0.6533	0.5033	0.3733	0.2633	0.1733	0.1033	0.0533			
0.5	0.5833	0.4433	0.3233	0.2233	0.1433	0.0833				

Case 25

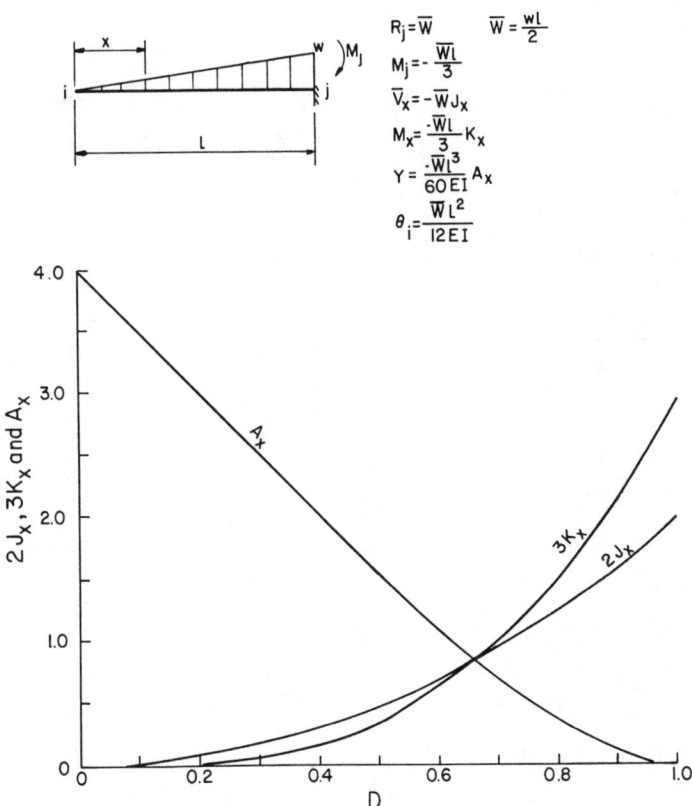

$R_j = \overline{W}$ $\overline{W} = \dfrac{wl}{2}$

$M_j = -\dfrac{\overline{W}l}{3}$

$\overline{V}_x = -\overline{W}J_x$

$M_x = \dfrac{-\overline{W}l}{3}K_x$

$Y = \dfrac{-\overline{W}l^3}{60EI}A_x$

$\theta_i = \dfrac{\overline{W}l^2}{12EI}$

D	0.0	0.1	0.2	0.3	0.4	0.5	0.6	0.7	0.8	0.9	1.0
J_x	0.0	0.01	0.04	0.09	0.16	0.25	0.36	0.49	0.64	0.81	1.00
K_x	0.0	0.001	0.008	0.027	0.064	0.125	0.216	0.343	0.512	0.729	1.00
A_x	4.0	3.5000	3.0003	2.5024	2.0102	1.5313	1.0778	0.6681	0.3277	0.0905	0

Case 26

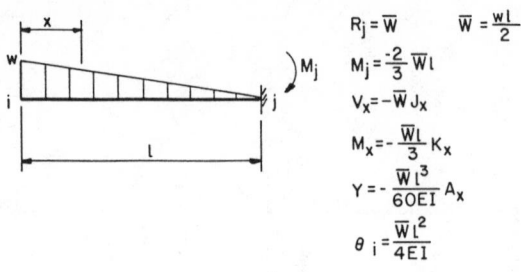

$$R_j = \overline{W} \qquad \overline{W} = \frac{wl}{2}$$

$$M_j = \frac{2}{3}\,\overline{W}l$$

$$V_x = -\overline{W}\,J_x$$

$$M_x = -\frac{\overline{W}l}{3}\,K_x$$

$$Y = -\frac{\overline{W}l^3}{60EI}\,A_x$$

$$\theta_i = \frac{\overline{W}l^2}{4EI}$$

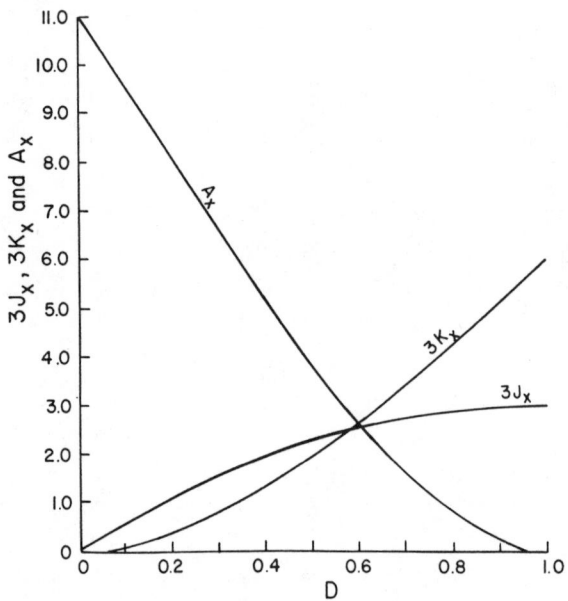

D	0.0	0.1	0.2	0.3	0.4	0.5	0.6	0.7	0.8	0.9	1.0
J_x	0.0	0.19	0.36	0.51	0.64	0.75	0.84	0.91	0.96	0.99	1.00
K_x	0.0	0.029	0.112	0.243	0.416	0.625	0.864	1.127	1.408	1.701	2.000
A_x	11.0000	9.5005	8.0077	6.5381	5.1178	3.7813	2.5702	1.5324	0.7203	0.1900	0.00

Chapter 3

Continuous Beams

METHODS OF ANALYSIS [1,2,3]

The most convenient method of solving statically indeterminate continuous beams is the moment distribution method discussed in "Statically Indeterminate Beams," Chapter 1.

The step-by-step procedures of moment distribution are summarized as follows:

Calculate Fixed End Moments

Figure 3-1 shows a continuous beam; $\overline{M}_{ij}$ and $\overline{M}_{ji}$ are the fixed end moments at ends i and j, respectively. Table 3-1 provides equations of fixed end moments for various loadings. The unbalanced moment at joint j is

$$M_j = -\overline{M}_{ij} + \overline{M}_{jk}$$

Calculate Distribution Factors

Stiffness factor of member ij $= I_1/\ell_1 = K_1$

Stiffness factor of member jk $= I_2/\ell_2 = K_2$

$\Sigma K = K_1 + K_2$

Distribution factor $DF_1 = K_1/\Sigma K$

Distribution factor $DF_2 = K_2/\Sigma K$

$DF_1 + DF_2 = 1.0$

If one end of a member is pinned, the value of the stiffness should be modified; the modified stiffness factor is $3K_i/4$, as mentioned in "Statically Indeterminate Beams," Chapter 1.

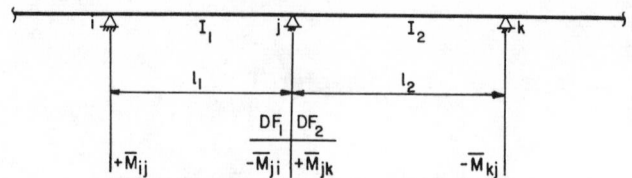

Figure 3-1. Distribution factor of a continuous beam.

Table 3-1
Fixed End Moments

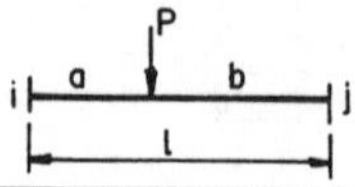 (beam with point load P at distance a, b; length l)	$\bar{M}_i = Pab^2/l^2$ $\bar{M}_j = Pa^2b/l^2$
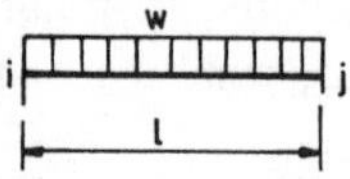 (uniform load w over length l)	$\bar{M}_i = \bar{M}_j = wl^2/12$
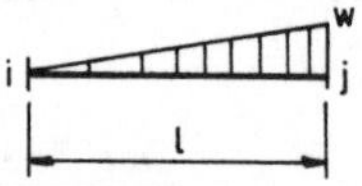 (triangular load w, length l)	$\bar{M}_i = wl^2/30$ $M_j = wl^2/20$
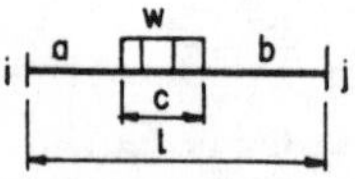 (partial uniform load w over length c, with a, b)	$\bar{M}_i = \dfrac{wc}{12\,l^2}\left[12ab(b+c)+6b^2c+4c^2(a+b)+c^3\right]$ $\bar{M}_j = \dfrac{wc}{12\,l^2}\left[12ab(a+c)+6a^2c+4c^2(a+b)+c^3\right]$
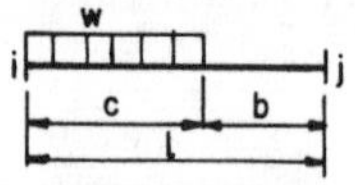 (uniform load w over length c from i, with b)	$\bar{M}_i = \dfrac{wc^2}{12\,l^2}(6b^2+4bc+c^2)$ $\bar{M}_j = \dfrac{wc^3}{12\,l^2}(4b+c)$
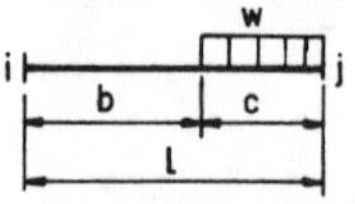 (uniform load w over length c from j, with b)	$\bar{M}_i = \dfrac{wc^3}{12\,l^2}(4a+c)$ $\bar{M}_j = \dfrac{wc^2}{12\,l^2}(6a^2+4ac+c^2)$
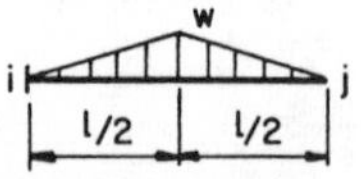 (triangular peak load w, $l/2$ and $l/2$)	$\bar{M}_i = \bar{M}_j = \dfrac{5wl^2}{96}$
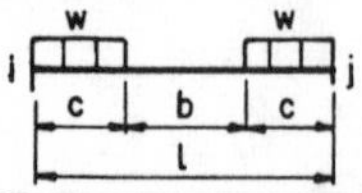 (two loads w over lengths c, b, c)	$\bar{M}_i = \bar{M}_j = \dfrac{wc^2}{6l}(3b+4c)$
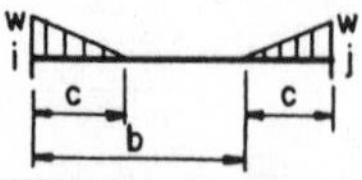 (loads w over lengths c, b, with c)	$\bar{M}_i = \bar{M}_j = \dfrac{wc^2}{12l}(2b+c)$
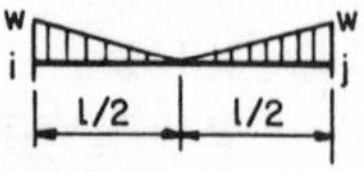 (load w, $l/2$ and $l/2$)	$\bar{M}_i = \bar{M}_j = \dfrac{wl^2}{32}$

Distribute Unbalanced Moments and Carry-over Moments

The sequence of distributions is first made at all joints as though all joints were simultaneously released. When the distribution is made, the carry-over moments occur simultaneously, then another complete distribution is made, and so forth.

Draw Shear and Moment Diagrams

The usual sign conventions for drawing the diagrams are that positive moment causes tension in the bottom fibers, and upward shear to the left is positive shear. Drawing shear and moment diagrams is a good way to check calculations and to obtain an overall picture of the stress condition in the beam.

Figure 3-2 shows a continuous beam. The fixed-end moments are calculated as follows:

$$\overline{M}_{12} = 10(7 \times 64 + 4 \times 121)/15^2 = +41.42 \text{ kip-ft}$$

$$\overline{M}_{21} = -10(11 \times 16 + 8 \times 49)/15^2 = -25.24 \text{ kip-ft}$$

$$\overline{M}_{23} = 3 \times 15^2/12 = +56.25 \text{ kip-ft}$$

$$\overline{M}_{32} = -3 \times 15^2/12 = -56.25 \text{ kip-ft}$$

Stiffness factor $K_{12} = I/15$

$$K_{23} = I/15$$

$$\Sigma K = 2I/15$$

Distribution factor $DF_{21} = K_{12}/\Sigma K = 0.5$

$$DF_{23} = K_{23}/\Sigma K = 0.5$$

The moment distribution and shear and moment diagrams are shown in Figure 3-2.

DATA FOR CONTINUOUS BEAMS

Equations and data for the end reactions, end moments, vertical shears, and the maximum positive bending moments of continuous beams are provided in this section. The beams are supported and loaded in various ways. The sign conventions for shearing force and bending moment are indicated in Figure 3-3. The forces shown are acting on an element cut out from a beam between two adjacent sections. The positive moment makes the top concave and the bottom convex.

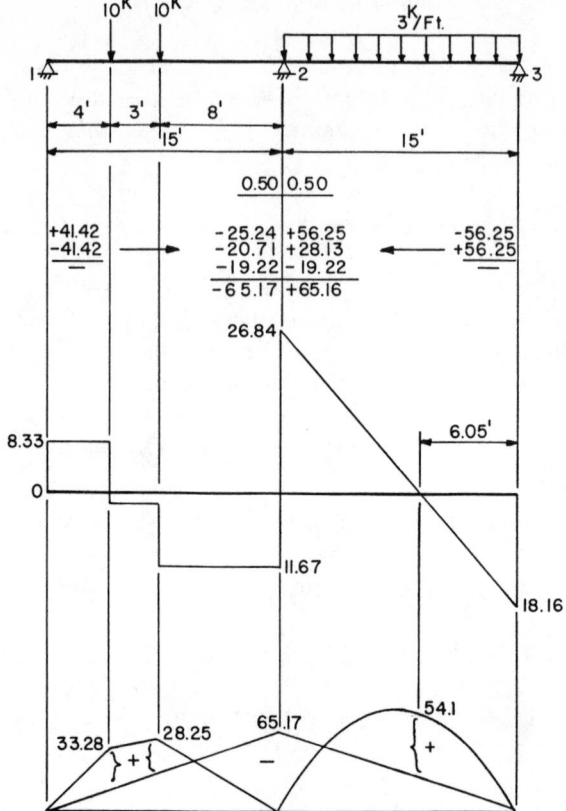

Figure 3-2. A continuous beam.

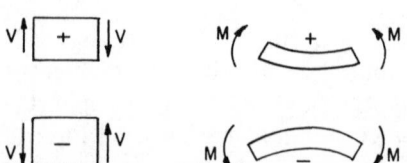

Figure 3-3. Sign conventions for shearing force and bending moment.

The following notations are used in this section:

w = unit load
P = concentrated load
R_i = reaction at joint i
M_i = moment at joint i
G_i = coefficient of reaction at joint i
K_i = coefficient of moment at joint i
K_p = coefficient of moment for M_p
M_p = positive moment or moment at loading point
β = span coefficient

The charts at the end of this chapter show reactions, shears, and moments of 27 cases of continuous beams. By superposition, the data can be applied to hundreds of loading cases, including combinations of uniformly distributed and concentrated loads.

Example 3-1

Calculate the reactions and moments of a continuous beam that is loaded as shown in Figure 3-4.

From Case 1,

$R_1 = 0.0625\ w\ell = .0625 \times 3 \times 15 = -2.8125$ kips
$R_2 = 0.625\ w\ell = .625 \times 3 \times 15 = 28.125$ kips
$R_3 = 0.4375\ w\ell = .4375 \times 3 \times 15 = 19.6875$ kips
$M_2 = -0.0625\ w\ell^2 = -.0625 \times 3 \times 15^2 = -42.1875$ kip-ft

From Case 3,

$a/\ell = 4/15 = 0.2667$
$G_1 = 2.687 \qquad G_2 = 0.7794$
$G_3 = 0.246 \qquad K_2 = 0.246$
$R_1 = PG_1/4 = 10 \times 2.685/4 = 6.7125$ kips
$R_2 = 10 \times 0.7794/2 = 3.897$ kips
$R_3 = -10 \times 0.246/4 = -0.615$ kips
$M_2 = P\ell\ K_2/4 = 10 \times 15 \times 0.246/4 = -9.225$ kip-ft

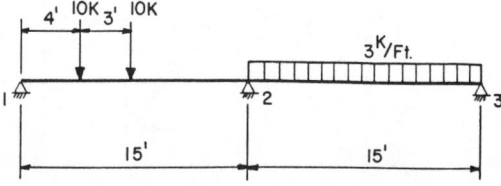

Figure 3-4. Continuous beam for Example 3-1.

From Case 3,

$a/\ell = 7/15 = 0.4667$
$G_1 = 1.770 \qquad G_2 = 1.2954$
$G_3 = 0.362 \qquad K_2 = 0.362$
$R_1 = 10 \times 1.77/4 = 4.425$ kips
$R_2 = 10 \times 1.2954/2 = 6.477$ kips
$R_3 = -10 \times 0.362/4 = -0.905$ kips
$M_2 = 10 \times 15 \times 0.362/4 = -13.575$ kip-ft
$\Sigma R_1 = -2.8125 + 6.7125 + 4.425 = 8.325$ kips
$\Sigma R_2 = 28.125 + 3.897 + 6.477 = 38.497$ kips

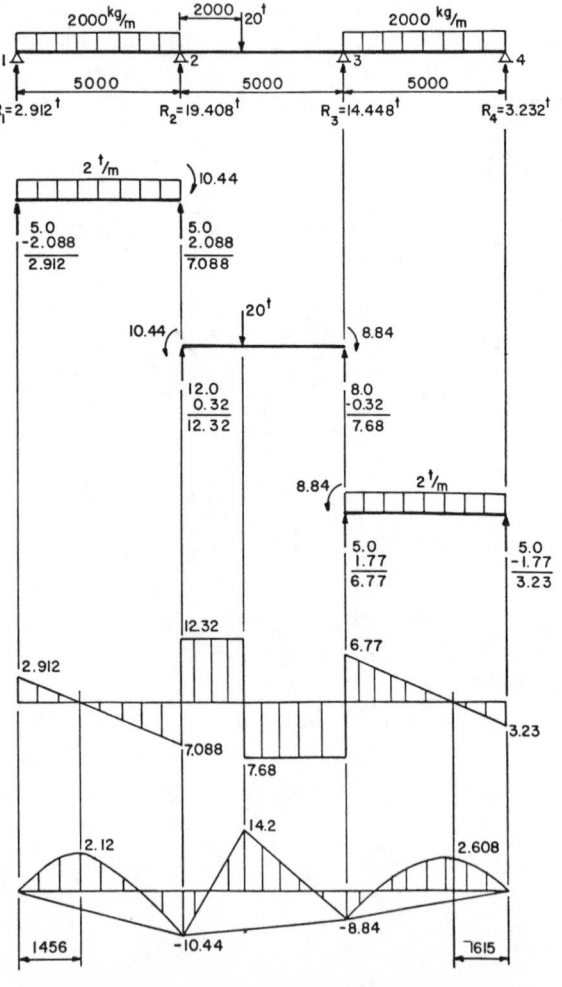

Figure 3-5. Continuous beam for Example 3-2.

$\Sigma R_3 = 19.6875 - 0.615 - 0.905 = 18.168$ kips
$\Sigma M_2 = -42.188 - 9.225 - 13.575 = -64.988$ kip-ft
$M_p = 8.325 \times 4 = 33.3$ kip-ft

$R_3 - 3x = 0 \quad x = 18.168/3 = 6.056$ ft

$M_p = 18.168 \times 6.056 - 3 \times 6.056^2/2 = 55.012$ kip-ft

Example 3-2

Calculate the reactions and moments of a continuous beam that is loaded as shown in Figure 3-5.

From Case 13,

$R_1 = 0.45 \ w\ell = 0.45 \times 2 \times 5 = 4.5$ t
$R_2 = 0.55 \ w\ell = 0.55 \times 2 \times 5 = 5.5$ t
$R_3 = 0.55 \ w\ell = 5.5$ t
$R_4 = 0.45 \ w\ell = 4.5$ t
$M_2 = -0.05 \ w\ell^2 = -0.05 \times 2 \times 25 = -2.5$ t-m
$M_3 = -0.05 \ w\ell^2 = -2.5$ t-m

From Case 16,

$a/\ell = 2/5 = 0.4$
$R_1 = -PG_1 = -20 \times 0.0794 = -1.588$ t
$R_2 = PG_2 = -20 \times 0.6954 = 13.908$ t
$R_3 = PG_3 = -20 \times 0.4474 = 8.948$ t
$R_4 = PG_4 = -20 \times 0.0634 = -1.268$ t
$M_2 = -P\ell \, K_2 = -20 \times 5 \times 0.0794 = -7.94$ t-m
$M_3 = -P\ell \, K_3 = -20 \times 5 \times 0.0634 = -6.34$ t-m
$\Sigma R_1 = 4.5 - 1.588 = 2.912$ t
$\Sigma R_2 = 5.5 + 13.908 = 19.408$ t
$\Sigma R_3 = 5.5 + 8.948 = 14.448$ t
$\Sigma R_4 = 4.5 - 1.268 = 3.232$ t
$\Sigma M_2 = -2.5 - 7.94 = -10.44$ t-m
$\Sigma M_3 = -2.5 - 6.34 = -8.84$ t-m

REFERENCES

1. Wang, P. C., *Numerical and Matrix Methods in Structural Mechanics,* John Wiley & Sons, Inc., New York, 1966.
2. McCormac, J. C. *Structural Analysis,* 2nd Edition, International Textbook Company, Scranton, Pennsylvania, 1967.
3. Norris, C. H. and Wilbur, J. B., *Elementary Structural Analysis,* 2nd Edition, McGraw-Hill Book Company, 1960.

Cases for Chapter 3

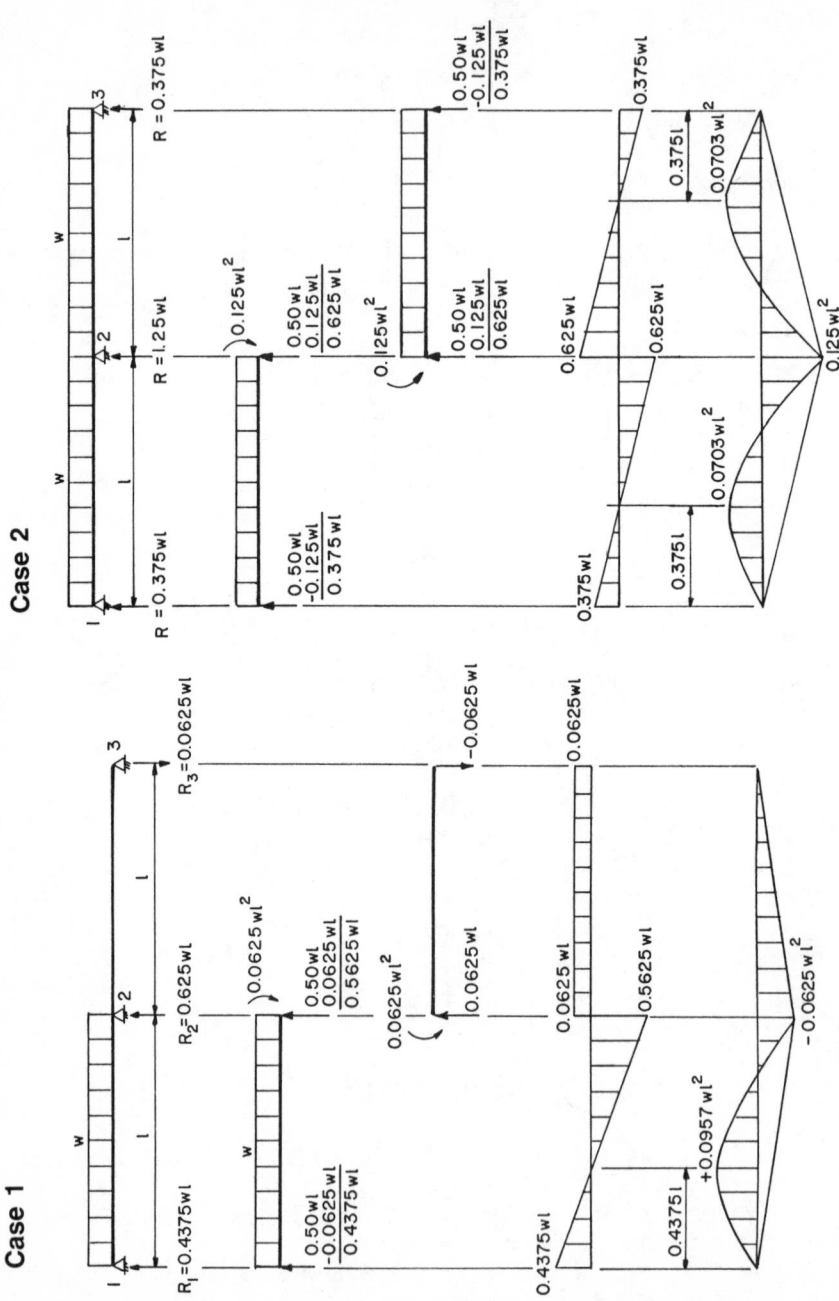

Case 2

Case 1

Case 3

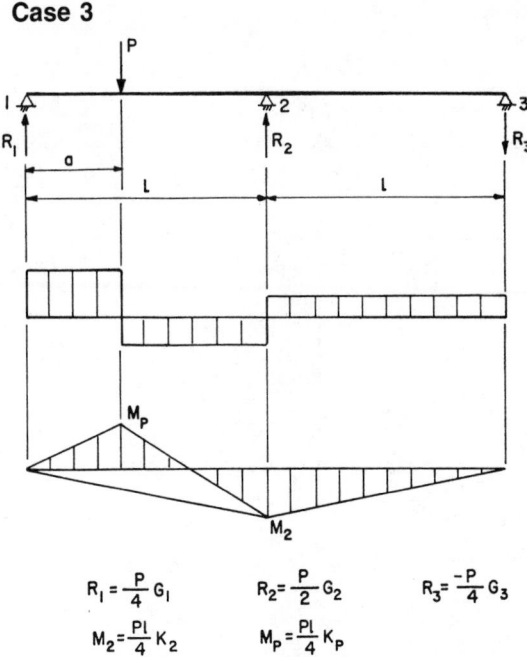

$$R_1 = \frac{P}{4} G_1 \qquad R_2 = \frac{P}{2} G_2 \qquad R_3 = \frac{-P}{4} G_3$$

$$M_2 = \frac{Pl}{4} K_2 \qquad M_P = \frac{Pl}{4} K_P$$

a/ℓ	0.10	0.20	0.30	0.40	0.50	0.60	0.70	0.80	0.90
G_1	3.501	3.008	2.527	2.064	1.625	1.216	0.843	0.512	0.229
G_2	0.299	0.592	0.873	1.136	1.375	1.584	1.757	1.888	1.971
G_3	0.099	0.192	0.273	0.336	0.375	0.384	0.357	0.288	0.171
K_2	0.099	0.192	0.273	0.336	0.375	0.384	0.357	0.288	0.171
K_p	0.350	0.602	0.758	0.826	0.813	0.730	0.590	0.410	0.206

Case 4

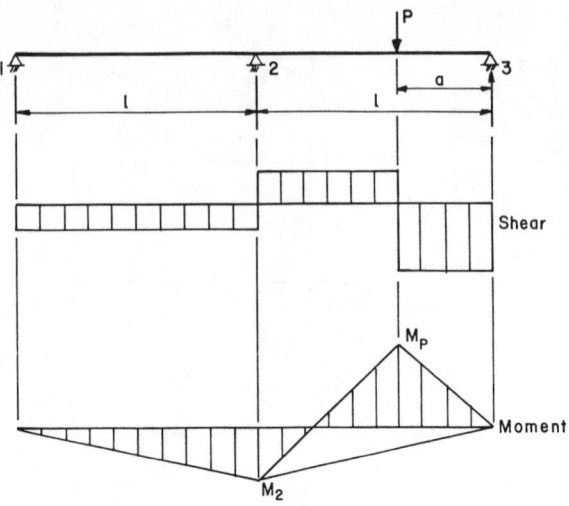

$$R_1 = -\frac{P}{4}G_1 \qquad R_2 = \frac{P}{2}G_2 \qquad R_3 = \frac{P}{4}G_3$$

$$M_2 = \frac{Pl}{4}K_2 \qquad M_P = \frac{Pl}{4}K_P$$

a/ℓ	0.10	0.20	0.30	0.40	0.50	0.60	0.70	0.80	0.90
G_1	0.099	0.192	0.273	0.336	0.375	0.384	0.357	0.288	0.171
G_2	0.299	0.592	0.873	1.136	1.375	1.584	1.757	1.888	1.971
G_3	3.501	3.008	2.527	2.064	1.625	1.216	0.843	0.512	0.229
K_2	0.099	0.192	0.273	0.336	0.375	0.384	0.357	0.288	0.171
K_P	0.350	0.602	0.758	0.826	0.813	0.730	0.590	0.410	0.206

Case 5

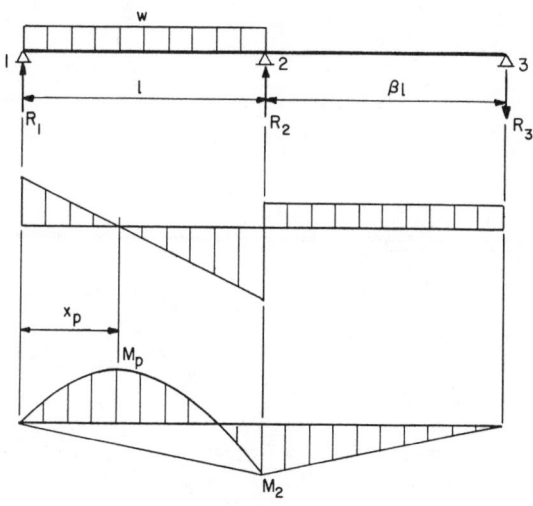

$$R_1 = wl\,G_1 \qquad R_2 = wl\,G_2 \qquad R_3 = -wl\,G_3$$

$$M_2 = wl^2 K_2 \qquad M_p = \frac{wl^2}{2} K_p \qquad x_p = G_1\, l$$

β	G_1	G_2	G_3	K_2	K_p
0.50	0.4166	0.6667	0.0834	0.0834	0.1736
0.60	0.4219	0.6563	0.0781	0.0781	0.1780
0.70	0.4265	0.6471	0.0735	0.0735	0.1819
0.80	0.4305	0.6389	0.0695	0.0695	0.1853
0.90	0.4342	0.6316	0.0658	0.0658	0.1885
1.00	0.4375	0.6250	0.0625	0.0625	0.1914
1.10	0.4405	0.6191	0.0595	0.0595	0.1940
1.20	0.4432	0.6136	0.0568	0.0568	0.1964
1.30	0.4456	0.6087	0.0544	0.0544	0.1986
1.40	0.4479	0.6042	0.0521	0.0521	0.2006
1.50	0.4500	0.6000	0.0500	0.0500	0.2025
1.60	0.4519	0.5962	0.0481	0.0481	0.2042
1.70	0.4532	0.5926	0.0468	0.0468	0.2054
1.80	0.4554	0.5893	0.0446	0.0446	0.2074
1.90	0.4569	0.5862	0.0431	0.0431	0.2088
2.00	0.4583	0.5833	0.0417	0.0417	0.2100

Case 6

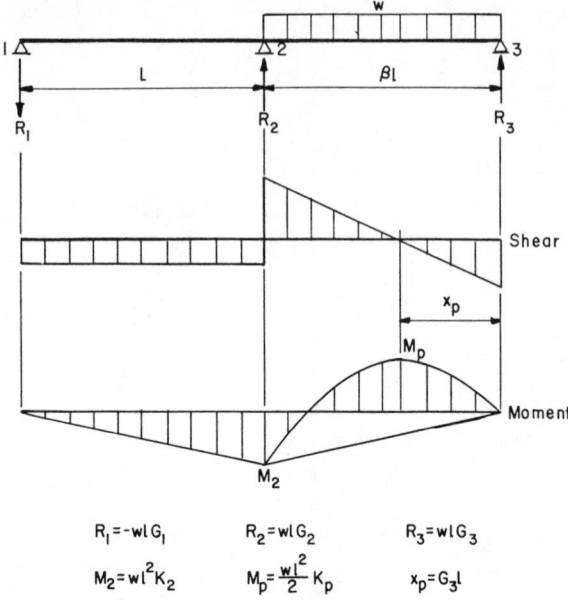

$$R_1 = -wl\,G_1 \qquad R_2 = wl\,G_2 \qquad R_3 = wl\,G_3$$

$$M_2 = wl^2 K_2 \qquad M_p = \frac{wl^2}{2} K_p \qquad x_p = G_3 l$$

β	G_1	G_2	G_3	K_2	K_p
0.50	0.0208	0.5417	0.4792	0.0208	0.2296
0.60	0.0281	0.5563	0.4719	0.0281	0.2227
0.70	0.0360	0.5721	0.4637	0.0360	0.2150
0.80	0.0444	0.5889	0.4556	0.0444	0.2076
0.90	0.0533	0.6066	0.4467	0.0533	0.1995
1.00	0.0625	0.6250	0.4375	0.0625	0.1914
1.10	0.0720	0.6441	0.4280	0.0720	0.1832
1.20	0.0818	0.6634	0.4182	0.0818	0.1749
1.30	0.0919	0.6837	0.4082	0.0919	0.1666
1.40	0.1021	0.7042	0.3979	0.1021	0.1583
1.50	0.1125	0.7250	0.3875	0.1125	0.1502
1.60	0.1231	0.7462	0.3769	0.1231	0.1421
1.70	0.1338	0.7676	0.3662	0.1338	0.1341
1.80	0.1446	0.7893	0.3554	0.1446	0.1263
1.90	0.1556	0.8112	0.3444	0.1556	0.1186
2.00	0.1667	0.8333	0.3333	0.1667	0.1111

Case 7

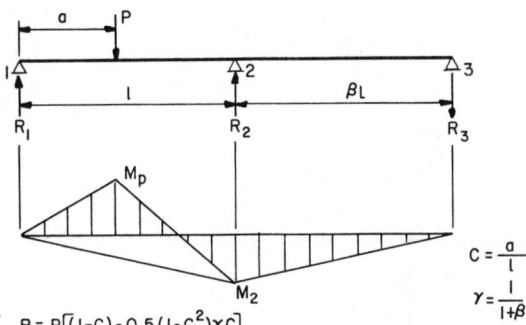

$$C = \frac{a}{l}$$

$$\gamma = \frac{1}{1+\beta}$$

check this

$$R_1 = P\left[(1-C) - 0.5(1-C^2)\gamma C\right]$$

$$R_2 = P\left[C + (1+C^2)\gamma C\right]$$

$$R_3 = -0.5P(1-C^2)\gamma C$$

$$M_2 = -0.5Pl(1-C^2)\gamma C$$

$$M_p = Pl\left[C(1-C) - 0.5(1-C^2)\gamma C^2\right]$$

Case 8

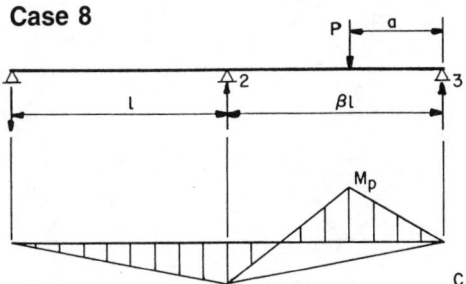

$$C = \frac{a}{\beta l}$$

$$a = \frac{\beta}{1+\beta}$$

$$R_1 = 0.5PC(1-C^2)a\beta$$

$$R_2 = PC\left[1 + 0.5(1-C^2)(1+\beta)a\right]$$

$$R_3 = P\left[(1-C) - 0.5C(1-C^2)a\right]$$

$$M_2 = -0.5P\beta lC(1-C^2)a$$

$$M_p = P\beta lC\left[(1-C) - 0.5C(1-C^2)a\right]$$

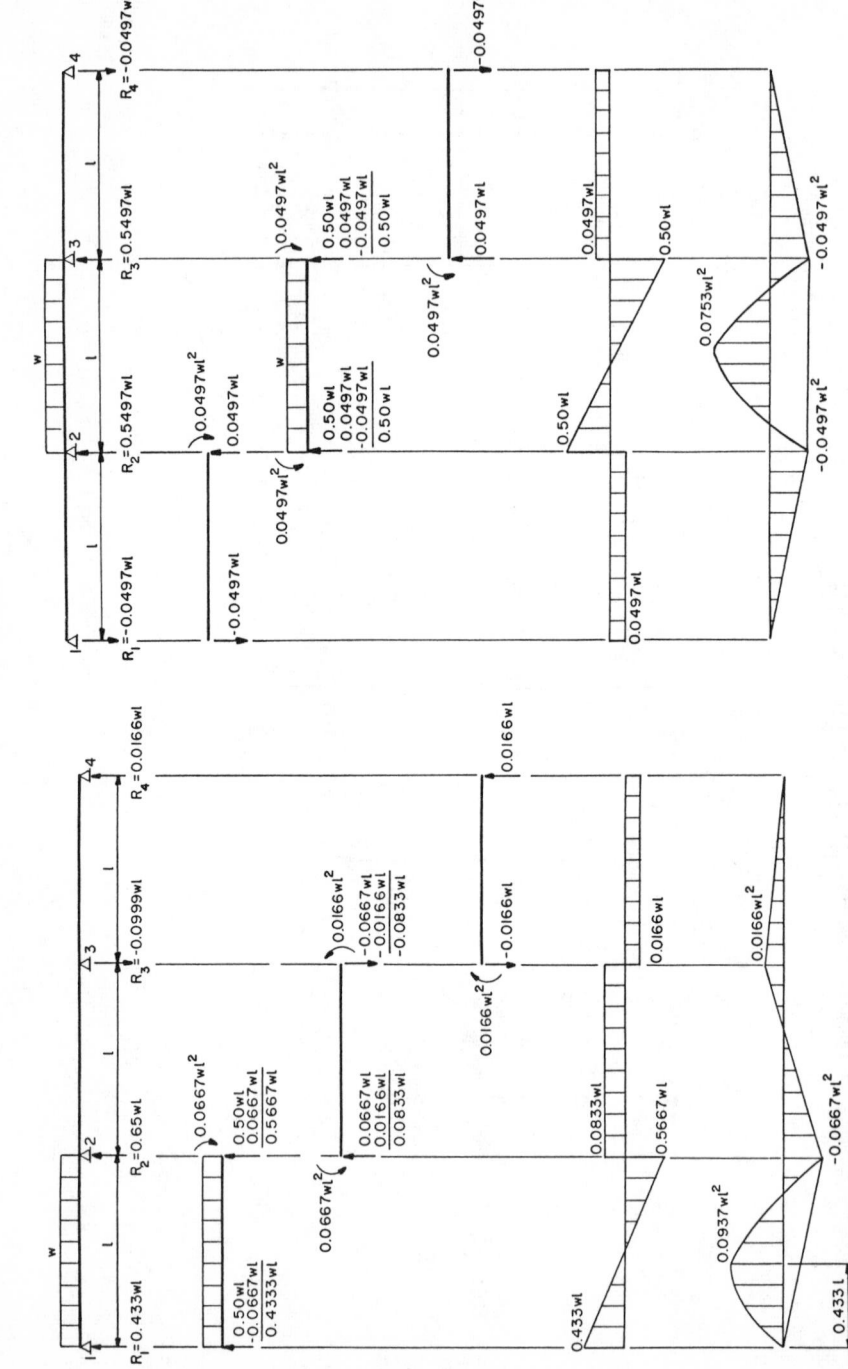

Case 9

Case 10

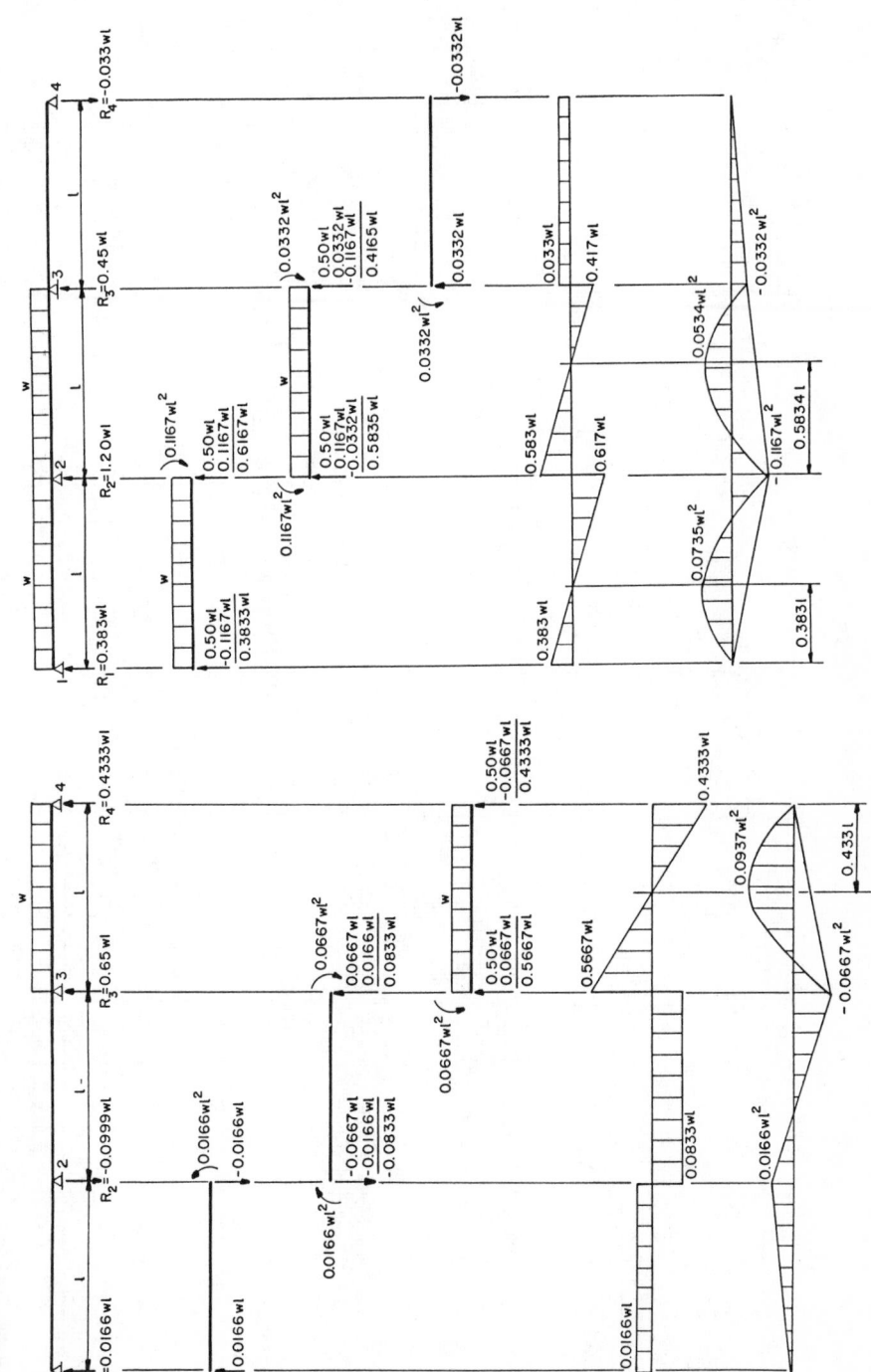

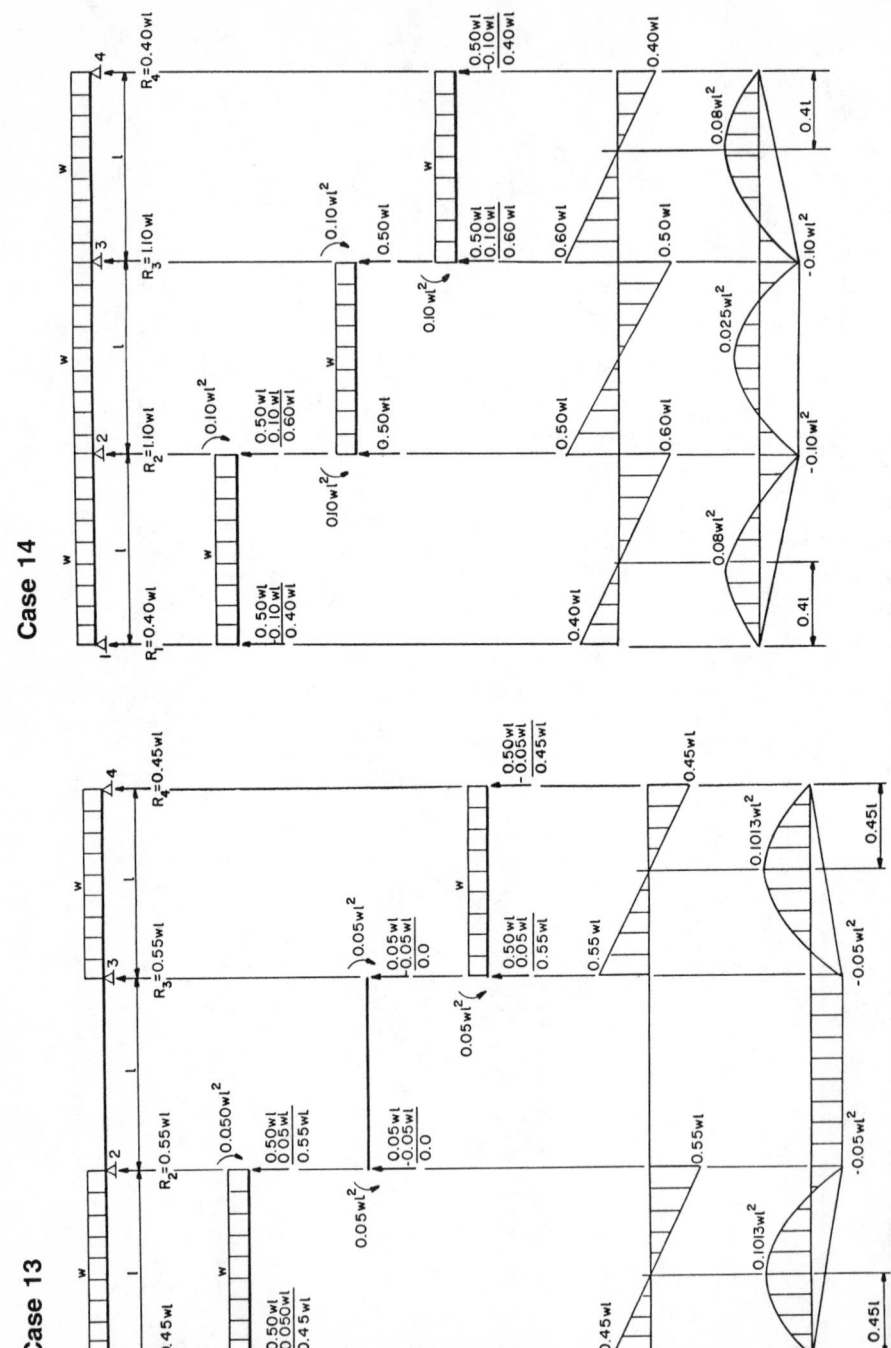

Case 14

Case 13

Case 15

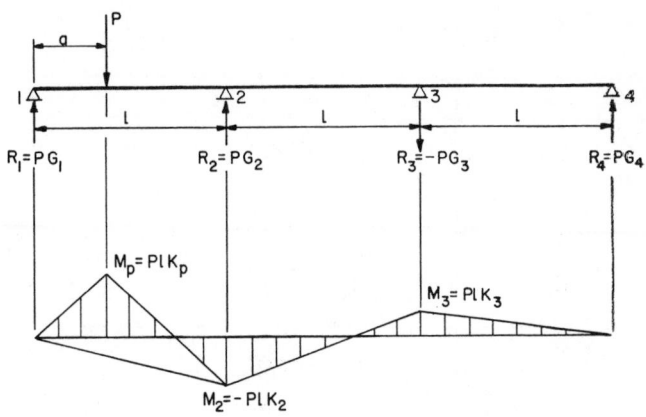

a/ℓ	0.10	0.20	0.30	0.40	0.50	0.60	0.70	0.80	0.90
G_1	0.8736	0.7488	0.6272	0.5104	0.4000	0.2976	0.2048	0.1232	0.0544
G_2	0.1594	0.3152	0.4637	0.6015	0.7249	0.8303	0.9141	0.9727	1.0026
G_3	0.0395	0.0367	0.1090	0.1341	0.1497	0.1533	0.1425	0.1150	0.0683
G_4	0.0066	0.0127	0.0181	0.0223	0.0248	0.0254	0.0237	0.0191	0.0113
K_p	0.0874	0.1498	0.1882	0.2042	0.2000	0.1786	0.1434	0.0986	0.0490
K_2	0.0264	0.0512	0.0728	0.0896	0.1000	0.1024	0.0952	0.0768	0.0456
K_3	0.0066	0.0127	0.0181	0.0223	0.0248	0.0254	0.0237	0.0191	0.0113

Case 16

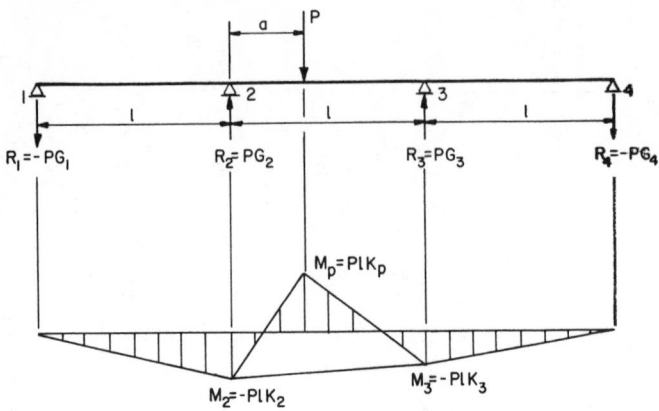

a/ℓ	0.10	0.20	0.30	0.40	0.50	0.60	0.70	0.80	0.90
G_1	0.0388	0.0636	0.0765	0.0794	0.0746	0.0634	0.0485	0.0316	0.0148
G_2	0.9628	0.8956	0.8045	0.6954	0.5746	0.4474	0.3205	0.1996	0.0908
G_3	0.0908	0.1996	0.3205	0.4474	0.5746	0.6954	0.8045	0.8956	0.9628
G_4	0.0148	0.0316	0.0485	0.0634	0.0744	0.0794	0.0765	0.0636	0.0388
K_2	0.0388	0.0636	0.0765	0.0794	0.0744	0.0634	0.0485	0.0316	0.0148
K_3	0.0148	0.0316	0.0485	0.0634	0.0744	0.0794	0.0765	0.0636	0.0388
K_p	0.0063	0.1124	0.1475	0.1686	0.1756	0.1686	0.1475	0.1124	0.0063

Case 17

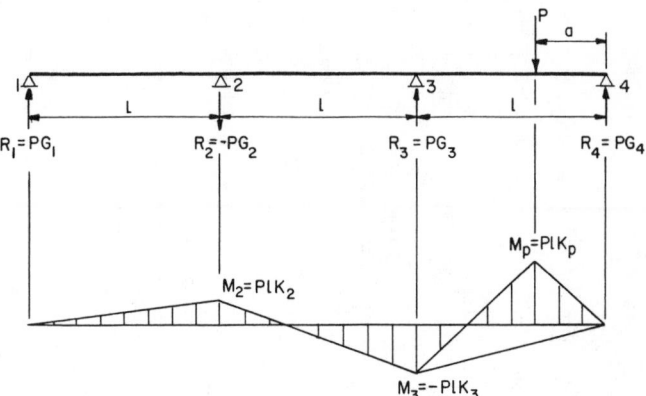

a/ℓ	0.10	0.20	0.30	0.40	0.50	0.60	0.70	0.80	0.90
G_1	0.0066	0.0127	0.0181	0.0223	0.0248	0.0254	0.0237	0.0191	0.0113
G_2	0.0395	0.0367	0.1090	0.1341	0.1497	0.1533	0.1425	0.1150	0.0683
G_3	0.1594	0.3152	0.4637	0.6015	0.7249	0.8303	0.9141	0.9727	1.0026
G_4	0.8736	0.7488	0.6272	0.5104	0.4000	0.2976	0.2048	0.1232	0.0544
K_p	0.0874	0.1498	0.1882	0.2042	0.2000	0.1786	0.1434	0.0986	0.0490
K_2	0.0066	0.0127	0.0181	0.0223	0.0248	0.0254	0.0237	0.0191	0.0113
K_3	0.0264	0.0512	0.0728	0.0896	0.1000	0.1024	0.0952	0.0768	0.0456

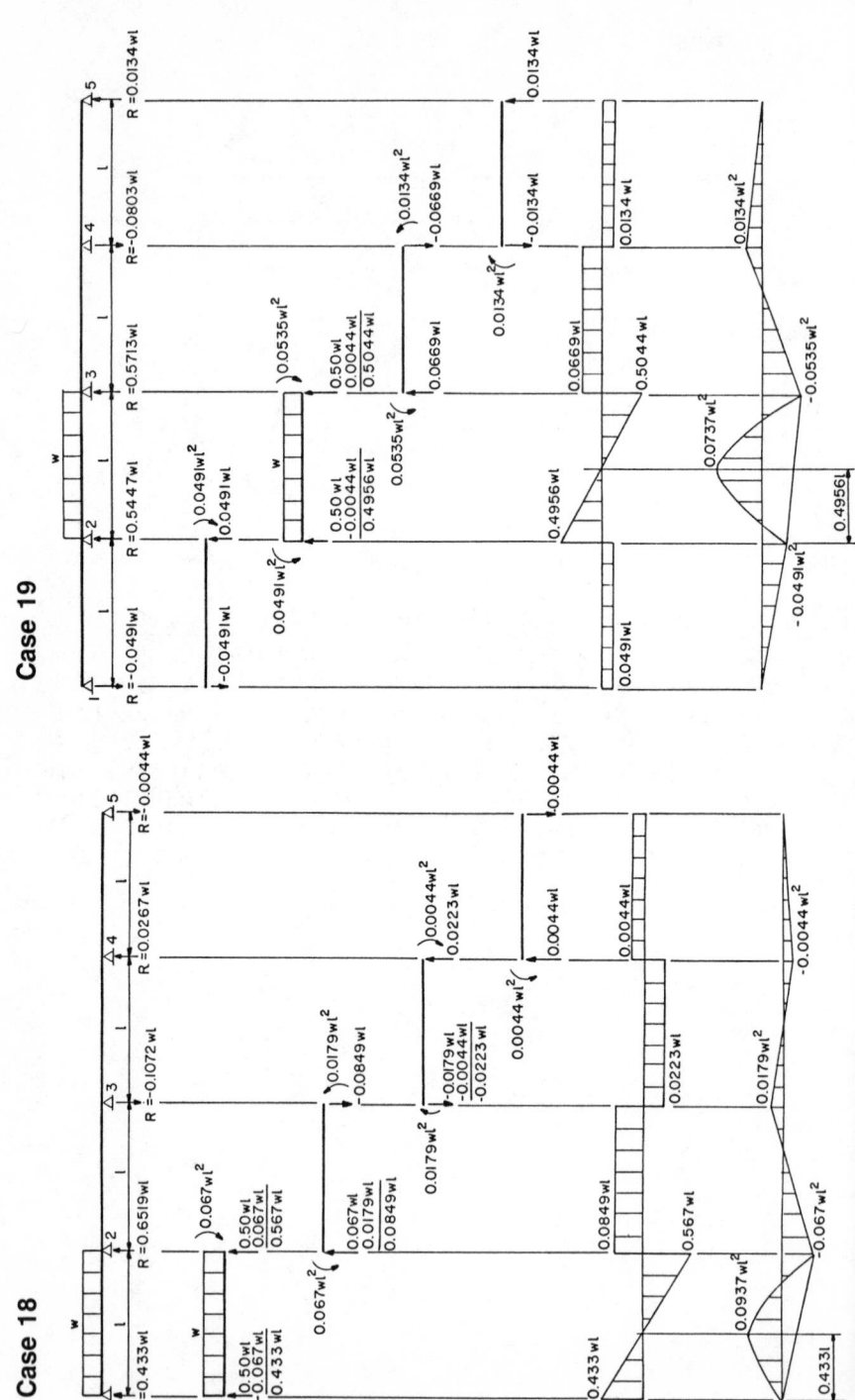

Case 18

Case 19

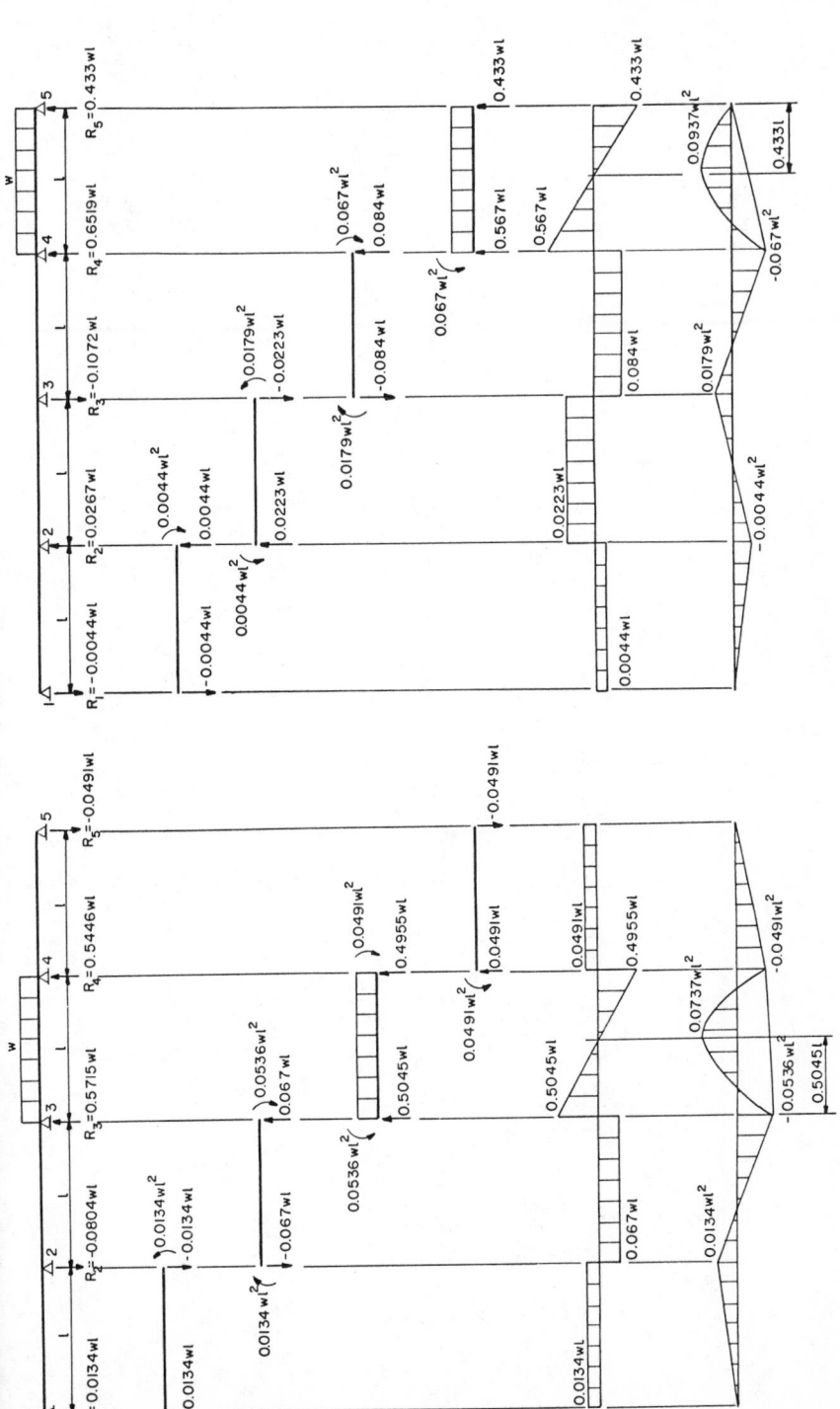

Case 22

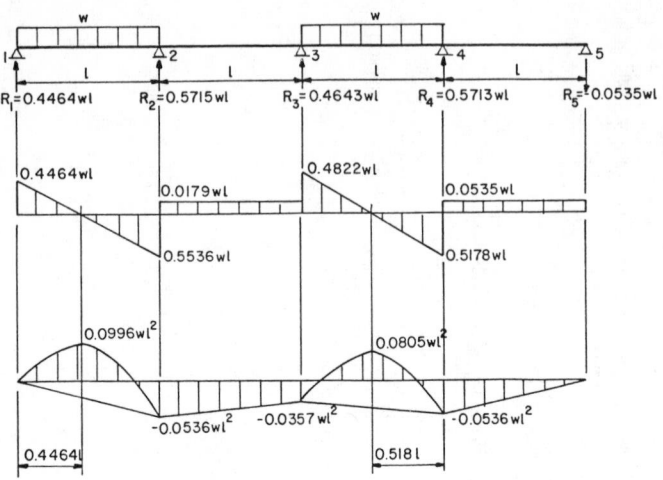

Case 23

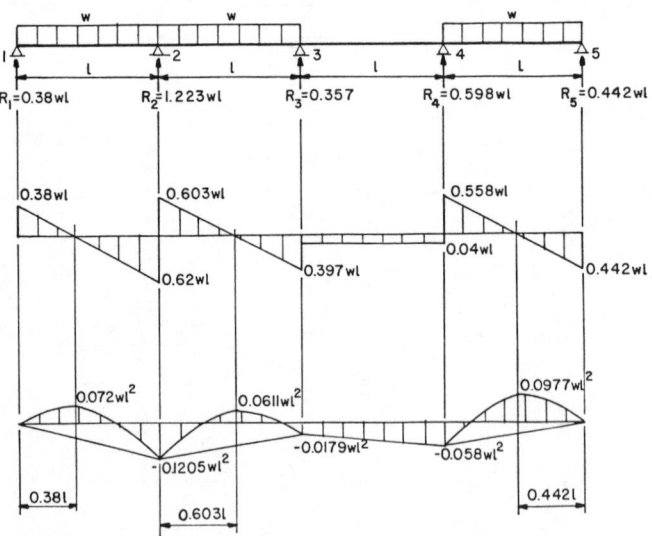

Case 24

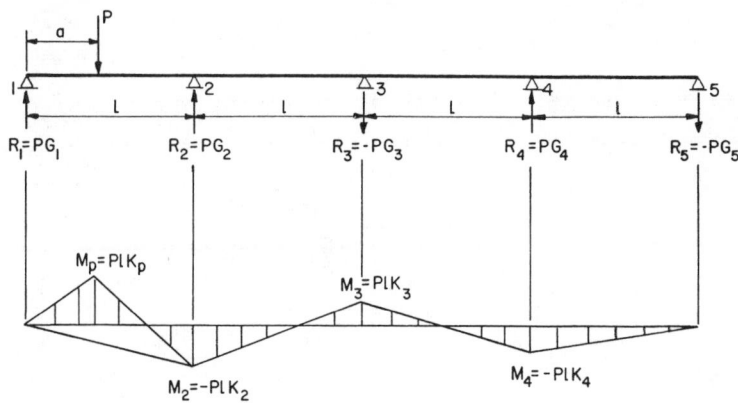

a/ℓ	0.10	0.20	0.30	0.40	0.50	0.60	0.70	0.80	0.90
G_1	0.8735	0.7486	0.6269	0.5100	0.3995	0.2971	0.2044	0.1228	0.0542
G_2	0.1601	0.3166	0.4658	0.6040	0.7277	0.8332	0.9168	0.9749	0.1004
G_3	0.0424	0.0823	0.1170	0.1440	0.1607	0.1646	0.1530	0.1235	0.0733
G_4	0.0106	0.0206	0.0292	0.0360	0.0401	0.0411	0.0382	0.0308	0.0183
G_5	0.0018	0.0034	0.0049	0.0060	0.0067	0.0068	0.0064	0.0051	0.0030
K_2	0.0265	0.0514	0.0731	0.0900	0.1005	0.1029	0.0956	0.0772	0.0458
K_3	0.0071	0.0137	0.0195	0.0240	0.0268	0.0274	0.0255	0.0206	0.0122
K_4	0.0018	0.0034	0.0049	0.0060	0.0067	0.0068	0.0064	0.0051	0.0030
K_p	0.0874	0.1497	0.1881	0.2040	0.1998	0.1782	0.1431	0.0983	0.0488

Case 25

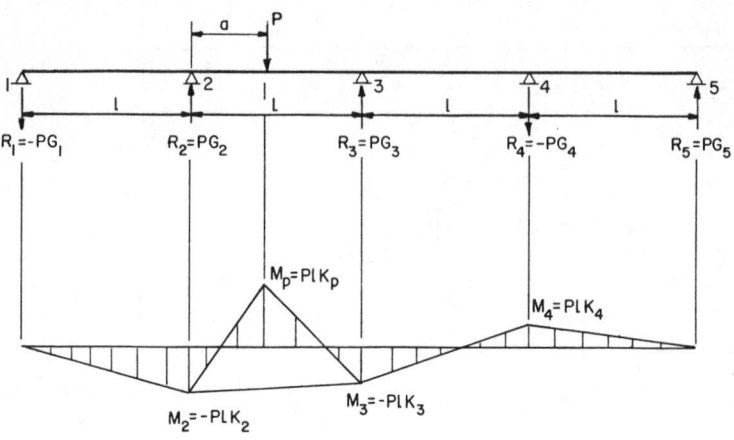

a/ℓ	0.10	0.20	0.30	0.40	0.50	0.60	0.70	0.80	0.90
G_1	0.0387	0.0634	0.0761	0.0788	0.0736	0.0625	0.0476	0.0308	0.0143
G_2	0.9614	0.8925	0.7997	0.6890	0.5669	0.4393	0.3127	0.1931	0.0868
G_3	0.0974	0.2137	0.3420	0.4754	0.6071	0.7303	0.8380	0.9234	0.9797
G_4	0.0241	0.0514	0.0787	0.1028	0.1204	0.1285	0.1236	0.1028	0.0626
G_5	0.0040	0.0086	0.0131	0.0171	0.0200	0.0214	0.0206	0.0171	0.0104
K_2	0.0387	0.0634	0.0761	0.0788	0.0736	0.0625	0.0476	0.0308	0.0143
K_3	0.0161	0.0343	0.0525	0.0686	0.0804	0.0857	0.0825	0.0686	0.0418
K_4	0.0040	0.0086	0.0131	0.0171	0.0200	0.0214	0.0206	0.0171	0.0104
K_p	0.0536	0.1024	0.1410	0.1653	0.1730	0.1636	0.1380	0.0990	0.0510

Case 26

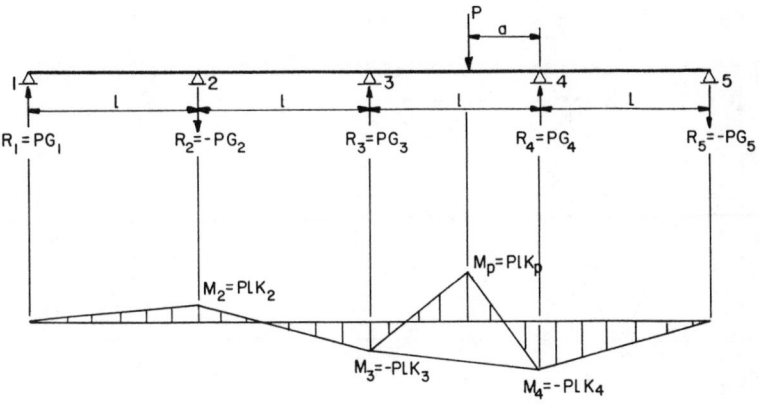

a/ℓ	0.10	0.20	0.30	0.40	0.50	0.60	0.70	0.80	0.90
G_1	0.0040	0.0086	0.0131	0.0171	0.0200	0.0214	0.0206	0.0171	0.0104
G_2	0.0241	0.0514	0.0787	0.1028	0.1204	0.1285	0.1236	0.1028	0.0626
G_3	0.0974	0.2137	0.3420	0.4754	0.6071	0.7303	0.8380	0.9234	0.9797
G_4	0.9614	0.8925	0.7997	0.6890	0.5669	0.4393	0.3127	0.1931	0.0868
G_5	0.0387	0.0634	0.0761	0.0788	0.0736	0.0625	0.0476	0.0308	0.0143
K_2	0.0040	0.0086	0.0131	0.0171	0.0200	0.0214	0.0206	0.0171	0.0104
K_3	0.0161	0.0343	0.0525	0.0686	0.0804	0.0857	0.0825	0.0686	0.0418
K_4	0.0387	0.0634	0.0761	0.0788	0.0736	0.0625	0.0476	0.0308	0.0143
K_p	0.0536	0.1024	0.1410	0.1653	0.1730	0.1636	0.1380	0.0990	0.0510

Case 27

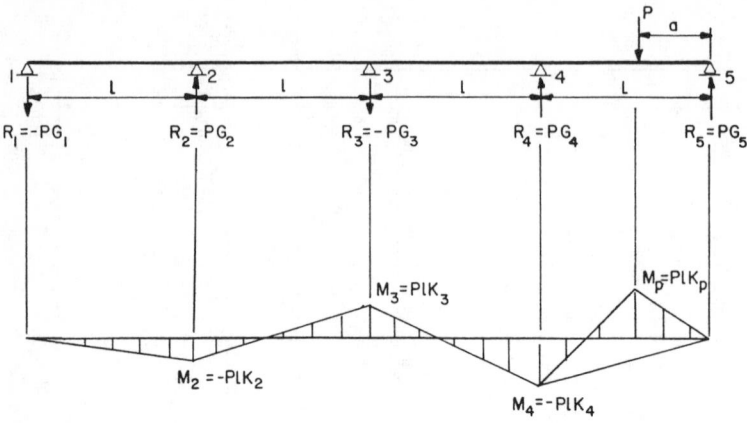

a/ℓ	0.10	0.20	0.30	0.40	0.50	0.60	0.70	0.80	0.90
G_1	0.0018	0.0034	0.0049	0.0060	0.0067	0.0068	0.0064	0.0051	0.0030
G_2	0.0106	0.0206	0.0292	0.0360	0.0401	0.0411	0.0382	0.0308	0.0183
G_3	0.0424	0.0823	0.1170	0.1440	0.1607	0.1646	0.1530	0.1235	0.0733
G_4	0.1601	0.3166	0.4658	0.6040	0.7277	0.8332	0.9168	0.9749	0.1004
G_5	0.8735	0.7486	0.6269	0.5100	0.3995	0.2971	0.2044	0.1228	0.0542
K_2	0.0018	0.0034	0.0049	0.0060	0.0067	0.0068	0.0064	0.0051	0.0030
K_3	0.0071	0.0137	0.0195	0.0240	0.0268	0.0274	0.0255	0.0206	0.0122
K_4	0.0265	0.0514	0.0731	0.0900	0.1005	0.1029	0.0956	0.0772	0.0458
K_p	0.0874	0.1497	0.1881	0.2040	0.1998	0.1782	0.1431	0.0983	0.0488

Chapter 4

Rigid Frames

BASIC CONCEPTS

When a frame is loaded, the joints of the frame undergo rotations and displacements and reestablish equilibrium. The end moments at the ends i and j of a member ij are designated by M_{ij} and M_{ji}, respectively. An end moment is positive if it acts clockwise. The rotation angles at ends i and j are designated by θ_i and θ_j. The rotation is positive if it rotates clockwise.

A member as shown in Figure 4-1a is rotated θ_i at joint i. The moment required to rotate θ_i is 4EK θ_i when j is fixed. Note that when $4EK\theta_i$ is applied at joint i, a moment of $2EK\theta_i$ will be induced at joint j. When joint j is pinned, the moment required to rotate θ_i is $3EK\theta_i$, as shown in Figure 4-1b. When one end of a member is settled δ, it induces a moment of $6EK\delta/\ell$ at both ends, as shown in Figure 4-2.

A structure will deform under a given loading, but we imagine that the joints are held by external forces and moments so that rotations and displacements are prevented. The fictitious external forces and moments that hold the joints are called *restraint forces* and *restraint moments*. After all fixed-end moments have been determined, the restraint moments of each joint may be calculated. The restraint moment $\overline{M}_i$ at a joint i is equal to the sum of the fixed-end moments of all members connecting at joint i.

$$\overline{M}_i = \Sigma \, \overline{M}_{ij}$$

Note that the restraint moment acts at the joint, while end moments act at the member ends.

KANI METHOD [1]

When the member ij of a loaded structure deforms, the final deformation is due to a superposition of the following four steps of deformation:

1. The member deforms under the given loading (fixed-end moment).
2. Joint i undergoes a rotation θ_i, while joint j is fixed.

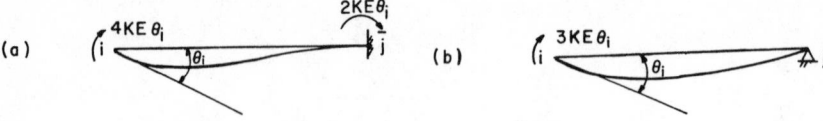

Figure 4-1. Moment induced by the rotation of a joint.

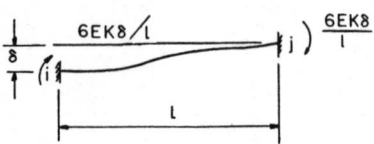

Figure 4-2. Moment induced by the settlement of a joint.

3. Joint j undergoes a rotation θ_j, while joint i is fixed.
4. The ends of the member have a relative displacement of δ, whereby the ends are not subjected to any additional rotation.

The end moment of the member is composed of four contributions: the fixed-end moment, the rotation of its own end, the rotation of the other end, and the displacement of the member. This may be expressed as

$$M_{ij} = \overline{M}_{ij} + 2M'_{ij} + M'_{ji} + M''_{ij} \tag{4-1}$$

From the condition of equilibrium at joint i ($\Sigma M_{ij} = 0$), one obtains

$$\Sigma M'_{ij} = -\frac{1}{2}\left[\Sigma \overline{M}_{ij} + \Sigma (M'_{ji} + M''_{ij})\right] \tag{4-2}$$

where $\overline{M}_{ij}$ = fixed end moment at joint i

M'_{ij} = the rotation contribution at joint i ($2KE\theta_i$)

M'_{ji} = the rotation contribution at joint j ($2KE\theta_j$)

M''_{ij} = linear displacement contribution

M_{ij} = the end moment at joint i

Let H_{ij} = the shear force of a column in the nth floor of a rigid frame, as shown in Figure 4-3. The shear force can be expressed as

$$H_{ij} = -\frac{M_{ij} + M_{ji}}{h_n} \tag{4-3}$$

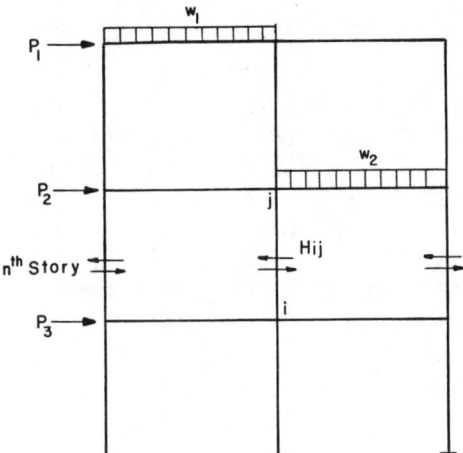

Figure 4-3. A rigid frame.

Let the total shear forces on the nth floor $= H_n$.

$$H_n = \Sigma H_{ij}$$

From Equations 4-1 and 4-3 we obtain

$$H_n = -\frac{1}{h_n} \Sigma (2M'_{ij} + M'_{ji} + M''_{ij} + 2M'_{ji} + M'_{ij} + M''_{ij})$$

$$H_n h_n = -\Sigma [3(M'_{ij} + M'_{ji}) + 2M''_{ij}]$$

The sum of the linear displacement contributions of all columns of the story n is

$$\Sigma M''_{ij} = \frac{-3}{2}\left[\frac{H_n h_n}{3} + \Sigma (M'_{ij} + M'_{ji})\right] \tag{4-4}$$

From Equations 4-2 and 4-4 we obtain

$$M'_{ij} = \alpha [\overline{M}_i + \Sigma (M'_{ji} + M''_{ij})] \tag{4-5}$$

$$M''_{ij} = \beta [\overline{M}_n + \Sigma (M'_{ij} + M'_{ji})] \tag{4-6}$$

where $\overline{M}_n = H_n h_n / 3$

$$\alpha = \frac{-1}{2} \frac{K_{ij}}{\Sigma K_{ij}}$$

$$\beta = \frac{-3}{2} \frac{K_{ij}}{\Sigma K_{ij}}$$

The rotation factor, represented by α, is obtained in such a way that at each joint the value -0.5 is distributed in proportion to the K values of the members connecting at the joint. Note that the sum of all rotation factors of a joint is -0.5. The displacement factor, β, is obtained by distributing the value -1.5 in proportion to the K values of all columns in the story. The calculations of Equations 4-5 and 4-6 are carried out one after the other. First calculate Equation 4-5 at each joint, and then calculate Equation 4-6 at each story, until all rotation and displacement contributions possess the desired accuracy. The final moments are obtained from Equation 4-1.

Example 4-1

Calculate the end moments of all joints of a two-story rigid frame that is loaded as shown in Figure 4-4.

1. Calculate fixed end moments and restraint moments.

   $w\ell^2/12 = 1 \times 12^2/12 = 12$ kip-ft

2. Calculate story shear and story moment.

   $H_1 = 3$ kips $\overline{M}_1 = H_1h_1/3 = 3 \times 10/3 = 10$ kip-ft

   $H_2 = 6$ kips $\overline{M}_2 = H_2h_2/3 = 6 \times 10/3 = 20$ kip-ft

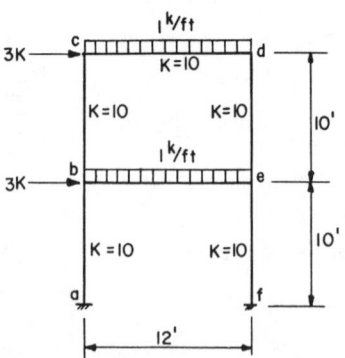

Figure 4-4. Rigid frame for Example 4-1.

3. Calculate rotation factor α and displacement factor β.

$$\alpha_{ij} = -0.5 \ K_{ij}/\Sigma K_{ij}$$

At joints c,d: $-0.5 \times 10/20 = -0.25$

At joints b,e: $-0.5 \times 10/30 = -0.1667$

$$\beta_{ij} = -1.5 \ K_{ij}/\Sigma K_{ij}$$

$$= -1.5 \times 10/20 = -0.75$$

4. Calculate the rotation contribution M_{ij}' at each joint using Equation 4-5.
5. Calculate the displacement contribution M_{ij}'' at each story using Equation 4-6.
6. Calculate the final end moment M_{ij} using Equation 4-1.

The detailed calculations of steps 4 and 5 are shown in Figure 4-5. Start from joint c, calculating the rotation contribution.

$$M_{cd}' = -0.25(-12) = 3.00 \text{ kip-ft}$$
$$M_{cb}' = -0.25(-12) = 3.00 \text{ kip-ft}$$
$$M_{dc}' = -0.25(12 + 3) = -3.75 \text{ kip-ft}$$
$$M_{de}' = -0.25(12 + 3) = -3.75 \text{ kip-ft}$$
$$M_{be}' = -0.1667(-12 + 3) = 1.50 \text{ kip-ft}$$
$$M_{bc}' = -0.1667(-12 + 3) = 1.50 \text{ kip-ft}$$
$$M_{ba}' = -0.1667(-12 + 3) = 1.50 \text{ kip-ft}$$
$$M_{eb}' = -0.1667(12 + 1.5 - 3.75) = -1.625 \text{ kip-ft}$$
$$M_{ed}' = -0.1667(12 + 1.5 - 3.75) = -1.625 \text{ kip-ft}$$
$$M_{ef}' = -0.1667(12 + 1.5 - 3.75) = -1.625 \text{ kip-ft}$$

Start from the upper story, calculating the displacement contribution.

$$-0.75 (10 + 3 + 1.5 - 3.75 - 1.625) = -6.844 \text{ kip-ft}$$
$$-0.75 (20 + 1.5 - 1.625) = -14.906 \text{ kip-ft}$$

This procedure is continued until the desired accuracy is obtained.
The final end moments are calculated using Equation 4-1 as follows:

$$M_{cd} = -12 + 2 \times 6.401 - 0.456 = 0.346 \text{ kip-ft}$$
$$M_{cb} = 0 + 2 \times 6.401 + 7.665 - 20.873 = -0.406 \text{ kip-ft}$$
$$M_{dc} = 12 + 2 (-0.456) + 6.401 = 17.498 \text{ kip-ft}$$
$$M_{de} = 0 + 2 (-0.456) + 4.22 - 20.873 = -17.565 \text{ kip-ft}$$

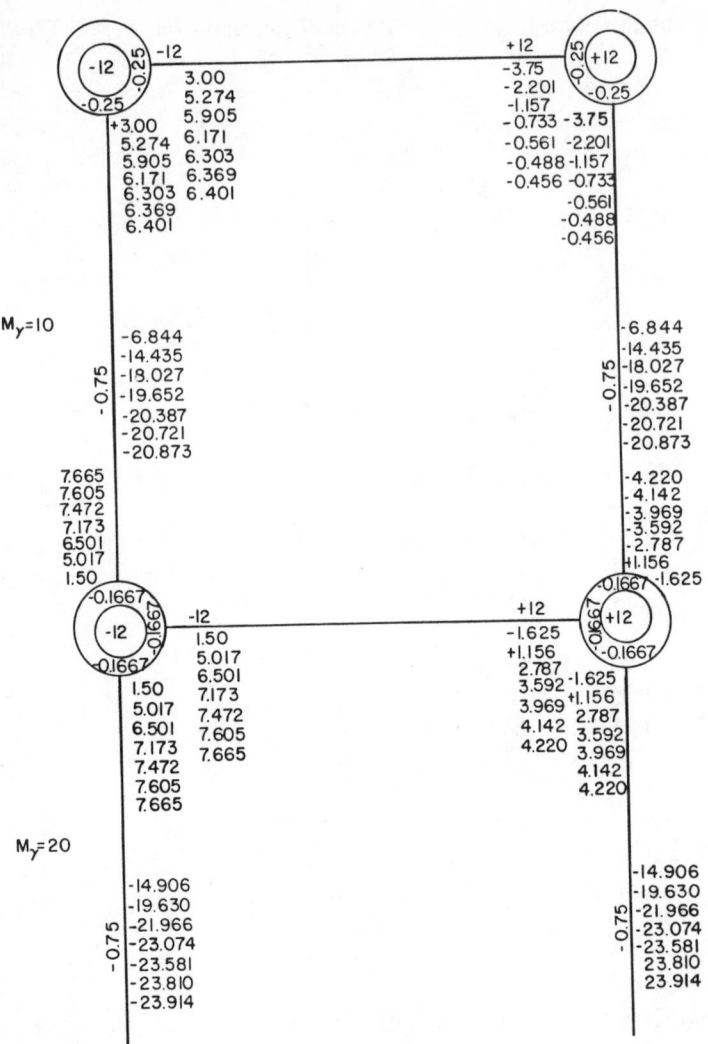

Figure 4-5.   Calculation of rigid frame for Example 4-1.

$M_{be} = -12 + 2 \times 7.665 + 4.22 = 7.55$ kip-ft
$M_{bc} = 0 + 2 \times 7.665 + 6.401 - 20.873 = 0.858$ kip-ft
$M_{ba} = 0 + 2 \times 7.665 - 23.914 = -8.584$ kip-ft
$M_{eb} = 12 + 2 \times 4.22 + 7.665 = 28.105$ kip-ft
$M_{ed} = 0 + 2 \times 4.22 - 0.456 - 20.873 = -12.889$ kip-ft
$M_{ef} = 0 + 2 \times 4.22 - 23.914 = -15.474$ kip-ft

MOMENT COEFFICIENT FOR FRAMES WITH CONSTANT STIFFNESS

The moment coefficients provided in this section involve frames with constant stiffness ratios; but for frames with varying stiffness, the coefficients are still applicable for design purposes. In the case of frames with varying stiffness ratios, the moment coefficients provide an approximate answer and are serviceable for preliminary designs. [2]

Note that the positive moment coefficients produce clockwise moments. The end moments may be calculated using the following equations:

$$M_{ij} = K_{ij} \frac{w\ell^2}{100} \text{ for vertical loads}$$

$$M_{ij} = K_{ij} \frac{Ph}{10} \text{ for horizontal loads}$$

where w = load per unit length
       ℓ = beam span
       P = concentrated horizontal load
       h = story height
       K_{ij} = moment coefficient

Moment coefficients for various frame loading cases are provided later in this section.

Example 4-2

Calculate the end moments of the frame shown in Figure 4-4 using moment coefficients.

From Case 5,

$$M_{cd} = -5.95 \times 1 \times 12^2/100 = -8.568 \text{ kip-ft}$$
$$M_{cb} = 5.95 \times 1 \times 12^2/100 = 8.568 \text{ kip-ft}$$
$$M_{be} = -7.142 \times 1 \times 12^2/100 = -10.284 \text{ kip-ft}$$
$$M_{bc} = 4.758 \times 1 \times 12^2/100 = 6.852 \text{ kip-ft}$$
$$M_{ba} = 2.383 \times 1 \times 12^2/100 = 3.432 \text{ kip-ft}$$
$$M_{dc} = 5.95 \times 1 \times 12^2/100 = 8.568 \text{ kip-ft}$$
$$M_{de} = -5.95 \times 1 \times 12^2/100 = -8.568 \text{ kip-ft}$$
$$M_{eb} = 7.142 \times 1 \times 12^2/100 = 10.284 \text{ kip-ft}$$

$M_{ed} = -4.758 \times 1 \times 12^2/100 = -6.852$ kip-ft
$M_{ef} = -2.383 \times 1 \times 12^2/100 = -3.432$ kip-ft
$M_a = 1.191 \times 1 \times 12^2/100 = 1.715$ kip-ft
$M_f = -1.715$ kip-ft

From Case 8,

$M_{cd} = M_{dc} = 3 \times 3 \times 10/10 = 9$ kip-ft
$M_{cb} = M_{de} = -3 \times 3 \times 10/10 = -9$ kip-ft
$M_{be} = M_{eb} = 6 \times 3 \times 10/10 = 18$ kip-ft
$M_{bc} = M_{ed} = -2 \times 3 \times 10/10 = -6$ kip-ft
$M_{ba} = M_{ef} = -4 \times 3 \times 10/10 = -12$ kip-ft
$M_a = M_f = -6 \times 3 \times 10/10 = -18$ kip-ft
$\Sigma M_{cd} = -8.568 + 9.0 = 0.432$ kip-ft
$\Sigma M_{cb} = 8.568 - 9 = -0.432$ kip-ft
$\Sigma M_{be} = -10.284 + 18 = 7.716$ kip-ft
$\Sigma M_{bc} = 6.852 - 6 = 0.852$ kip-ft
$\Sigma M_{ba} = 3.432 - 12 = -8.568$ kip-ft
$\Sigma M_a = 1.715 - 18 = -16.285$ kip-ft
$\Sigma M_{dc} = -8.568 + 9.0 = 17.568$ kip-ft
$\Sigma M_{de} = -8.568 - 9 = -17.568$ kip-ft
$\Sigma M_{eb} = 10.284 + 18 = 28.284$ kip-ft
$\Sigma M_{ed} = -6.852 - 6 = -12.852$ kip-ft
$\Sigma M_{ef} = -3.432 - 12 = -15.432$ kip-ft
$\Sigma M_f = -1.715 - 18 = -19.715$ kip-ft

REFERENCES

1. Kani, G., *Analysis of Multistory Frames,* translated from *Die Berechnung Mehrstockiger Rahmen,* 5th German Edition, by C. J. Hyman, Frederick Ungar Publishing Company, New York, 1957.
2. Takabeya, F. *Multistory Frames,* Wilhelm Ernst & Sohn, Berlin/Munich, 1965.

Cases for Chapter 4

Case 1

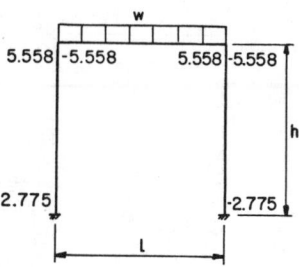

Case 2

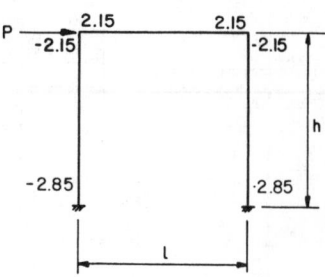

Case 3

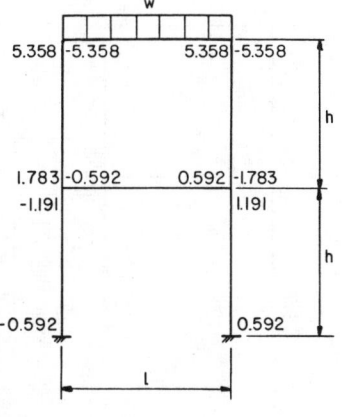

Case 4

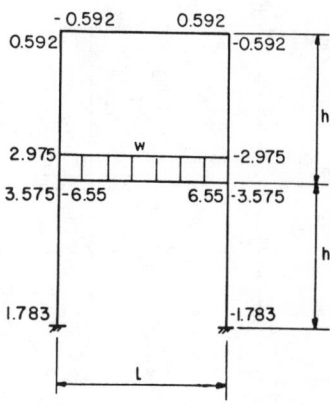

Case 5

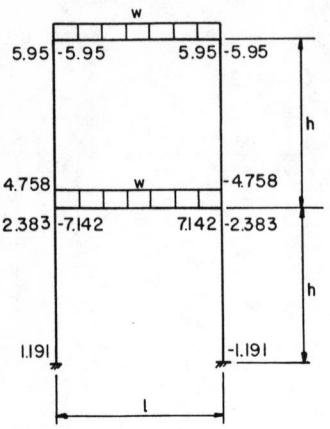

Case 6

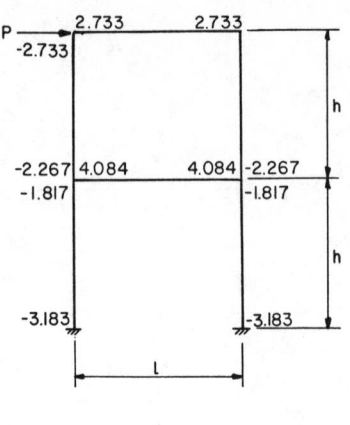

Case 7

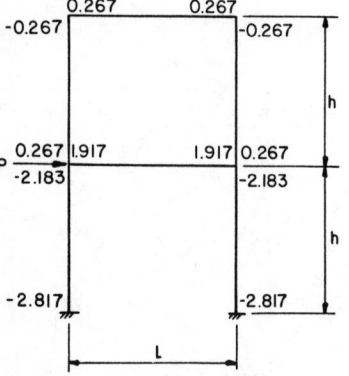

Case 8

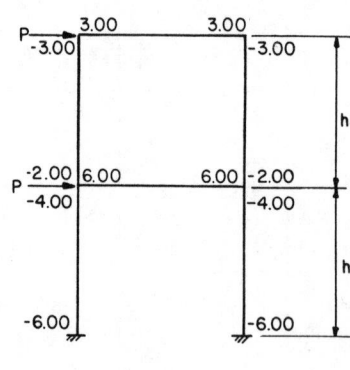

Case 9

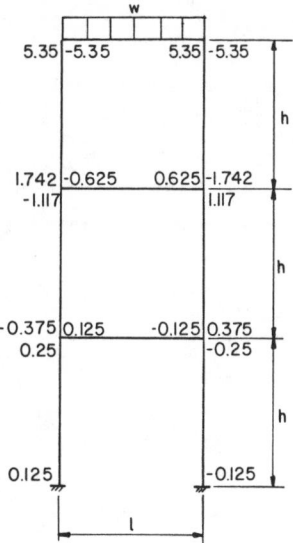

Case 10

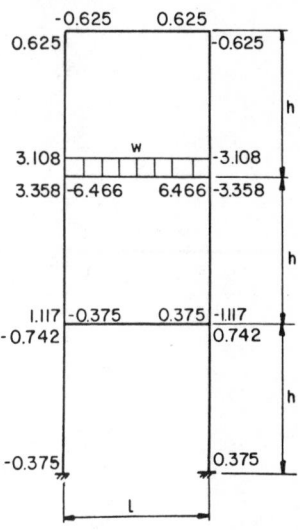

Case 11

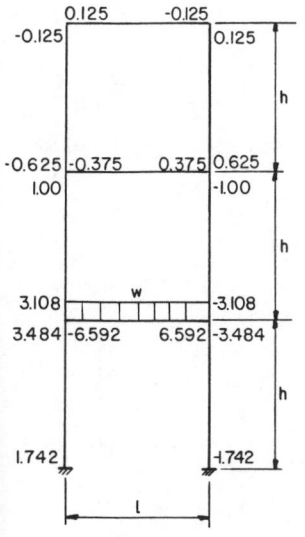

Case 12

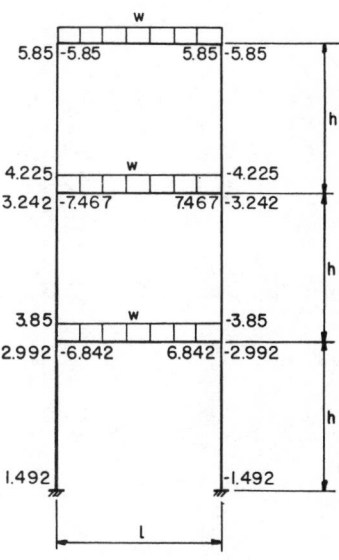

Case 13

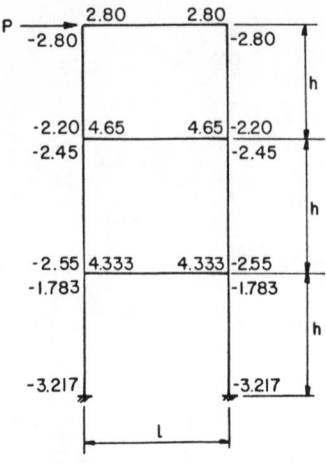

Case 14

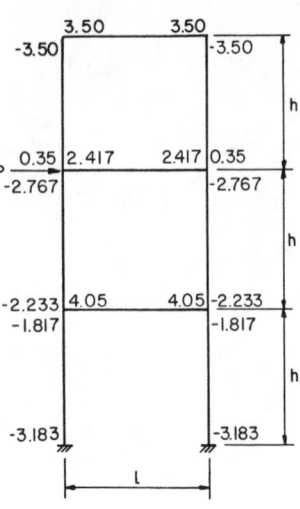

Case 15

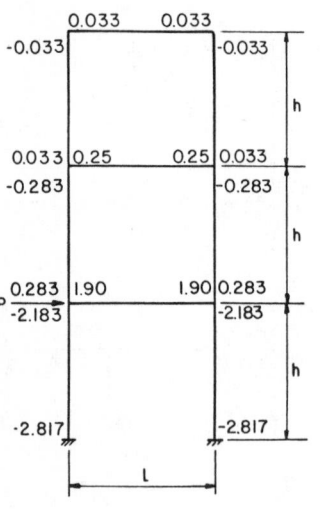

Case 16

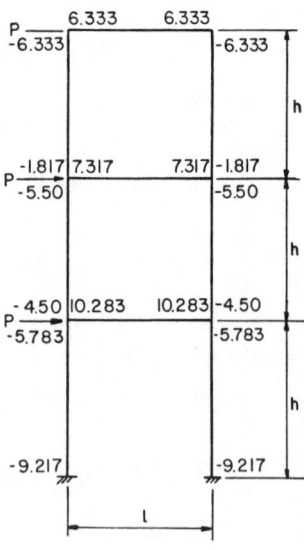

Case 17

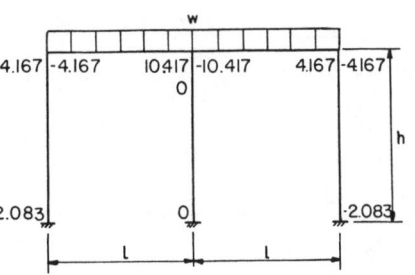

Case 18

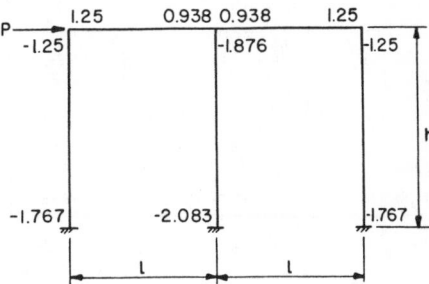

Case 19

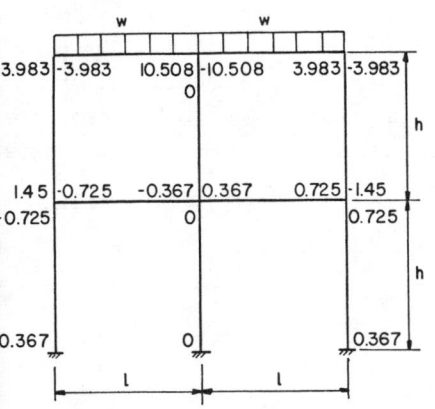

Case 20

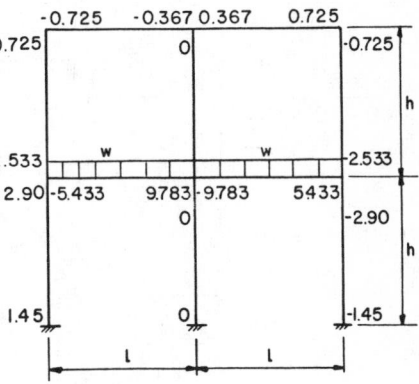

Case 21

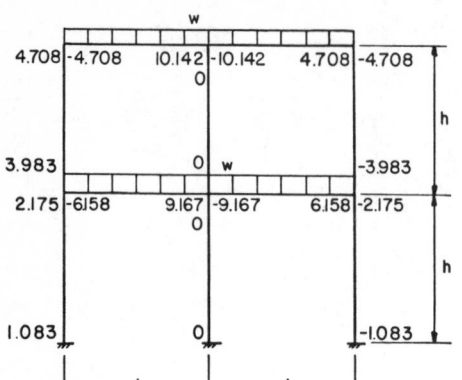

Case 22

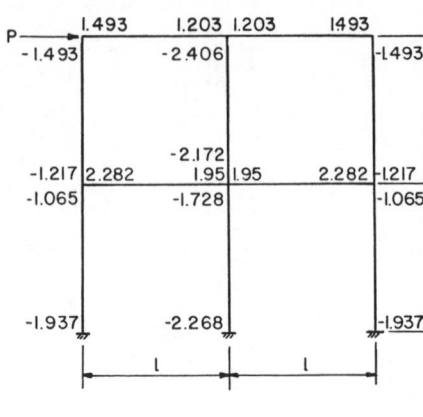

Case 23

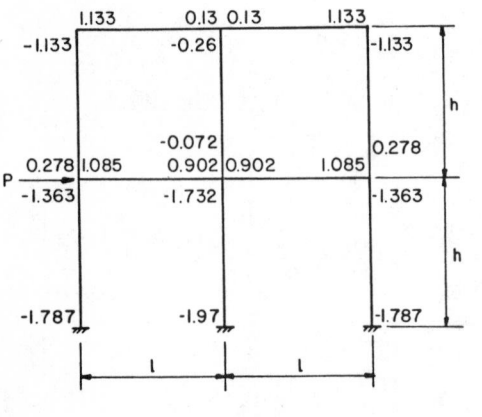

Case 24

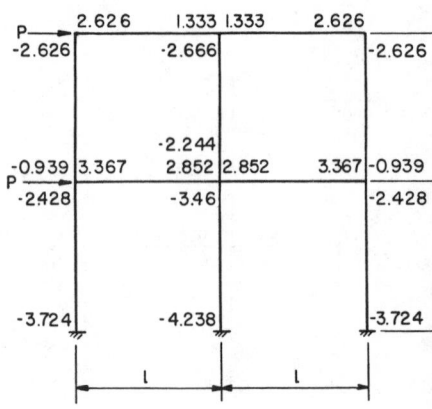

Chapter 5

Columns

A column may be defined as a compression member whose length is considerably greater than its cross-sectional dimensions. An ideal column is

1. Perfectly straight
2. Loaded at its centroid
3. Free of residual stresses
4. Made of isotropic material
5. Impervious to local buckling

A sufficiently slender column will fail by elastic instability, and the maximum stress is less than the proportional limit of the material. The column strength depends on the modulus of elasticity, the slenderness ratio, and the end conditions. For a given material and given end conditions, there is a certain slenderness ratio that marks the dividing point between columns failing by elastic stability and those yielding stress. This is called the *critical slenderness ratio.*

For an ideal column the buckling load is known as the *Euler load* and is given by

$$P_e = \frac{C\pi^2 EI}{L^2} \tag{5-1}$$

where P_e = Euler load
E = modulus of elasticity
I = moment of inertia of the cross section
L = length of the column
C = coefficient of constraint

The coefficient C depends on the column-end conditions. The ideal column, as shown in Figure 5-1, is fixed at the lower end and free at the upper end. When P is less than the critical value, the column remains

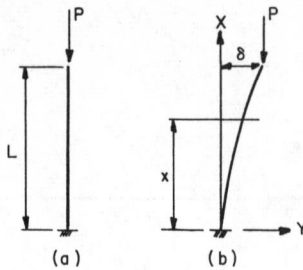

Figure 5-1. An ideal column.

straight and undergoes only axial compression. When P reaches the critical value, the column becomes unstable.

The moment at any cross section is

$$M = -P(\delta - y) \tag{5-2}$$

Equation 5-2 can be expressed as

$$EI \frac{d^2y}{dx^2} = P(\delta - y)$$

let $k^2 = \dfrac{P}{EI}$

$$\frac{d^2y}{dx^2} + k^2y = k^2\delta \tag{5-3}$$

The solution of Equation 5-3 is

$$y = \delta(1 - \cos kx)$$

and cos kL = 0, since y = δ at x = L.

$$kL = (2n-1)\frac{\pi}{2} \tag{5-4}$$

The smallest value of kL that satisfies Equation 5-4 is n = 1, and the smallest critical load can be obtained.

$$P_e = \frac{\pi^2 EI}{4L^2} \tag{5-5}$$

In this case the coefficient C is $1/4$. If both ends of the column are pinned, C = 1, and for fixed ends, C = 4.

Equation 5-1 can be expressed as

$$P_e = \frac{\pi^2 EI}{(kL)^2} = \frac{\pi^2 E \, A}{(kL/r)^2}$$

$$F_c = \frac{P_e}{A} = \frac{\pi^2 E}{(kL/r)^2} \tag{5-6}$$

where F_c = buckling stress
$\quad\quad\;\; k$ = effective length factor
$\quad\quad\;\; r$ = radius of gyration

kL/r is the effective slenderness ratio. Equation 5-6 indicates that when $kL/r < \sqrt{\pi^2 E/F_y}$, the column buckles inelastically at F_y. The effective slenderness ratio kL/r for which the Euler buckling stress equals the yield stress is $\sqrt{\pi^2 E/F_y}$.

STRENGTH OF ACTUAL COLUMNS

An actual column always presents residual stresses, initial curvature, and eccentricity of the applied load. The strength of an actual column is less than that of an ideal column.

Initial Curvature [3,4]

For a pinned-end column with initial curvature as shown in Figure 5-2, the curvature is treated as a cosine curve, and the following equation holds true:

$$y = \frac{y_0}{1 - P/P_e} \cos \frac{\pi x}{L} \tag{5-7}$$

where y_0 is the initial out-of-straightness at midheight. The final midheight out-of-straightness y, after P is applied, is expressed by

$$y = y_0 \left(\frac{P_e}{P_e - P} \right) \tag{5-8}$$

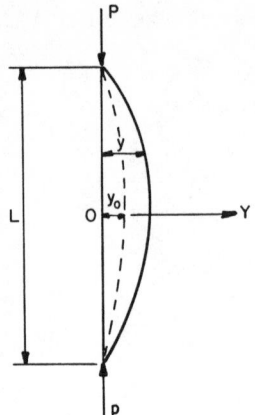

Figure 5-2. Column with initial curvature.

The parenthetic term in Equation 5-8 is the *amplification factor.* The maximum stress in the column is $P/A + Mc/I$ and is expressed as

$$F_m = \frac{P}{A}\left[1 + \frac{y_0 c}{r^2}\left(\frac{P_e}{P_e - P}\right)\right] \tag{5-9}$$

The initial maximum out-of-straightness of a column is often given as a function of the column length. The American Institute of Steel Construction (AISC) permits a maximum value of $y_0/L = 1/1{,}000$. The load that causes initial yielding in the outer fibers of the column is calculated from Equation 5-9 by substituting F_y for F_m.

Eccentrically Applied Load [1,4]

If the load at each end of an initially straight pinned-end column is eccentrically applied from the centroid, the out-of-straightness at the midheight is

$$y = e \sec \frac{\pi}{2}\sqrt{\frac{P}{P_e}} \tag{5-10}$$

The y value in Equation 5-10 approaches infinity as P approaches P_e. The maximum compressive stress is defined by the secant formula:

$$F_m = \frac{P}{A}\left(1 + \frac{ec}{r^2} \sec \frac{\pi}{2}\sqrt{\frac{P}{P_e}}\right) \tag{5-11}$$

In the past, an equivalent eccentricity of 0.25 was assumed for ec/r^2 to account for initial curvature and accidental eccentricity. The maximum column strength is reached when F_m reaches F_y.

Residual Stress

Stresses present in the cross section of a member before the application of external load are known as *residual stresses*. Residual stresses may be due to cooling after hot-rolling or welding, or to fabrication operation such as cold-bending or cambering. Research has shown that the influence of residual stresses on column strength is often greater than that of initial crookedness and unintentional eccentricity.

A column cross section containing residual stresses will have certain fibers yield even though the average applied stress is less than the yield stress. The buckling strength of a column that contains residual stresses may be determined by

$$F_c = \frac{\pi^2 EI_e}{(kL)^2 \, A} = \frac{\pi^2 \, E}{(kL/r)^2} \frac{I_e}{I} \tag{5-12}$$

where I_e is the moment of inertia of the elastic portion of the cross section. Note that the I in the Euler equation is replaced by I_e, since the stiffness of the cross section is limited to the stiffness of the elastic portion EI_e. To use Equation 5-12 the relationship between F_c and I_e must be established.

COLUMN STRENGTH OF HOT-ROLLED WIDE FLANGE SHAPES [1,4]

The buckling strength of a column containing residual stress is a function of the stiffness of the elastic portion EI_e. The relations between I_e/I and E_T/E can be mathematically established. E_T is the tangent modulus of the column cross section. It can be determined from a stub-column test. The tangent modulus is determined from the average stress-strain curve. The I_e/I ratio for a wide-flange shape can be approximately expressed as

$$I_e/I = E_T/E \quad \text{for the strong axis}$$

$$I_e/I = (E_T/E)^3 \quad \text{for the weak axis}$$

An approximate value for F_c is given by

$$F_c = \frac{\pi^2 E}{(kL/r)^2} \left(\frac{E_T}{E}\right) = \frac{\pi^2 E_T}{(kL/r)^2} \quad \text{(strong axis)} \qquad (5\text{-}13a)$$

$$F_c = \frac{\pi^2 E}{(kL/r)^2} \left(\frac{E_T}{E}\right)^3 \quad \text{(weak axis)} \qquad (5\text{-}13b)$$

Column curves for wide-flange shapes can be prepared by plotting Equations 5-13a and 5-13b. This shows that straight lines and parabolic curves give satisfactory predictions for column strength in the weak and strong axes, respectively. Test results also indicate the efficacy of the straight-line and parabolic assumptions. A set of column curves are plotted for $F_y = 36$ kips/in.^{2}, as shown in Figure 5-3. The maximum residual stress of $0.3\,F_y$ is a typical value for carbon steel and a conservative value for the other structural steels. As shown in Figure 5-3, when F_c is less than 25.2 kips/in.^{2} ($0.7\,F_y$), no localized yielding occurs. However, when F_c is greater than 25.2 kips/in.^{2}, the weak axis bending strength is less than the strong axis bending strength of a column with the same slenderness parameter.

Figure 5-3 also shows the curve suggested by the Column Research Council (CRC) [5]. The CRC basic column curve is an average curve that is used for bending about both axes. It serves as a single *basic*

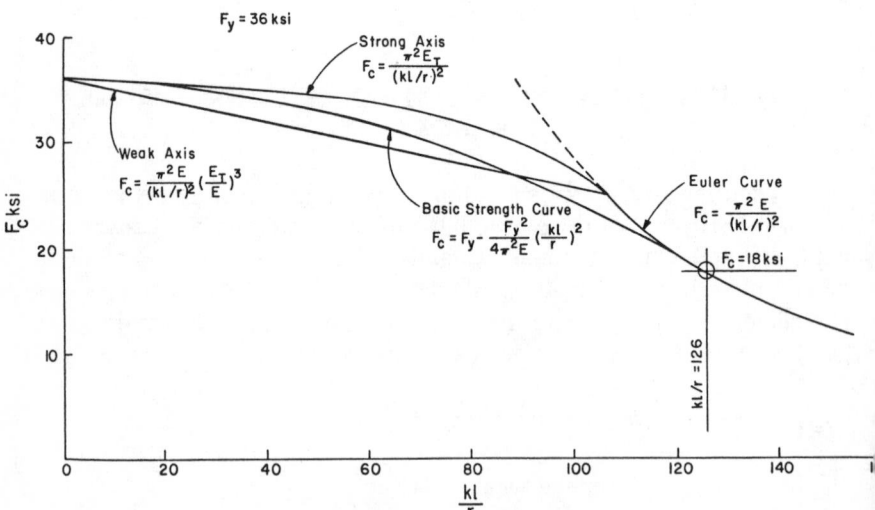

Figure 5-3. Column curves.

strength curve for hot-rolled wide-flange shapes. The basic strength curve is defined by the equation

$$F_c = F_y - \frac{F_y^2}{4\pi^2 E}\left(\frac{K\,L}{r}\right)^2 \quad \text{when } kL/r \leqslant C_c \tag{5-14}$$

and by the Euler equation:

$$F_c = \frac{\pi^2 E}{(kL/r)^2} \quad \text{when } kL/r > C_c$$

C_c corresponds to the point of tangency on the Euler curve; it is the kL/r value when $F_c = F_y/2$.

$$C_c = \sqrt{\frac{2\pi^2 E}{F_y}} \tag{5-15}$$

C_c curves for various steels are plotted in Figure 5-4.

$$\text{Let } \beta = \frac{kL/r}{C_c}$$

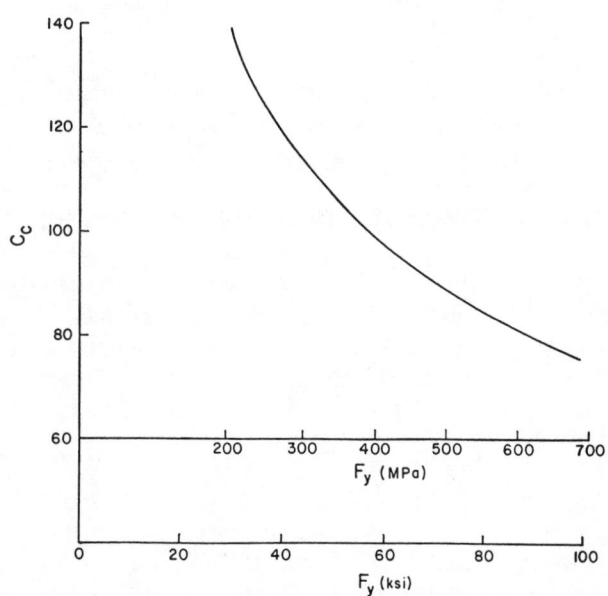

Figure 5-4. C_c values for steel columns.

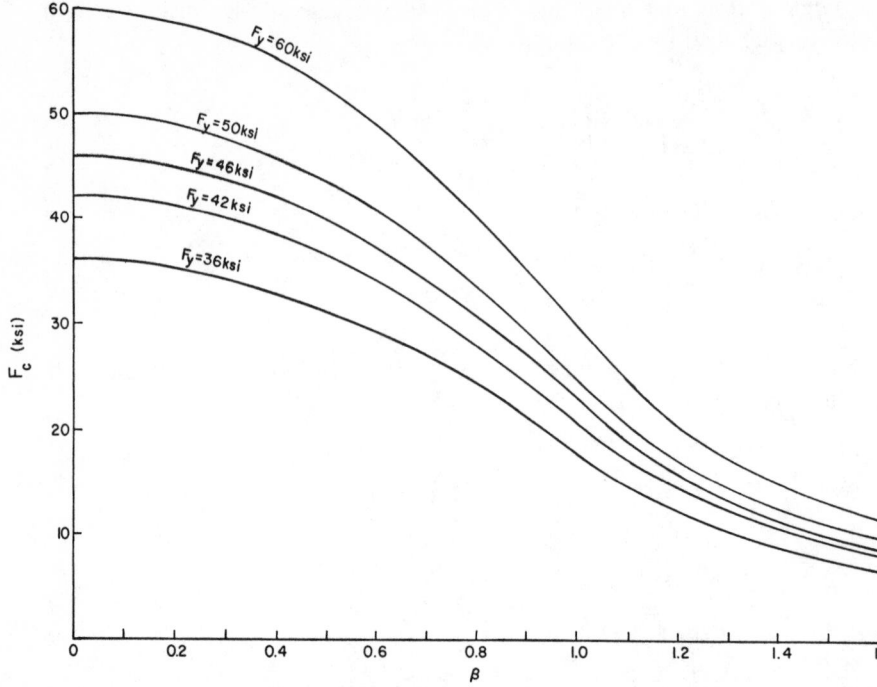

Figure 5-5. Buckling stresses for steel columns.

The basic strength curves for structural steels with specified F_y are plotted in Figure 5-5. Allowable stresses for design purposes can be obtained by dividing the strength F_c by an appropriate factor of safety (F.S.).

AISC ALLOWABLE COLUMN STRESSES [6]

The CRC basic column strength curve is used as the basic equation for the ultimate strength of a column, and is divided by a factor of safety to give the equation for allowable axial stress. For columns with $kL/r \leqslant C_c$, allowable stresses are calculated by

$$F_a = \frac{F_y \left[1 - \frac{F_y}{4\pi^2 E}\left(\frac{kL}{r}\right)^2\right]}{\text{F.S.}} \tag{5-16}$$

$$\text{F.S.} = \frac{5}{3} + \frac{3(kL/r)}{8\,C_c} - \frac{(kL/r)^3}{8\,C_c^{\,3}} \tag{5-17}$$

When the slenderness ratio kL/r $> C_c$, allowable stresses are calculated by dividing the Euler equation by a safety factor of 1.917.

$$F_a = \frac{12\pi^2 E}{23(kL/r)^2} \qquad (5\text{-}18)$$

Allowable stresses for columns with yield stresses of 36–65 kips/in.^{2} are calculated according to the AISC specification and presented in Tables 5-1A through 5-1F.

REFERENCES

1. Beedle, L. S., Blackmon, J. H., et al. *Structural Steel Design*, The Ronald Press Company, New York, 1964.
2. Timoshenko, S. P. and Gere, J. M., *Theory of Elastic Stability*, McGraw-Hill Book Company, New York, 1961.
3. Roark, R. J. , *Formulas for Stress and Strain*, 4th Edition, McGraw-Hill Book Company, New York, 1965.
4. Brockenbrough, R. L., Johnston, B. G., *Steel Design Manual*, U.S. Steel Corporation, Pittsburgh, Pennsylvania, 1968.
5. Column Research Council, *Guide to Design Criteria for Metal Compression Members*, 2nd Edition, John Wiley & Sons, Publisher.
6. American Institute of Steel Construction, Inc., *Manual of Steel Construction*, AISC, 1980.

Table 5-1A
Allowable Stresses for Columns with 36 kips/in.² Yield Stress

$\frac{kL}{r}$	F_a		$\frac{kL}{r}$	F_a	
	kips/in.²	MPa		kips/in.²	MPa
2	21.52	148.38	62	17.24	118.87
4	21.44	147.82	64	17.04	117.49
6	21.35	147.20	66	16.84	116.11
8	21.25	146.51	68	16.64	114.73
10	21.16	145.89	70	16.43	113.28
12	21.05	145.13	72	16.22	111.83
14	20.95	144.45	74	16.01	110.38
16	20.83	143.62	76	15.79	108.87
18	20.72	142.86	78	15.58	107.42
20	20.60	142.03	80	15.36	105.90
22	20.48	141.20	82	15.13	104.32
24	20.35	140.31	84	14.90	102.73
26	20.22	139.41	86	14.67	101.15
28	20.08	138.45	88	14.44	99.56
30	19.94	137.48	90	14.20	97.91
32	19.80	136.52	92	13.97	96.32
34	19.65	135.48	94	13.72	94.60
36	19.50	134.45	96	13.48	92.94
38	19.35	133.41	98	13.23	91.22
40	19.19	132.31	100	12.98	89.49
42	19.03	131.21	102	12.72	87.70
44	18.86	130.04	104	12.47	85.98
46	18.70	128.93	106	12.20	84.12
48	18.53	127.76	108	11.94	82.32
50	18.35	126.52	110	11.67	80.46
52	18.17	125.28	112	11.40	78.60
54	17.99	124.04	114	11.13	76.74
56	17.81	122.80	116	10.85	74.81
58	17.62	121.49	118	10.57	72.88
60	17.43	120.18	120	10.28	70.88

Table 5-1B
lowable Stresses for Columns with 42 kips/in.² Yield Stress

F_a		$\frac{kL}{r}$	F_a	
kips/in.²	MPa	r	kips/in.²	MPa
25.10	173.06	62	19.53	134.65
24.99	172.30	64	19.27	132.86
24.88	171.54	66	19.01	131.07
24.76	170.71	68	18.75	129.28
24.63	169.82	70	18.48	127.42
24.50	168.92	72	18.20	125.48
24.36	167.96	74	17.92	123.55
24.22	166.99	76	17.64	121.62
24.07	165.96	78	17.35	119.62
23.92	164.92	80	17.06	117.62
23.76	163.82	82	16.77	115.63
23.59	162.65	84	16.47	113.56
23.42	161.48	86	16.17	111.49
23.24	160.23	88	15.86	109.35
23.06	158.99	90	15.55	107.21
22.88	157.75	92	15.23	105.00
22.69	156.44	94	14.91	102.80
22.49	155.06	96	14.59	100.59
22.29	153.68	98	14.26	98.32
22.08	152.24	100	13.93	96.04
21.87	150.79	102	13.59	93.70
21.66	149.34	104	13.25	91.36
21.44	147.82	106	12.90	88.94
21.22	146.31	108	12.55	86.53
20.99	144.72	110	12.19	84.05
20.76	143.14	112	11.83	81.56
20.52	141.48	114	11.47	79.08
20.28	139.83	116	11.10	76.53
20.03	138.10	118	10.70	73.77
19.79	136.45	120	10.35	71.36

Table 5-1C
Allowable Stresses for Columns with 45 kips/in.^{2} Yield Stress

$\frac{kL}{r}$	F_a kips/in.^{2}	MPa	$\frac{kL}{r}$	F_a kips/in.^{2}	MPa
2	26.89	185.40	62	20.63	142.24
4	26.77	184.57	64	20.34	140.24
6	26.64	183.68	66	20.04	138.17
8	26.51	182.78	68	19.74	136.10
10	26.37	181.81	70	19.43	133.97
12	26.22	180.78	72	19.12	131.83
14	26.07	179.75	74	18.81	129.69
16	25.91	178.64	76	18.49	127.48
18	25.74	177.47	78	18.17	125.28
20	25.57	176.30	80	17.84	123.00
22	25.39	175.06	82	17.51	120.73
24	25.20	173.75	84	17.17	118.38
26	25.01	172.44	86	16.82	115.97
28	24.81	171.06	88	16.48	113.63
30	24.61	169.68	90	16.12	111.14
32	24.40	168.23	92	15.77	108.73
34	24.18	166.72	94	15.40	106.18
36	23.96	165.20	96	15.04	103.70
38	23.74	163.68	98	14.66	101.08
40	23.51	162.10	100	14.28	98.46
42	23.27	160.44	102	13.90	95.84
44	23.03	158.79	104	13.51	93.15
46	22.78	157.06	106	13.12	90.46
48	22.53	155.34	108	12.72	87.70
50	22.27	153.55	110	12.31	84.87
52	22.01	151.75	112	11.90	82.05
54	21.74	149.89	114	11.49	79.22
56	21.47	148.03	116	11.10	76.53
58	21.19	146.03	118	10.72	73.91
60	20.91	144.17	120	10.37	71.50

Table 5-1D
llowable Stresses for Columns with 50 kips/in.² Yield Stress

L	Fₐ		kL	Fₐ	
	kips/in.²	MPa	r	kips/in.²	MPa
2	29.87	205.95	62	22.37	154.24
4	29.73	204.98	64	22.02	151.82
6	29.58	203.95	66	21.67	149.41
8	29.42	202.84	68	21.31	146.93
0	29.26	201.74	70	20.94	144.38
2	29.08	200.50	72	20.56	141.76
4	28.90	199.26	74	20.19	139.21
6	28.71	197.95	76	19.80	136.52
8	28.51	196.57	78	19.41	133.83
0	28.30	195.12	80	19.01	131.07
2	28.08	193.60	82	18.61	128.31
4	27.86	192.09	84	18.20	125.48
6	27.63	190.50	86	17.79	122.66
8	27.40	188.92	88	17.37	119.76
0	27.15	187.19	90	16.94	116.80
2	26.90	185.47	92	16.50	113.76
4	26.64	183.68	94	16.06	110.73
6	26.38	181.88	96	15.62	107.70
8	26.11	180.02	98	15.17	104.59
0	25.83	178.09	100	14.71	101.42
2	25.55	176.16	102	14.24	98.18
4	25.26	174.16	104	13.77	94.94
6	24.96	172.09	106	13.29	91.63
8	24.66	170.02	108	12.80	88.25
0	24.35	167.89	110	12.34	85.08
2	24.04	165.75	112	11.90	82.05
4	23.72	163.54	114	11.49	79.22
6	23.39	161.27	116	11.10	76.53
8	23.06	158.99	118	10.72	73.91
0	22.72	156.65	120	10.37	71.50

Table 5-1E
Allowable Stresses for Columns with 60 kips/in.^{2} Yield Stress

$\frac{kL}{r}$	F_a		$\frac{kL}{r}$	F_a	
	kips/in.^{2}	MPa		kips/in.^{2}	MPa
2	35.83	247.04	62	25.58	176.37
4	35.64	245.73	64	25.10	173.06
6	35.44	244.35	66	24.61	169.68
8	35.23	242.90	68	24.11	166.23
10	35.01	241.39	70	23.60	162.72
12	34.77	239.73	72	23.08	159.13
14	34.52	238.00	74	22.56	155.55
16	34.27	236.28	76	22.02	151.82
18	34.00	234.42	78	21.48	148.10
20	33.71	232.42	80	20.93	144.31
22	33.42	230.42	82	20.37	140.45
24	33.12	228.35	84	19.80	136.52
26	32.81	226.22	86	19.22	132.52
28	32.48	223.94	88	18.63	128.45
30	32.15	221.67	90	18.04	124.38
32	31.81	219.32	92	17.43	120.18
34	31.45	216.84	94	16.81	115.90
36	31.09	214.36	96	16.19	111.63
38	30.72	211.81	98	15.55	107.21
40	30.34	209.19	100	14.93	102.94
42	29.95	206.50	102	14.35	98.94
44	29.55	203.74	104	13.81	95.22
46	29.15	200.98	106	13.29	91.63
48	28.73	198.09	108	12.80	88.25
50	28.31	195.19	110	12.34	85.08
52	27.87	192.16	112	11.90	82.05
54	27.43	189.12	114	11.49	79.22
56	26.98	186.02	116	11.10	76.53
58	26.53	182.92	118	10.72	73.91
60	26.06	179.68	120	10.37	71.50

Table 5-1F
Allowable Stresses for Columns with 65 kips/in.² Yield Stress

kL/r	F_a kips/in.²	MPa	kL/r	F_a kips/in.²	MPa
2	38.81	267.59	62	27.05	186.50
4	38.59	266.07	64	26.50	182.71
6	38.37	263.86	66	25.93	178.78
8	38.13	262.90	68	25.35	174.78
10	37.87	261.10	70	24.76	170.71
12	37.61	259.31	72	24.17	166.65
14	37.32	257.31	74	23.56	162.44
16	37.03	255.31	76	22.94	158.17
18	36.72	253.18	78	22.32	153.89
20	36.40	250.97	80	21.68	149.48
22	36.06	248.62	82	21.03	145.00
24	35.71	246.21	84	20.37	140.45
26	35.36	243.80	86	19.70	135.83
28	34.99	241.25	88	19.02	131.14
30	34.60	238.56	90	18.32	126.31
32	34.21	235.87	92	17.62	121.49
34	33.81	233.11	94	16.90	116.52
36	33.39	230.22	96	16.20	111.70
38	32.96	227.25	98	15.55	107.21
40	32.53	224.29	100	14.93	102.94
42	32.08	221.18	102	14.35	98.94
44	31.62	218.01	104	13.81	95.22
46	31.35	216.15	106	13.29	91.63
48	30.68	211.53	108	12.80	88.25
50	30.19	208.15	110	12.34	85.08
52	29.69	204.71	112	11.90	82.05
54	29.18	201.19	114	11.49	79.22
56	28.66	197.60	116	11.10	76.53
58	28.14	194.02	118	10.72	73.91
60	27.60	190.30	120	10.37	71.50

Chapter 6

Beam-Columns

Members that are subjected to a combination of axial compression and bending are known as beam-columns. The bending may be caused by end moments, lateral loading, or eccentric application of the axial load. The analysis of beam-columns is not a straightforward matter. In beam-column analysis, the portion of the bending due to the axial force multiplied by the deflection cannot be ignored.

The beam shown in Figure 6-1 is subjected to only lateral loads, and the small deflection has an insignificant effect on the moment and shear forces. The beam shown in Figure 6-2 is subjected to axial and lateral loads simultaneously. The bending moment and shear forces are dependent upon the magnitude of the deflection produced. The moment Py cannot be ignored in the analysis, and this moment cannot be found until the deflections are determined. Therefore, the beam-column is statically indeterminate.

The relations among load, shear force, and bending moment can be obtained from the equilibrium of an element; the following relations are obtained:

$$\frac{M}{EI} = \frac{d^2y}{dx^2} \tag{6-1}$$

$$V = EI \frac{d^3y}{dx^3} + P \frac{dy}{dx} \tag{6-2}$$

$$-w = EI \frac{d^4y}{dx^4} + P \frac{d^2y}{dx^2} \tag{6-3}$$

Note that when $P = 0$, Equations 6-1, 6-2, and 6-3 reduce to the usual equations for bending by lateral loads only.

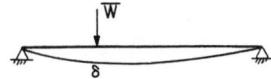

Figure 6-1. Beam subjected to lateral load.

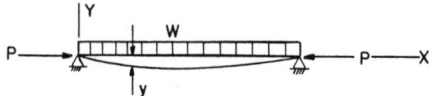

Figure 6-2. Beam subjected to axial and lateral loads.

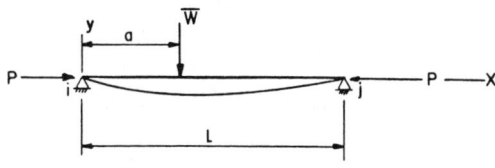

Figure 6-3. A beam-column.

BEAM-COLUMNS WITH CONCENTRATED LATERAL LOAD [1,2]

To analyze the beam-column shown in Figure 6-3, it is necessary to solve the differential equation of the deflection curve. The bending moments are expressed as follows:

$$M = \frac{W(\ell - a)}{\ell} x + Py \qquad x \leqslant a$$

$$M = \frac{Wa}{\ell} (\ell - x) + Py \qquad a \leqslant x < \ell$$

From Equation 6-1

$$\frac{d^2y}{dx^2} = \frac{1}{EI} \frac{W(\ell - a)}{\ell} x + \frac{P}{EI} y \qquad x \leqslant a \tag{6-4}$$

$$\frac{d^2y}{dx^2} = \frac{1}{EI} \frac{Wa(\ell - x)}{\ell} + \frac{P}{EI} y \qquad a \leqslant x < \ell \tag{6-5}$$

By letting $k^2 = P/EI$ and solving Equations 6-4 and 6-5, the following equations of deflection curves are obtained:

$$y = -\frac{W \sin k(\ell - a)}{Pk \sin k\ell} \sin kx + \frac{W(\ell - a)}{P\ell} x \qquad x \le a \qquad (6\text{-}6)$$

$$y = -\frac{W \sin ka}{Pk \sin k\ell} \sin k(\ell - x) + \frac{Wa(\ell - x)}{P\ell} \qquad a \le x < \ell \qquad (6\text{-}7)$$

The following equations are obtained by differentiation:

$$y' = \frac{-W \sin k(\ell - a)}{P \sin k\ell} \cos kx + \frac{W(\ell - a)}{P\ell} \qquad x \le a \qquad (6\text{-}8)$$

$$y' = \frac{W \sin ka}{P \sin k\ell} \cos k(\ell - x) - \frac{Wa}{P\ell} \qquad a \le x < \ell \qquad (6\text{-}9)$$

$$y'' = \frac{Wk \sin k(\ell - a)}{P \sin k\ell} \sin kx \qquad x \le a \qquad (6\text{-}10)$$

$$y'' = \frac{Wk \sin ka}{P \sin k\ell} \sin k(\ell - x) \qquad a \le x < \ell \qquad (6\text{-}11)$$

When the load W is applied at the midspan, the deflection curve is symmetrical. Let $u = k\ell$; the maximum deflection is calculated as follows:

$$y_{\ell/2} = \frac{-W \sin (k\ell/2)}{Pk \sin k\ell} \sin (k\ell/2) + \frac{W\ell}{4P}$$

$$\delta_{max} = \frac{-W}{2Pk} \left[\tan (k\ell/2) - \frac{k\ell}{2} \right]$$

$$= \frac{-W}{2EI} \frac{\ell^3}{u^3} \left(\tan \frac{u}{2} - \frac{u}{2} \right)$$

$$= \frac{W\ell^3}{48EI} \frac{24\left(\tan \frac{u}{2} - \frac{u}{2} \right)}{u^3}$$

$$\delta_{max} = \frac{-W\ell^3}{48EI} A_u \tag{6-12}$$

where $A_u = \dfrac{24\left(\tan\dfrac{u}{2} - \dfrac{u}{2}\right)}{u^3}$

Note that when $u = \pi$, A_u approaches infinity, and we can find the critical value of P as follows:

$$u = \pi = k\ell = \sqrt{\frac{P_{cr}\ell^2}{EI}}$$

$$P_{cr} = \frac{\pi^2 EI}{\ell^2}$$

When P approaches P_{cr}, even a small lateral load will cause considerable deflection.

Similarly, we can find the maximum bending moment as follows:

$$M_{max} = EI(y'')_{x = \frac{\ell}{2}} = \frac{EIW \sin (k\ell/2)}{EIk \sin k\ell} \sin (k\ell/2)$$

$$M_{max} = \frac{W\ell \tan (u/2)}{2u} = \frac{W\ell}{4} \frac{2 \tan (u/2)}{u}$$

$$M_{max} = \frac{W\ell}{4} K_u \tag{6-13}$$

$$K_u = \frac{2 \tan (u/2)}{u}$$

The factors of K_u in Equation 6-13 and A_u in Equation 6-12 approach unity as the compressive force P becomes smaller and smaller. In this case the deflection and moment reduce to that caused by lateral load only.

BEAM-COLUMNS WITH END MOMENTS

If Wa in Figure 6-3 remains finite and we let $M_i = Wa$, we have a beam-column with moment at the left end. When a approaches zero, sin ka = ka and Equation 6-7 becomes

$$y_1 = \frac{M_i}{P}\left[-\frac{\sin k(\ell - x)}{\sin k\ell} + \left(1 - \frac{x}{\ell}\right)\right] \tag{6-14}$$

Similarly, if $W(\ell - a)$ is finite, when a increases and approaches ℓ, sin $k(\ell - a) = k(\ell - a)$. Let $W(\ell - a) = M_j$; then we have a beam-column with moment at the right end, and Equation 6-6 becomes

$$y_2 = \frac{M_j}{P}\left(-\frac{\sin kx}{\sin k\ell} + \frac{x}{\ell}\right) \tag{6-15}$$

When a beam-column has moments at ends as shown in Figure 6-4, the deflection curve is given by

$$y = \frac{M_i}{P}\left[-\frac{\sin k(\ell - x)}{\sin k\ell} + \left(1 - \frac{x}{\ell}\right)\right] + \frac{M_j}{P}\left(-\frac{\sin kx}{\sin k\ell} + \frac{x}{\ell}\right)$$

If $M_i = M_j = M_0$,

$$y = \frac{M_0}{P}\left[1 - \frac{\sin kx + \sin k(\ell - x)}{\sin k\ell}\right]$$

Figure 6-4. Beam column with end moments.

$$y = \frac{M_0}{P}\left[1 - \frac{\cos (k\ell/2 - kx)}{\cos (k\ell/2)}\right]$$ (6-16)

Let $u = k\ell$, and Equation 6-16 simplifies to

$$y = \frac{M_0\ell^2}{EI}\frac{1}{u^2 \cos (u/2)}\left[-\cos(u/2 - ux/\ell) + \cos \frac{u}{2}\right]$$

The maximum moment at the midspan is obtained by

$$M_{max} = EI(y'') = M_0 \sec(u/2)$$ (6-17)

DESIGN OF BEAM-COLUMNS BY SECANT FORMULA [2,6]

The secant formula method specifies that the maximum fiber stress in the beam-column may not exceed the yield stress of the material. The beam-column shown in Figure 6-4 will be used to illustrate the secant formula approach. The maximum moment which occurs at the middle of the beam-column is obtained by Equation 6-17:

$$M_{max} = M_0 \sec(u/2)$$

where $u = \sqrt{P\ell^2/EI}$

The maximum fiber stress at the middle cross section is equal to

$$f_{max} = \frac{P}{A} + \frac{M_{max}c}{I}$$ (6-18)

Equation 6-18 can be extended to the well-known secant formula by replacing $M_0 = Pe$:

$$f_{max} = \frac{P}{A}\left(1 + \frac{ec}{r^2} \sec \frac{\ell}{2r}\sqrt{\frac{P}{AE}}\right)$$ (6-19)

The ratio r^2/c in Equation 6-19 is called the radius of the core; it defines the core of the cross section within which a compressive force can act on a short column without causing tensile stress.

By using $M = Pe$ and introducing an initial eccentricity ratio $(ec/r^2)_{init.}$ of 0.25, the following equation is obtained for the average axial stress P/A:

$$\frac{P}{A} = \frac{f_{max}}{1 + (0.25 + ec/r^2) \sec \left(\frac{\ell}{2r} \sqrt{\frac{P}{AE}}\right)}$$

(6-20)

Equation 6-20 represents the relation between the maximum compressive fiber stress f_{max} and the average compressive stress P/A. We can use the secant formula 6-20 directly in designing eccentrically loaded columns. If P denotes the safe load and F.S. is the required factor of safety, then P(F.S.) is the load on the column at which the maximum fiber stress equals the yield stress. Equation 6-20 becomes

$$F_a = \frac{P}{A} = \frac{F_y/F.S.}{1 + (0.25 + ec/r^2) \sec \left(\frac{\ell}{2r} \sqrt{\frac{P(F.S.)}{AE}}\right)}$$

(6-21)

Equation 6-21 can be solved by trial and error. We can obtain the allowable stress F_a for a given column. Let $(F.S.)P/A = (F_c)_y$, and plot $(F_c)_y$ against ℓ/r. The average compressive stress at which yielding begins can be obtained directly from the curve. The allowable compressive stress can be obtained by dividing the $(F_c)_y$ of the curves by the desired factor of safety. Figure 6-5 is a plot for A36 steel for eccentricity ratios of 0.25, 0.5, 0.75 and 1.00. The desired allowable stress can be obtained by simply dividing the ordinates of the curve by the desired factor of safety.

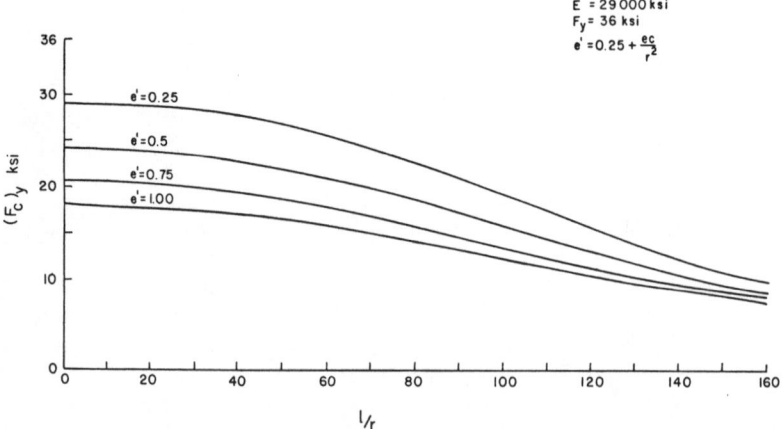

Figure 6-5. Allowable compressive stress of a beam column.

Example 6-1

A column in the living quarters of a production platform has an effective length $k\ell = 25$ ft. The column is to be designed to support 150 kips load and provide a factor of safety of 2.3 against failure. If an 8 in.^{2} tubular section with a yielding point of 36 kips/in.^{2} is used for the column, what thickness of tube is required?

Try $\frac{1}{2}$ in. wall thickness.

$A = (64 - 49) = 15$ in.^{2}

$I = (8^4 - 7^4)/12 = 199.67$ in.^{4}

$r = \sqrt{199.67/15} = 3.648$ in.

$k\ell/r = 25 \times 12/3.648 = 82.2$

From Figure 6-5 (use $e' = 0.25$)

$(F_c)_y = 23.5$ kips/in.^{2}

$(P)_y = 23.5 \times 15 = 352.5$ kips

F.S. $= 352.5/150 = 2.35 > 2.3$

DESIGN OF BEAM-COLUMNS BY INTERACTION EQUATIONS [2,3]

Interaction equations relate the interaction between the axial load, the bending moments, and the beam-column geometry at the allowable working stresses by means of simple mathematical expressions. They are frequently used because of their simplicity and versatility. The AISC specification gives the following interaction equations for combined stresses in beam-columns:

$$\frac{f_a}{F_a} + \frac{C_{mx}f_{bx}}{(1 - f_a/F'_{ex})F_{bx}} + \frac{C_{my}f_{by}}{(1 - f_a/F'_{ey})F_{by}} \leqslant 1.0 \qquad (6\text{-}22)$$

$$\frac{f_a}{0.6F_y} + \frac{f_{bx}}{F_{bx}} + \frac{f_{by}}{F_{by}} \leqslant 1.0 \qquad (6\text{-}23)$$

When $f_a/F_a \leqslant 0.15$, the following formula may be used in lieu of 6-22:

$$\frac{f_a}{F_a} + \frac{f_{bx}}{F_{bx}} + \frac{f_{by}}{F_{by}} \leqslant 1.0 \qquad (6\text{-}24)$$

Equation 6-24 neglects the bending caused by the axial load when the computed axial stress f_a is less than 15 percent of the allowable axial stress F_a.

where F_a = axial compressive stress that would be permitted if axial load alone existed.

F_b = compressive bending stress that would be permitted if bending moment alone existed.

$F_e{}' = \dfrac{12\pi^2 E}{23(kL_b/r_b)^2}$ (the Euler stress divided by a factor of safety)

f_a = computed axial stress

f_b = computed bending stress

C_m = a reduction factor (the C_m factor is used to modify the amplication factor $1/(1 - f_b/F_e{}')$ so that it cannot overestimate the extent of the secondary moment)

AISC specification recommends that C_m values depend on two conditions of the beam-column: the stability against sidesway and the presence or absence of transverse loading between points of support in the plane of bending. For compression members in frames subject to sidesway, the value of C_m can be taken as

$$C_m = 1 - 0.18\, f_a/F_e{}' = 0.85$$

For restrained compression members in frames braced against joint translation and not subject to transverse loading, the greatest amplification occurs when the end moments M_1 and M_2 are numerically equal and cause single curvature. Amplification is least when M_1 and M_2 are numerically equal and cause reverse curvature. For proper evaluation of the relationship between end moment and amplified moment, AISC recommends the following approximate value for C_m:

$$C_m = 0.6 + 0.4(M_1/M_2) \geqslant 0.4$$

where M_1/M_2 is the ratio of the smaller to larger moments at the ends of that portion of the member unbraced in the plane of bending under consideration. M_1/M_2 is positive when the member is bent in single curvature and negative when it is bent in reverse curvature.

For compression members braced against joint translation and subjected to transverse loading between the supports, the value of C_m can be approximated using the equation

$$C_m = 1 + \psi \, \frac{f_a}{F_e{}'}$$

Values of C_m in this case may be determined by rational analysis. However, in lieu of such analysis AISC recommends using the following values:

$C_m = 0.85$ for members with restrained ends

$C_m = 1.0$ for members with unrestrained ends

It should be noted that in the AISC interaction equations F_a is governed by the maximum slenderness ratio, regardless of the plane of bending. $F_e{}'$ is always governed by the slenderness ratio in the plane of bending. Thus two different values of slenderness ratio may be involved even when the flexure is about the strong axis only.

Example 6-2

The beam-column shown in Figure 6-6 is 20 ft long and is subjected to equal end moments of 65 kip-ft which cause single curvature about the strong axis of the member. If the axial load is 250 kips and the member is braced against buckling, design the member by AISC specification (use $k_x = k_y = 0.85$).

Try W12 × 96 A36 steel.

$A = 28.2$ in.^{2} $r_x = 5.44$ in.

$S = 131$ in.^{3} $r_y = 3.09$ in.

$k\ell/r_x = 0.85 \times 20 \times 12/5.44 = 37.5$

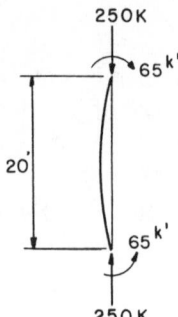

Figure 6-6. Beam-column for Example 6-2.

$k\ell/r_y = 0.85 \times 20 \times 12/3.09 = 66.0$

$F_a = 16.84 \text{ kips/in.}^2 \quad F_e' = 106.25 \text{ kips/in.}^2$

$f_a = 250/28.2 = 8.86 \text{ kips/in.}^2$

$f_b = 65 \times 12/131 = 5.95 \text{ kips/in.}^2$

Reduction factor $C_m = 0.6 + 0.4 = 1.0$

Amplification factor $= 1 - f_a/F_e' = 1 - 7.80/106.25 = 0.927$

$$\frac{f_a}{F_a} + \frac{C_m f_b}{(1 - f_a/F_e')F_b} = \frac{8.86}{16.84} + \frac{5.95}{0.927 \times 22} = 0.817 < 1.0$$

$$\frac{f_a}{0.6F_y} + \frac{f_b}{F_b} = \frac{8.86}{22} + \frac{5.95}{22} = 0.673 < 1.0$$

AXIAL TENSION AND BENDING

When members are subject to both tension and bending stresses, the axial tension tends to reduce the bending stress, since the secondary moment is opposite in sense to the applied moments. Thus the secondary moment, which is the product of the axial tension and the deflection, diminishes, rather than amplifies, the primary moment. Members subject to axial tension and bending usually fail by excessive yielding after plastification of the full cross section at the location of maximum combined stress, but the member could fail through lateral instability if bending stresses are predominant.

For the design of a member subject to both axial tension and bending stresses, AISC specification requires proportion at all points along the member's length to satisfy Equation 6-23, where f_b is the computed bending tensile stress and F_b is the allowable tensile bending stress.

DATA FOR BEAMS UNDER BOTH AXIAL AND LATERAL LOADING [4,5,7]

Data and formulas for members simultaneously subject to axial and transverse loading are provided in this section. These data and formulas can be used to compute the maximum moments, deflections, and slopes. The equations are expressed in the following forms:

$$M = M' K_u$$

$$y = y' A_u$$

$$\theta = \theta' B_u$$

where M', y', and θ' are the moment, deflection, and slope of the laterally loaded beam without the axial load. K_u, A_u and B_u are factors which give the influence of the axial force. Numerical values of the factors for various values of u are provided. By using these data, the maximum moment, deflection, and slope of a beam under simultaneous axial and lateral loading can be calculated. It should be noted that the principle of superposition, which is widely used when lateral loads act alone on a beam, can also be applied with a simple modification. If several lateral loads are acting on a beam-column, the resultant deflection (moments) can be obtained by superposing the separate deflections (moments) produced by the lateral load acting in combination with the total axial load.

Example 6-3

Calculate the maximum moment, deflection, and the end slopes of the beam-column (W14 × 120) shown in Figure 6-7.

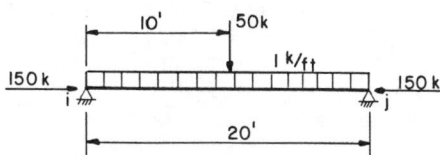

Figure 6-7. Beam-column for Example 6-3.

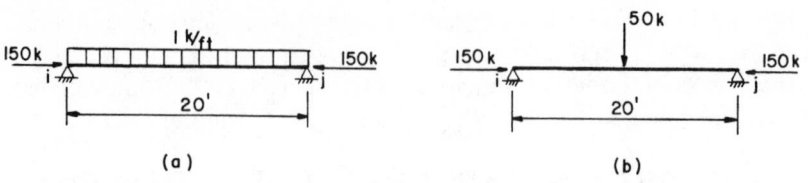

Figure 6-8. Superposition of a beam-column for Example 6-3.

The beam-column is a superposition of the two beam-columns shown in Figure 6-8. From Cases 3 and 4 the maximum moment is at the midspan and is superposed from the two individual lateral loads in combination with the axial force.

$$M_{max} = \frac{W\ell^2}{8} K_u' + \frac{W\ell}{4} K_u''$$

$$u = k\ell = \sqrt{P/EI} \; \ell = 240 \times \sqrt{150/(29,000 \times 1,380)} = 0.4646$$

$$M_{max} = \frac{400}{8} (1.02326) + \frac{50 \times 20}{4} (1.0186) = 305.81 \; \text{kip-ft}$$

$$-y_{max} = \frac{5w\ell^4}{384EI} A_u' + \frac{W\ell^3}{48EI} A_u''$$

$$= \frac{5 \times 20^4 \times 12^3 \times 1.0227}{384 \times 29,000 \times 1,380} + \frac{50 \times 20^3 \times 12^3 \times 1.02233}{48 \times 29,000 \times 1,380}$$

$$= 0.4599 \; \text{in.}$$

$$\theta_i = \theta_j = \frac{w\ell^3}{24EI} B_u' + \frac{W\ell^2}{16EI} B_u''$$

$$= \frac{1 \times 20^3 \times 12^2 \times 1.022326}{24 \times 29,000 \times 1,380} + \frac{50 \times 20^2 \times 12^2 \times 1.02326}{16 \times 29,000 \times 1,380}$$

$$= 0.005829 \; \text{radian}$$

REFERENCES

1. Timoshenko, S. and Gere, J., *Theory of Elastic Stability*, McGraw-Hill Book Company, New York, 1961.
2. Beedle, L., Blackmon, J., et al., *Structural Steel Design*, The Ronald Press Company, New York, 1964.
3. American Institute of Steel Construction, "Specification for the Design, Fabrication, and Erection of Structural Steel for Buildings," 1980.
4. Roark, R., *Formulas for Stress and Strain*, Fourth Edition, McGraw-Hill Book Company, New York, 1965.
5. Roark, R. and Young, W., *Formulas for Stress and Strain*, Fifth Edition, Mc-Graw-Hill Book Company, New York, 1975.
6. Brockenbrough, R. and Johnston, B., *Steel Design Manual*, U.S. Steel Corporation, 1968.
7. Johnston, B., Editor. *Guide to Stability Design Criteria for Metal Structures*, Third Edition, John Wiley & Sons, New York, 1976.

Cases for Chapter 6

Case 1

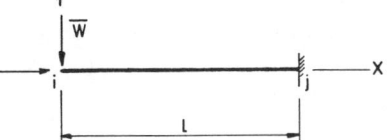

$$M_j = M_{max} = -\overline{W} l\, K_u$$

$$Y_i = Y_{max} = -\frac{\overline{W} l^3}{3EI}\, A_u$$

$$\theta_i = \frac{\overline{W} l^2}{2EI}\, B_u$$

u	K_u	A_u	B_u
0.1	1.00335	1.00402	1.00418
0.2	1.01355	1.01626	1.01694
0.3	1.03112	1.03736	1.03892
0.4	1.05698	1.06843	1.07131
0.5	1.09260	1.11126	1.11595
0.6	1.14023	1.16857	1.17571
0.7	1.20327	1.24450	1.25494
0.8	1.28705	1.34554	1.36039
0.9	1.40018	1.48213	1.50303
1.0	1.55741	1.67222	1.70163
1.1	1.78615	1.94912	1.99108
1.2	2.14346	2.38221	2.44033
1.3	2.77085	3.14352	3.24063
1.4	4.14135	4.80818	4.98315

Case 2

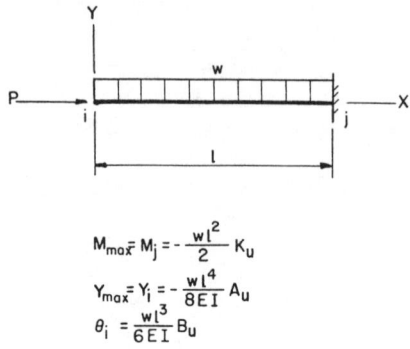

$$M_{max} = M_j = -\frac{wl^2}{2} K_u$$

$$Y_{max} = Y_i = -\frac{wl^4}{8EI} A_u$$

$$\theta_i = \frac{wl^3}{6EI} B_u$$

u	K_u	A_u	B_u
0.1	1.00251	1.00390	1.00452
0.2	1.01016	1.01581	1.01830
0.3	1.02332	1.03632	1.04205
0.4	1.04266	1.06652	1.07705
0.5	1.06926	1.10814	1.12533
0.6	1.10474	1.16381	1.19000
0.7	1.15160	1.23757	1.27580
0.8	1.21371	1.33568	1.39009
0.9	1.29732	1.46827	1.54481
1.0	1.41318	1.65274	1.76045
1.1	1.58121	1.92135	2.07500
1.2	1.84289	2.34135	2.56768
1.3	2.30106	3.07944	3.43486
1.4	3.29954	4.69293	5.33310

Case 3

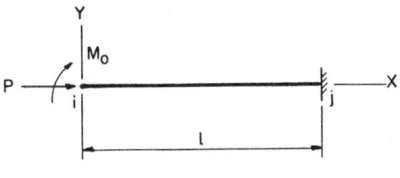

$$M_j = M_o\, K_u$$
$$Y_{max} = \frac{M_o\, l^2}{2\,EI}\, A_u$$
$$\theta_i = -\frac{M_o\, l}{EI}\, B_u$$

u	K_u	A_u	B_u
0.1	1.00502	1.00418	1.00335
0.2	1.02034	1.01694	1.01355
0.3	1.04675	1.03892	1.03112
0.4	1.08570	1.07131	1.05698
0.5	1.13949	1.11595	1.09260
0.6	1.21163	1.17571	1.14023
0.7	1.30746	1.25494	1.20327
0.8	1.43532	1.36039	1.28705
0.9	1.60873	1.50303	1.40018
1.0	1.85082	1.70163	1.55741
1.1	2.20460	1.99108	1.78615
1.2	2.75970	2.44403	2.14346
1.3	3.73833	3.24063	2.77085
1.4	5.88349	4.98315	4.14135
1.5	14.13683	11.67719	9.40094

Case 4

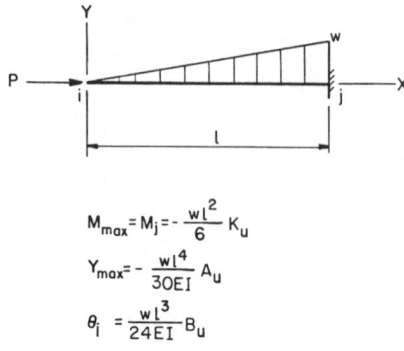

$$M_{max} = M_j = -\frac{wl^2}{6} K_u$$

$$Y_{max} = -\frac{wl^4}{30EI} A_u$$

$$\theta_i = \frac{wl^3}{24EI} B_u$$

u	K_u	A_u	B_u
0.1	1.00201	1.00387	1.00462
0.2	1.00812	1.01549	1.01899
0.3	1.01864	1.03558	1.04362
0.4	1.03408	1.06515	1.07993
0.5	1.05530	1.10591	1.13004
0.6	1.08355	1.16042	1.19718
0.7	1.12080	1.23263	1.28629
0.8	1.17007	1.32865	1.40505
0.9	1.23626	1.45841	1.56591
1.0	1.32778	1.63888	1.79021
1.1	1.46020	1.90163	2.11758
1.2	1.66596	2.31237	2.63059
1.3	2.02551	3.03405	3.53398
1.4	2.80767	4.61141	5.51227
1.5	5.80032	10.66737	13.11812

Case 5

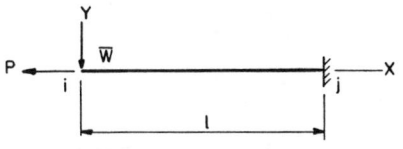

$$M_{max} = M_j = -\overline{W} l \, K_u$$
$$Y_{max} = -\frac{\overline{W} l^3}{3EI} A_u$$
$$\theta_i = \frac{\overline{W} l^2}{2EI} B_u$$

u	K_u	A_u	B_u
0.1	0.99668	0.99600	0.99585
0.2	0.98688	0.98426	0.98360
0.3	0.97104	0.96527	0.96382
0.4	0.94987	0.93989	0.93741
0.5	0.92423	0.90919	0.90545
0.6	0.89508	0.87431	0.86916
0.7	0.86338	0.83643	0.82977
0.8	0.83005	0.79666	0.78844
0.9	0.79589	0.75598	0.74619
1.0	0.76159	0.71522	0.70389
1.5	0.60343	0.52876	0.51103
2.0	0.48201	0.38849	0.36710
2.5	0.39465	0.29057	0.26782
3.0	0.33168	0.22277	0.20015
3.5	0.28519	0.17505	0.15341
4.0	0.24983	0.14066	0.12042
4.5	0.22217	0.11523	0.09657
5.0	0.19998	0.09600	0.07892

Case 6

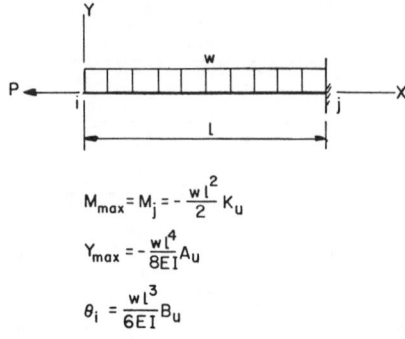

$$M_{max} = M_j = -\frac{w\,l^2}{2}\,K_u$$

$$Y_{max} = -\frac{w\,l^4}{8EI}A_u$$

$$\theta_i = \frac{w\,l^3}{6EI}B_u$$

u	K_u	A_u	B_u
0.1	0.99751	0.99608	0.99552
0.2	0.99015	0.98469	0.98229
0.3	0.97826	0.96624	0.96094
0.4	0.96234	0.94158	0.93243
0.5	0.94302	0.91168	0.89797
0.6	0.92100	0.87775	0.85887
0.7	0.89699	0.84089	0.81645
0.8	0.87165	0.80211	0.77199
0.9	0.84559	0.76251	0.72661
1.0	0.81930	0.72281	0.68124
1.5	0.69584	0.54073	0.47556
2.0	0.59693	0.40307	0.32432
2.5	0.52147	0.30626	0.22231
3.0	0.46322	0.23857	0.15491
3.5	0.41697	0.19038	0.11013
4.0	0.37924	0.15519	0.08000
4.5	0.34776	0.12884	0.05925
5.0	0.32104	0.10863	0.04476

Case 7

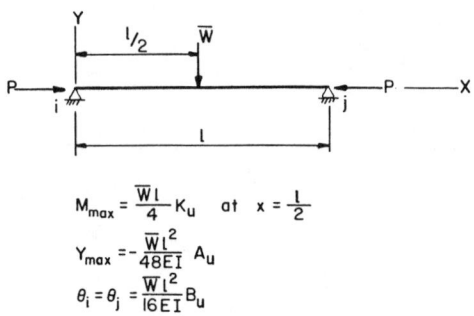

$$M_{max} = \frac{\overline{W}l}{4} K_u \quad at \quad x = \frac{l}{2}$$

$$Y_{max} = -\frac{\overline{W}l^2}{48EI} A_u$$

$$\theta_i = \theta_j = \frac{\overline{W}l^2}{16EI} B_u$$

u	K_u	A_u	B_u	u	K_u	A_u	B_u
0.1	1.00083	1.00100	1.00104	1.6	1.28705	1.34554	1.36039
0.2	1.00335	1.00402	1.00418	1.7	1.33921	1.40851	1.42613
0.3	1.00757	1.00908	1.00946	1.8	1.40018	1.48213	1.50303
0.4	1.01355	1.01626	1.01694	1.9	1.47198	1.56891	1.59368
0.5	1.02137	1.02565	1.02672	2.0	1.55741	1.67222	1.70163
0.6	1.03112	1.03736	1.03892	2.1	1.66030	1.79674	1.83177
0.7	1.04294	1.05156	1.05372	2.2	1.78615	1.94912	1.99108
0.8	1.05698	1.06843	1.07131	2.3	1.94304	2.13922	2.18988
0.9	1.07346	1.08823	1.09194	2.4	2.14346	2.38221	2.44403
1.0	1.09260	1.11126	1.11595	2.5	2.40766	2.70270	2.77934
1.1	1.11474	1.13789	1.14371	2.6	2.77085	3.14352	3.24063
1.2	1.14023	1.16857	1.17571	2.7	3.300164	3.78628	3.91338
1.3	1.16954	1.20387	1.21254	2.8	4.14135	4.80818	4.98315
1.4	1.20327	1.24450	1.25494	2.9	5.68144	6.67982	6.94275
1.5	1.24213	1.29135	1.30383	3.0	9.40095	11.20126	11.67718

Case 8

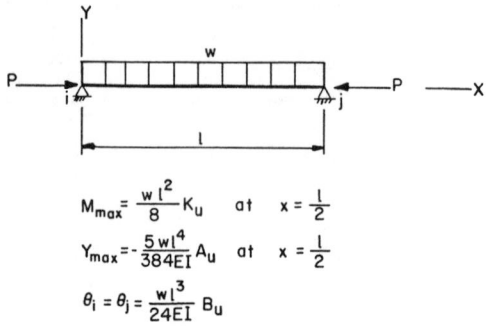

$$M_{max} = \frac{w l^2}{8} K_u \quad at \quad x = \frac{l}{2}$$

$$Y_{max} = -\frac{5 w l^4}{384 EI} A_u \quad at \quad x = \frac{l}{2}$$

$$\theta_i = \theta_j = \frac{w l^3}{24 EI} B_u$$

u	K_u	A_u	B_u	u	K_u	A_u	B_u
0.1	1.00104	1.00116	1.00100	1.6	1.36039	1.35146	1.34554
0.2	1.00418	1.00408	1.00416	1.7	1.42613	1.41553	1.40851
0.3	1.00946	1.00923	1.00908	1.8	1.50303	1.49045	1.48213
0.4	1.01694	1.01653	1.01626	1.9	1.59368	1.57877	1.56891
0.5	1.02672	1.02608	1.02565	2.0	1.70163	1.68392	1.67222
0.6	1.03892	1.03800	1.03736	2.1	1.83177	1.81066	1.79674
0.7	1.05372	1.05242	1.05156	2.2	1.99108	1.96578	1.94912
0.8	1.07131	1.06958	1.06843	2.3	2.18988	2.15932	2.13922
0.9	1.09194	1.08972	1.08823	2.4	2.44403	2.40672	2.38221
1.0	1.11595	1.11313	1.11126	2.5	2.77934	2.73306	2.70270
1.1	1.14371	1.14021	1.13789	2.6	3.24063	3.18196	3.14352
1.2	1.17571	1.17142	1.16857	2.7	3.91338	3.83655	3.78628
1.3	1.21254	1.20733	1.20387	2.8	4.98315	4.87733	4.80818
1.4	1.25494	1.24866	1.24450	2.9	6.94275	6.78364	6.67982
1.5	1.30383	1.29633	1.29135	3.0	11.67718	11.38900	11.20126

Case 9

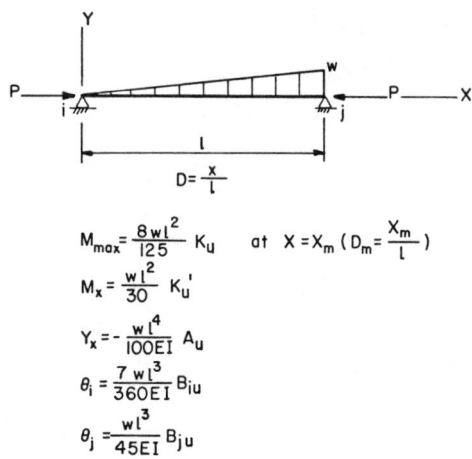

$$M_{max} = \frac{8wl^2}{125} K_u \quad \text{at} \quad X = X_m \left(D_m = \frac{X_m}{l}\right)$$

$$M_x = \frac{wl^2}{30} K_u'$$

$$Y_x = -\frac{wl^4}{100EI} A_u$$

$$\theta_i = \frac{7wl^3}{360EI} B_{iu}$$

$$\theta_j = \frac{wl^3}{45EI} B_{ju}$$

u	D_m	K_u	B_{iu}	B_{ju}
0.2	0.5771	1.00641	1.00427	1.00380
0.4	0.5763	1.01866	1.01715	1.01548
0.6	0.5750	1.03981	1.03978	1.03557
0.8	0.5732	1.07099	1.07208	1.06512
1.0	0.5708	1.11390	1.11746	1.10583
1.2	0.5678	1.17161	1.17807	1.16025
1.4	0.5642	1.24806	1.25846	1.23229
1.6	0.5600	1.34990	1.36557	1.32801
1.8	0.5551	1.48781	1.51057	1.45725
2.0	0.5494	1.70535	1.71268	1.63682
2.2	0.5429	1.96077	2.00755	1.89800
2.4	0.5356	2.40071	2.46949	2.30584
2.6	0.5273	3.17576	3.28279	3.02166
2.8	0.5182	4.87399	5.06360	4.58469
3.0	0.5079	11.40696	11.91016	10.58098

Case 9 continued

u	K'$_u$				A$_u$			
	D = 0.2	D = 0.4	D = 0.6	D = 0.8	D = 0.2	D = 0.4	D = 0.6	D = 0.8
0.2	0.96442	1.68733	1.92761	1.44490	0.36848	0.61105	0.63402	0.40800
0.4	0.97791	1.70970	1.95080	1.45982	0.37319	0.61872	0.64175	0.41281
0.6	1.00118	1.74825	1.99075	1.48548	0.38131	0.63195	0.65510	0.42113
0.8	1.03551	1.80507	2.04955	1.52321	0.39327	0.65143	0.67475	0.43378
1.0	1.08293	1.88349	2.13055	1.57507	0.40977	0.67829	0.70185	0.45025
1.2	1.14656	1.98855	2.23885	1.64425	0.43186	0.71423	0.73807	0.47280
1.4	1.23116	2.12800	2.38221	1.73557	0.46116	0.76190	0.78607	0.50267
1.6	1.34415	2.31387	2.57275	1.85654	0.50019	0.82535	0.84993	0.54237
1.8	1.49754	2.56567	2.83002	2.01929	0.55303	0.91118	0.93624	0.59598
2.0	1.71197	2.91684	3.18758	2.24461	0.62664	1.03070	1.05631	0.67051
2.2	2.02580	3.42954	3.70766	2.57102	0.73402	1.20492	1.23117	0.77894
2.4	2.51900	4.23325	4.52000	3.07871	0.90220	1.47758	1.50457	0.94833
2.6	3.38999	5.64915	5.94562	3.96637	1.19822	1.95718	1.98502	1.24574
2.8	5.30237	8.75113	9.05892	5.89792	1.84625	3.00643	3.03526	1.89538
3.0	12.67050	20.68195	21.00282	13.28815	4.33722	7.03776	7.06771	4.38820

Case 10

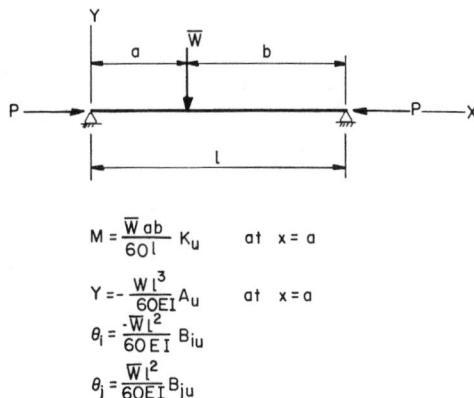

$$M = \frac{\overline{W} ab}{60 l} K_u \quad \text{at} \quad x = a$$

$$Y = -\frac{W l^3}{60 EI} A_u \quad \text{at} \quad x = a$$

$$\theta_i = \frac{-\overline{W} l^2}{60 EI} B_{iu}$$

$$\theta_j = \frac{\overline{W} l^2}{60 EI} B_{ju}$$

	K_u				A_u			
u	C = 0.2	C = 0.4	C = 0.6	C = 0.8	C = 0.2	C = 0.4	C = 0.6	C = 0.8
0.2	1.00214	1.00321	1.00321	1.00214	0.51381	1.15656	1.15656	0.51381
0.4	1.00866	1.01300	1.01300	1.00866	0.51932	1.17048	1.17048	0.51932
0.6	1.01983	1.02986	1.02986	1.01983	0.52881	1.19446	1.19446	0.52881
0.8	1.03618	1.05466	1.05466	1.03618	0.54277	1.22976	1.22976	0.54277
1.0	1.05854	1.08878	1.08878	1.05854	0.56197	1.27841	1.27841	0.56197
1.2	1.08814	1.13435	1.13435	1.08814	0.58759	1.34349	1.34349	0.58759
1.4	1.12688	1.19460	1.19460	1.12688	0.62144	1.42969	1.42969	0.62144
1.6	1.17769	1.27455	1.27455	1.17769	0.66633	1.54433	1.54433	0.66633
1.8	1.24529	1.38233	1.38233	1.24529	0.72680	1.69925	1.69925	0.72680
2.0	1.33775	1.53187	1.53187	1.33775	0.81060	1.91473	1.91473	0.81060
2.2	1.46996	1.74901	1.74901	1.46996	0.93216	2.22845	2.22845	0.93216
2.4	1.67288	2.08754	2.08754	1.67288	1.12147	2.71886	2.71886	1.12147
2.6	2.02306	2.68084	2.68084	2.02306	1.45287	3.58049	3.58049	1.45287
2.8	2.77607	3.97464	3.97464	2.77607	2.17478	5.46362	5.46362	2.17478
3.0	5.63048	8.93314	8.93314	5.63048	4.93918	12.69302	12.69302	4.93918

	B_{iu}				B_{ju}			
u	C = 0.2	C = 0.4	C = 0.6	C = 0.8	C = 0.2	C = 0.4	C = 0.6	C = 0.8
0.2	2.88979	3.85522	3.37467	1.92884	1.92884	3.37467	3.85522	2.88979
0.4	2.91963	3.90161	3.41940	1.95583	1.95583	3.41940	3.90161	2.91963
0.6	2.97096	3.98150	3.49650	2.00236	2.00236	3.49650	3.98150	2.97096
0.8	3.04642	4.09910	3.61015	2.07101	2.07101	3.61015	4.09910	3.04642
1.0	3.15015	4.26110	3.76697	2.16586	2.16586	3.76697	4.26110	3.15015
1.2	3.28850	4.47769	3.97710	2.29313	2.29313	3.97710	4.47769	3.28850
1.4	3.47114	4.76442	4.25600	2.46232	2.46233	4.25600	4.76442	3.47114
1.6	3.71307	5.14549	4.62774	2.68830	2.68830	4.62774	5.14549	3.71307
1.8	4.03859	5.66005	5.13134	2.99508	2.99508	5.13134	5.66005	4.03859
2.0	4.48922	6.37515	5.83369	3.42394	3.42394	5.83369	6.37515	4.48922
2.2	5.14204	7.41531	6.85908	4.05160	4.05160	6.85908	7.41531	5.14204
2.4	6.15742	9.03979	8.46651	5.03800	5.03800	8.46651	9.03979	6.15742
2.6	7.93274	11.89124	11.29831	6.78000	6.78000	11.29831	11.89124	7.93274
2.8	11.79585	18.11784	17.50226	10.60474	10.60474	17.50226	18.11784	11.79585
3.0	26.57630	42.00565	41.36390	25.34100	25.34100	41.36390	42.00565	26.57630

Case 11

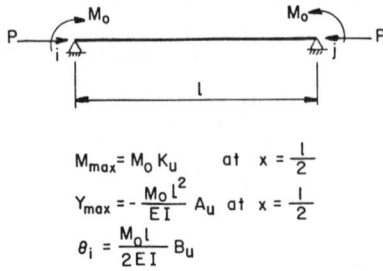

$$M_{max} = M_0 K_u \quad at \quad x = \frac{l}{2}$$

$$Y_{max} = -\frac{M_0 l^2}{EI} A_u \quad at \quad x = \frac{l}{2}$$

$$\theta_i = \frac{M_0 l}{2EI} B_u$$

u	K_u	A_u	B_u	u	K_u	A_u	B_u
0.1	1.00125	0.12513	1.00084	1.6	1.43532	0.17005	1.28705
0.2	1.00502	0.12552	1.00335	1.7	1.51519	0.17827	1.33935
0.3	1.01136	0.12618	1.00757	1.8	1.60873	0.18788	1.40018
0.4	1.02034	0.12712	1.01355	1.9	1.71915	0.19921	1.47198
0.5	1.03209	0.12834	1.02137	2.0	1.85082	0.21270	1.55741
0.6	1.04675	0.12987	1.03112	2.1	2.00976	0.22897	1.66030
0.7	1.06454	0.13171	1.04294	2.2	2.20460	0.24885	1.78615
0.8	1.08570	0.13391	1.05698	2.3	2.44806	0.27373	1.94304
0.9	1.11056	0.13649	1.07346	2.4	2.75970	0.30550	2.14346
1.0	1.13949	0.13949	1.09260	2.5	3.17136	0.34742	2.40766
1.1	1.17299	0.14296	1.11474	2.6	3.73833	0.40508	2.77085
1.2	1.21163	0.14696	1.14023	2.7	4.56607	0.48917	3.30016
1.3	1.25615	0.15157	1.16955	2.8	5.88349	0.62289	4.14135
1.4	1.30746	0.15687	1.20327	2.9	8.29856	0.86784	5.68144
1.5	1.36670	0.16298	1.24213	3.0	14.13683	1.45965	9.40094

Case 12

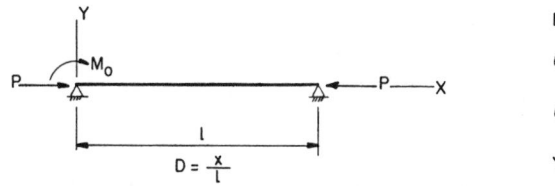

$$M_x = M_0 K_u$$
$$\theta_i = -\frac{M_0 l}{3EI} B_{iu}$$
$$\theta_j = \frac{M_0 l}{6EI} B_{ju}$$
$$Y = \frac{-M_0 l^2}{EI} A_u$$

			K_u			
u	B_{iu}	B_{ju}	D = 0.2	D = 0.4	D = 0.6	D = 0.8
0.2	1.00268	1.00469	0.80193	0.60257	0.40225	0.20129
0.4	1.01083	1.01899	0.80779	0.61040	0.40912	0.20522
0.6	1.02485	1.04366	0.81783	0.62389	0.42098	0.21201
0.8	1.04545	1.08006	0.83250	0.64372	0.43851	0.22209
1.0	1.07372	1.13037	0.85250	0.67102	0.46278	0.23610
1.2	1.11138	1.19792	0.87892	0.70746	0.49545	0.25504
1.4	1.16102	1.28777	0.91339	0.75564	0.53903	0.28044
1.6	1.22665	1.40785	0.95842	0.81954	0.59745	0.31470
1.8	1.31477	1.57100	1.01808	0.90564	0.67709	0.36173
2.0	1.43649	1.79925	1.09928	1.02501	0.78891	0.42826
2.2	1.61242	2.13360	1.21479	1.19817	0.95330	0.52683
2.4	1.88544	2.65950	1.39111	1.46782	1.21278	0.68365
2.6	2.36176	3.58902	1.69375	1.93975	1.67294	0.96388
2.8	3.39626	5.63151	2.34132	2.96740	2.68696	1.58569
3.0	7.34859	13.50566	4.78645	6.90085	6.60459	4.00115

	A_u			
u	D = 0.2	D = 0.4	D = 0.6	D = 0.8
0.2	0.04816	0.06425	0.05624	0.03215
0.4	0.04866	0.06502	0.05699	0.03260
0.6	0.04952	0.06636	0.05828	0.03337
0.8	0.05077	0.06832	0.06017	0.03452
1.0	0.05250	0.07102	0.06278	0.03610
1.2	0.05481	0.07463	0.06629	0.03822
1.4	0.05785	0.07941	0.07093	0.04104
1.6	0.06188	0.08576	0.07713	0.04481
1.8	0.06731	0.09433	0.08552	0.04918
2.0	0.07482	0.10625	0.09723	0.05707
2.2	0.08570	0.12359	0.11432	0.06753
2.4	0.10262	0.15066	0.14111	0.08397
2.6	0.13221	0.19819	0.18831	0.11300
2.8	0.19660	0.30196	0.29170	0.17675
3.0	0.44293	0.70009	0.68940	0.42235

Case 13

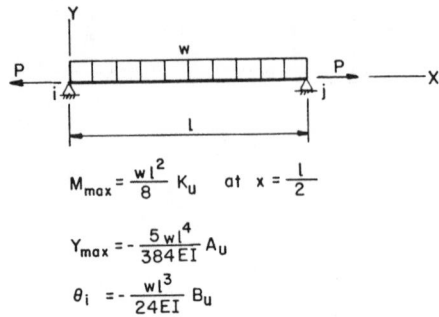

$$M_{max} = \frac{w l^2}{8} K_u \quad at \quad x = \frac{l}{2}$$

$$Y_{max} = -\frac{5 w l^4}{384 EI} A_u$$

$$\theta_i = -\frac{w l^3}{24 EI} B_u$$

u	K_u	A_u	B_u
0.1	0.99896	0.99886	0.99901
0.2	0.99585	0.99595	0.99601
0.3	0.99071	0.99093	0.99108
0.4	0.98360	0.98399	0.98426
0.5	0.97460	0.97521	0.97562
0.6	0.96382	0.96470	0.96527
0.7	0.95138	0.95250	0.95332
0.8	0.93741	0.93890	0.93989
0.9	0.92205	0.92390	0.92513
1.0	0.90545	0.90769	0.90919
1.5	0.80928	0.81373	0.81672
2.0	0.70389	0.71066	0.71522
2.5	0.60068	0.61335	0.61704
3.0	0.51103	0.52157	0.52876
3.5	0.43274	0.44454	0.45264
4.0	0.36710	0.37974	0.38849
4.5	0.31280	0.32578	0.33501
5.0	0.26782	0.28116	0.29057

Case 14

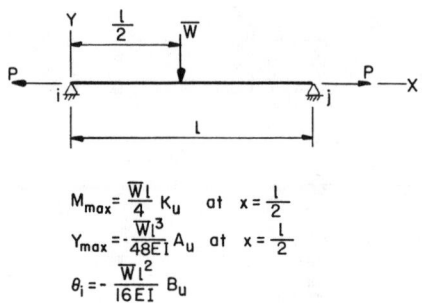

$$M_{max} = \frac{\overline{W}l}{4} K_u \quad at \quad x = \frac{l}{2}$$

$$Y_{max} = -\frac{\overline{W}l^3}{48EI} A_u \quad at \quad x = \frac{l}{2}$$

$$\theta_i = -\frac{\overline{W}l^2}{16EI} B_u$$

u	K_u	A_u	B_u
0.1	0.99917	0.99888	0.99896
0.2	0.99668	0.99600	0.99585
0.3	0.99257	0.99111	0.99071
0.4	0.98688	0.98426	0.98360
0.5	0.97967	0.97562	0.97461
0.6	0.97104	0.96527	0.96385
0.7	0.96107	0.95332	0.95139
0.8	0.94987	0.93989	0.93741
0.9	0.93755	0.92514	0.92205
1.0	0.92423	0.90919	0.90545
1.5	0.84687	0.81672	0.80928
2.0	0.76159	0.71522	0.70389
2.5	0.67863	0.61704	0.60219
3.0	0.60343	0.52876	0.51103
3.5	0.53793	0.45264	0.43274
4.0	0.48201	0.38849	0.36710
4.5	0.43419	0.33530	0.31260
5.0	0.39465	0.29057	0.26782

Case 15

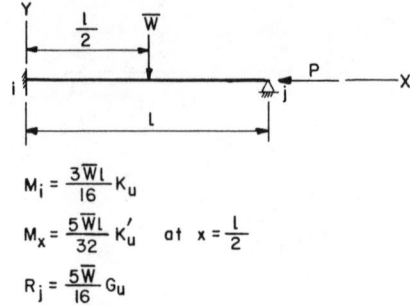

$$M_i = \frac{3\overline{W}L}{16} K_u$$

$$M_x = \frac{5\overline{W}L}{32} K'_u \quad \text{at} \quad x = \frac{L}{2}$$

$$R_j = \frac{5\overline{W}}{16} G_u$$

u	K_u	K'_u	G_u
0.2	1.00150	1.00144	0.99916
0.4	1.00604	1.00578	0.99637
0.6	1.01373	1.01312	0.99177
0.8	1.02474	1.02364	0.98516
1.0	1.03933	1.03758	0.97640
1.2	1.05788	1.05531	0.96527
1.4	1.08089	1.07730	0.95147
1.6	1.10903	1.10419	0.93458
1.8	1.14319	1.13683	0.91409
2.0	1.18458	1.17639	0.88925
2.2	1.23484	1.22443	0.85910
2.4	1.29627	1.28315	0.82224
2.6	1.37213	1.35568	0.77672
2.8	1.46725	1.44664	0.71535
3.0	1.58904	1.56313	0.64249

Case 16

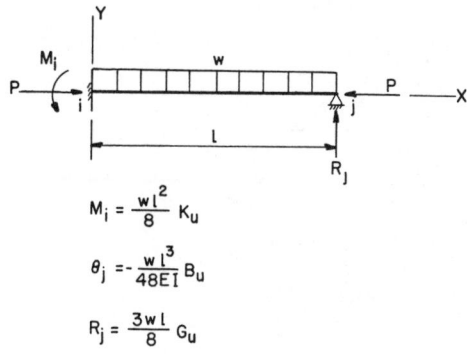

$$M_i = \frac{wl^2}{8} K_u$$

$$\theta_j = -\frac{wl^3}{48EI} B_u$$

$$R_j = \frac{3wl}{8} G_u$$

u	K_u	B_u	G_u
0.2	1.00134	1.00210	0.99955
0.4	1.00537	1.00806	0.99821
0.6	1.01220	1.01833	0.99593
0.8	1.02199	1.03306	0.99267
1.0	1.03496	1.05263	0.98835
1.2	1.05145	1.07758	0.98285
1.4	1.07191	1.10864	0.97603
1.6	1.09692	1.14677	0.96769
1.8	1.12730	1.19328	0.95757
2.0	1.16411	1.24993	0.94530
2.2	1.20882	1.31911	0.93039
2.4	1.26348	1.40421	0.91217
2.6	1.33101	1.51002	0.88966
2.8	1.41573	1.64369	0.86142
3.0	1.52427	1.81618	0.82524

Case 17

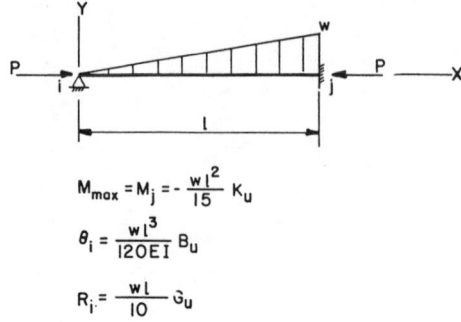

$$M_{max} = M_j = -\frac{w\,l^2}{15}\,K_u$$

$$\theta_i = \frac{w\,l^3}{120EI}\,B_u$$

$$R_i = \frac{w\,l}{10}\,G_u$$

u	K_u	B_u	G_u	u	K_u	B_u	G_u
0.1	1.00011	1.00029	0.99984	1.6	1.08264	1.15407	0.94491
0.2	1.00115	1.00209	0.99923	1.7	1.09489	1.17733	0.93674
0.3	1.00258	1.00474	0.99828	1.8	1.10837	1.20301	0.92775
0.4	1.00460	1.00845	0.99693	1.9	1.12318	1.23136	0.91788
0.5	1.00722	1.01326	0.99518	2.0	1.13946	1.26268	0.90702
0.6	1.01045	1.01921	0.99303	2.1	1.15738	1.29730	0.89508
0.7	1.01431	1.02632	0.99046	2.2	1.17711	1.33564	0.88193
0.8	1.01882	1.03465	0.98745	2.3	1.19889	1.37818	0.86741
0.9	1.02401	1.04425	0.98399	2.4	1.22297	1.42550	0.85135
1.0	1.02991	1.05518	0.98006	2.5	1.24968	1.47829	0.83355
1.1	1.03655	1.06752	0.97563	2.6	1.27941	1.53739	0.81373
1.2	1.04397	1.08137	0.97069	2.7	1.31263	1.60385	0.79158
1.3	1.05223	1.09682	0.96518	2.8	1.34992	1.67895	0.76672
1.4	1.06139	1.11399	0.95908	2.9	1.39202	1.76429	0.73865
1.5	1.07150	1.13302	0.95234	3.0	1.43987	1.86191	0.70676

Case 18

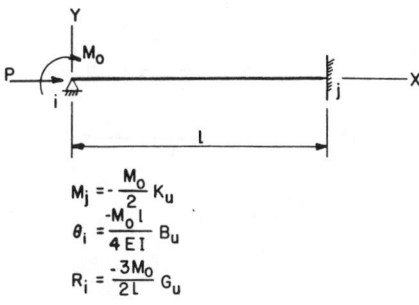

$$M_j = -\frac{M_o}{2} K_u$$

$$\theta_i = \frac{-M_o l}{4EI} B_u$$

$$R_i = \frac{-3M_o}{2l} G_u$$

u	K_u	B_u	G_u
0.2	1.00200	1.00134	1.00067
0.4	1.00807	1.00537	1.00269
0.6	1.01835	1.01220	1.00612
0.8	1.03311	1.02199	1.01104
1.0	1.05276	1.03496	1.01759
1.2	1.07786	1.05145	1.02595
1.4	1.10917	1.07191	1.03639
1.6	1.14772	1.09692	1.04924
1.8	1.19489	1.12730	1.06496
2.0	1.25254	1.16411	1.08418
2.2	1.32323	1.20882	1.10774
2.4	1.41054	1.26348	1.13685
2.6	1.51964	1.33101	1.17321
2.8	1.65815	1.41572	1.21938
3.0	1.83786	1.52427	1.27929

Case 19

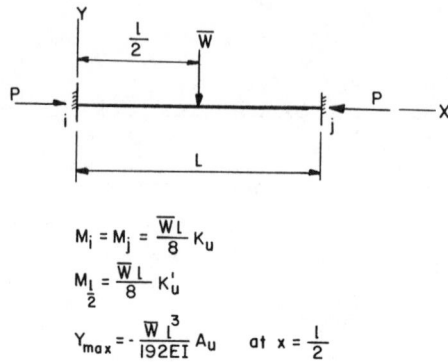

$$M_i = M_j = \frac{\overline{W}L}{8} K_u$$

$$M_{\frac{L}{2}} = \frac{\overline{W}L}{8} K_u'$$

$$Y_{max} = -\frac{\overline{W}L^3}{192EI} A_u \quad \text{at} \ x = \frac{L}{2}$$

u	K_u	K'_u	A_u	u	K_u	K_u'	A_u
0.1	1.00021	1.00021	1.00025	1.6	1.05698	1.05698	1.06843
0.2	1.00083	1.00083	1.00100	1.7	1.06490	1.06490	1.07795
0.3	1.00188	1.00188	1.00225	1.8	1.07346	1.07346	1.08823
0.4	1.00335	1.00335	1.00402	1.9	1.08268	1.08268	1.09932
0.5	1.00524	1.00524	1.00629	2.0	1.09261	1.09261	1.11126
0.6	1.00757	1.00757	1.00908	2.1	1.10328	1.10328	1.12410
0.7	1.01034	1.01034	1.01240	2.2	1.11474	1.11474	1.13789
0.8	1.01355	1.01355	1.01626	2.3	1.12704	1.12704	1.15269
0.9	1.01722	1.01722	1.02067	2.4	1.14023	1.14023	1.16857
1.0	1.02137	1.02137	1.02565	2.5	1.15438	1.15438	1.18560
1.1	1.02599	1.02599	1.03121	2.6	1.16955	1.16955	1.20387
1.2	1.03112	1.03112	1.03736	2.7	1.18581	1.18581	1.22347
1.3	1.03676	1.03676	1.04414	2.8	1.20327	1.20327	1.24450
1.4	1.04294	1.04294	1.05156	2.9	1.22200	1.22200	1.26709
1.5	1.04967	1.04967	1.05965	3.0	1.24213	1.24213	1.29135

Case 20

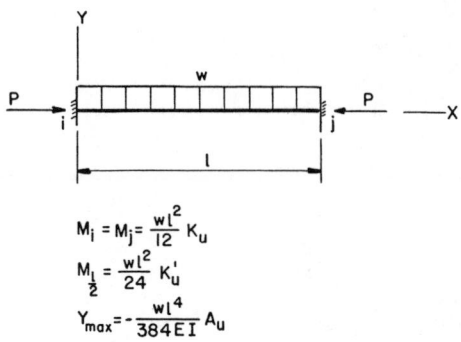

$$M_i = M_j = \frac{wl^2}{12} K_u$$

$$M_{\frac{l}{2}} = \frac{wl^2}{24} K_u'$$

$$Y_{max} = -\frac{wl^4}{384EI} A_u$$

u	K_u	K_u'	A_u	u	K_u	K_u'	A_u
0.1	1.00017	1.00029	1.00016	1.6	1.04545	1.08006	1.06843
0.2	1.00067	1.00117	1.00101	1.7	1.05174	1.09123	1.07795
0.3	1.00150	1.00263	1.00226	1.8	1.05853	1.10330	1.08823
0.4	1.00268	1.00469	1.00416	1.9	1.06585	1.11633	1.09932
0.5	1.00419	1.00734	1.00629	2.0	1.07372	1.13037	1.11126
0.6	1.00605	1.01060	1.00908	2.1	1.08218	1.14548	1.12410
0.7	1.00826	1.01448	1.01240	2.2	1.09124	1.16172	1.13789
0.8	1.01083	1.01899	1.01626	2.3	1.10097	1.17917	1.15269
0.9	1.01377	1.02414	1.02067	2.4	1.11138	1.19792	1.16857
1.0	1.01707	1.02996	1.02565	2.5	1.12254	1.21804	1.18560
1.1	1.02077	1.03645	1.03121	2.6	1.13450	1.23964	1.20387
1.2	1.02485	1.04366	1.03736	2.7	1.14730	1.26284	1.22347
1.3	1.02935	1.05159	1.04414	2.8	1.16102	1.28777	1.24451
1.4	1.03427	1.06028	1.05156	2.9	1.17573	1.31456	1.26709
1.5	1.03963	1.06976	1.05965	3.0	1.19150	1.34338	1.29135

Case 21

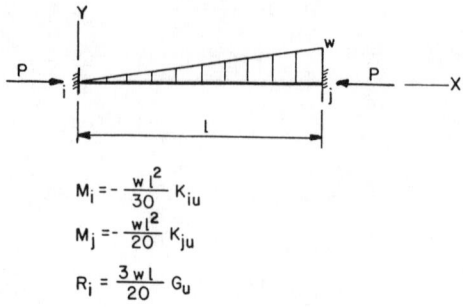

$$M_i = -\frac{w l^2}{30} K_{iu}$$

$$M_j = -\frac{w l^2}{20} K_{ju}$$

$$R_i = \frac{3 w l}{20} G_u$$

u	K_{iu}	K_{ju}	G_u
0.2	1.00074	1.00121	0.99979
0.4	1.00306	1.00241	0.99988
0.6	1.00692	1.00548	0.99971
0.8	1.01239	1.00979	0.99949
1.0	1.01954	1.01543	0.99920
1.2	1.02845	1.02245	0.99884
1.4	1.03926	1.03094	0.99841
1.6	1.05210	1.04101	0.99791
1.8	1.06716	1.05278	0.99733
2.0	1.08468	1.06642	0.99668
2.2	1.10492	1.08213	0.99594
2.4	1.12824	1.10015	0.99511
2.6	1.15506	1.12079	0.99420
2.8	1.18593	1.14441	0.99318
3.0	1.22151	1.17150	0.99206

Case 22

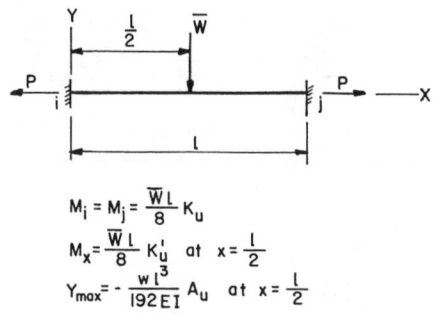

$$M_i = M_j = \frac{\overline{W} L}{8} K_u$$

$$M_x = \frac{\overline{W} L}{8} K_u' \quad \text{at} \quad x = \frac{l}{2}$$

$$Y_{max} = -\frac{w\,l^3}{192\,EI} A_u \quad \text{at} \quad x = \frac{l}{2}$$

u	K_u	K'_u	A_u
0.25	0.99867	0.99867	0.99950
0.50	0.99482	0.99482	0.99379
0.75	0.98844	0.98844	0.98609
1.00	0.97967	0.97967	0.97562
1.25	0.96867	0.96867	0.96242
1.50	0.95562	0.95562	0.94678
1.75	0.94073	0.94073	0.92894
2.00	0.92423	0.92423	0.90919
2.50	0.88736	0.88736	0.86508
3.00	0.84687	0.84687	0.81672
3.50	0.80446	0.80446	0.76618
4.00	0.76159	0.76159	0.71522
4.50	0.71938	0.71938	0.66518
5.00	0.67863	0.67863	0.61704
5.50	0.63987	0.63987	0.57144
5.00	0.60343	0.60343	0.52876

Case 23

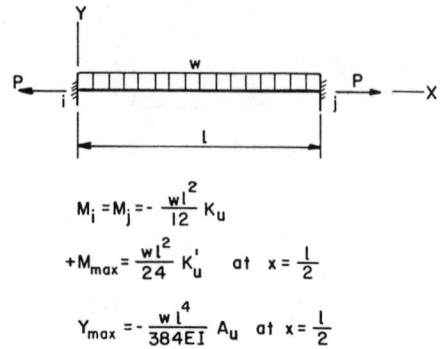

$$M_i = M_j = -\frac{wl^2}{12} K_u$$

$$+M_{max} = \frac{wl^2}{24} K'_u \quad \text{at} \quad x = \frac{l}{2}$$

$$Y_{max} = -\frac{wl^4}{384EI} A_u \quad \text{at} \quad x = \frac{l}{2}$$

u	K_u	K'_u	A_u
0.25	0.99896	0.99826	0.99822
0.50	0.99586	0.99276	0.99379
0.75	0.99075	0.98384	0.98614
1.00	0.98372	0.97158	0.97562
1.25	0.97489	0.95623	0.96242
1.50	0.96440	0.93806	0.94678
1.75	0.95242	0.91736	0.92894
2.00	0.93911	0.89449	0.90919
2.50	0.90924	0.84359	0.86508
3.00	0.87625	0.78810	0.81672
3.50	0.84145	0.73049	0.76618
4.00	0.80597	0.67284	0.71522
4.50	0.77070	0.61674	0.66518
5.00	0.73628	0.56332	0.61704
5.50	0.70317	0.51329	0.57144
6.00	0.67164	0.46702	0.52876

Chapter 7

Plates

A plane x-y is the middle plane of a plate before bending, and w denotes the small deflections of the plate in the z direction. The plate undergoes small deflections during bending, and the x-y plane forms the middle surface of the plate. We assume that:

1. The middle surface of the plate is the neutral surface of the plate.
2. The middle surface of the bent plate is a developable surface and undergoes no stretching during bending.
3. Deflections of the plate, w, are small in comparison with its thickness, t.

The curvature of the surface in a plane parallel to the x-z plane can be expressed by

$$\rho_x = -\frac{\partial^2 w}{\partial x^2}$$

and the curvature in the plane parallel to the y-z plane is expressed by

$$\rho_y = -\frac{\partial^2 w}{\partial y^2}$$

The twist of the surface with respect to the x and y axes is expressed by

$$\rho_{xy} = \frac{\partial^2 w}{\partial x \partial y}$$

A curvature is considered positive when it is convex downward. Let M_x and M_y be bending moments per unit length acting on the axes paral-

159

lel to the y and x axes, respectively, as shown in Figure 7-1. The bending moments can be expressed by

$$M_x = -D\left(\frac{\partial^2 w}{\partial x^2} + \nu \frac{\partial^2 w}{\partial y^2}\right) = D\left(\frac{1}{r_x} + \nu \frac{1}{r_y}\right) \tag{7-1}$$

$$M_y = -D\left(\frac{\partial^2 w}{\partial y^2} + \nu \frac{\partial^2 w}{\partial x^2}\right) = D\left(\frac{1}{r_y} + \nu \frac{1}{r_x}\right) \tag{7-2}$$

$$M_{xy} = D(1 - \nu)\frac{\partial^2 w}{\partial x \partial y} \tag{7-3}$$

where ν is Poisson's ratio, D is the flexural rigidity of the plate, and w denotes small deflections of the plate in the z direction. Note that the D quantity takes the place of the quantity EI in the case of beams and is expressed by

$$D = \frac{Et^3}{12(1 - \nu^2)} \tag{7-4}$$

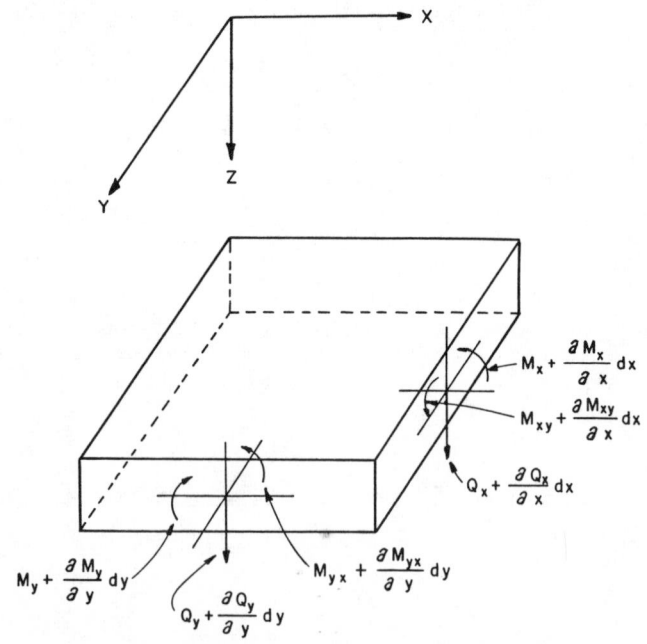

Figure 7-1. Moments and shear forces on a plate element.

The set of Equations 7-1 through 7-3 is equivalent to the following single equation in the beam theory:

$$M = EI \frac{d^2y}{dx^2}$$

The moments in Equations 7-1 and 7-2 are considered positive when they produce compression in the upper surface and tension in the lower surface of the plate.

Next we can derive the equivalent of the beam equation

$$V = \frac{dM}{dx} = EI \frac{d^3y}{dx^3}$$

for flat plate. If the shear forces per unit length are denoted by Q_x and Q_y, we can obtain the expression of unit shear forces for flat plate from the requirement of torque equilibrium. From Figure 7-1 we can see that the moments M_y and M_{xy} create an unbalanced torque of

$$\Delta T = \left(\frac{\partial M_y}{\partial y} + \frac{\partial M_{xy}}{\partial x} \right) dx \, dy$$

ΔT will be held in equilibrium by the moment Q_x dxdy, produced by the vertical shear force Q_x dx. Equating this to the unbalanced torque gives the following equation:

$$Q_x = \frac{\partial M_x}{\partial x} + \frac{\partial M_{yx}}{\partial y} = -D \frac{\partial}{\partial x} \left(\frac{\partial^2 w}{\partial x^2} + \frac{\partial^2 w}{\partial y^2} \right) \tag{7-5}$$

Similarly, we can obtain

$$Q_y = \frac{\partial M_y}{\partial y} - \frac{\partial M_{xy}}{\partial x} = -D \frac{\partial}{\partial y} \left(\frac{\partial^2 w}{\partial x^2} - \frac{\partial^2 w}{\partial y^2} \right) \tag{7-6}$$

Finally, we can derive the equivalent of the beam formula

$$q = \frac{dV}{dx} = EI \frac{d^4y}{dx^4}$$

for flat plate. From the vertical force equilibrium of the plate element we can obtain

$$\frac{\partial Q_x}{\partial y} + \frac{\partial Q_y}{\partial x} = q$$

Substituting this into Equations 7-1 and 7-3 we obtain

$$\frac{\partial^4 w}{\partial x^4} + 2 \frac{\partial^4 w}{\partial x^2 \partial y^2} + \frac{\partial^4 w}{\partial y^4} = \frac{q}{D} \tag{7-7}$$

With Laplace operator ∇^2, Equations 7-5 through 7-7 can be written as

$$Q_x = D\frac{\partial}{\partial y} (\nabla^2 w) \tag{7-5a}$$

$$Q_y = D\frac{\partial}{\partial x} (\nabla^2 w) \tag{7-6a}$$

$$\nabla^4 w = \frac{q}{D} \tag{7-7a}$$

The problem of plate bending is to find a solution of Equation 7-7a that satisfies all boundary conditions. When such a solution is found, the unit shear forces can be obtained from Equations 7-5a and 7-6a, and the unit moments from Equations 7-1 through 7-3. Because Equation 7-7 is linear, the most powerful method for solving the problem is the Fourier series. If enough miscellaneous solutions are known, one can build up others by superposition.

A particular case of plate bending is pure bending; that is, $q = 0$. If $M_x = M_y = M$ uniformly distributed along the boundary, the plate is bent to a spherical surface with a curvature of

$$\rho = \frac{M}{D(1 + \nu)}$$

If the plate is bent by moments uniformly distributed along the edges parallel to the y axis, then $\partial^2 w/\partial y^2 = 0$, and the plate is bent to a cylindrical surface with its longitudinal axis parallel to the y axis. From Equations 7-1 and 7-2 we obtain

$$M_x = -D \frac{\partial^2 w}{\partial x^2} \tag{7-8}$$

$$M_y = -\nu D \frac{\partial^2 w}{\partial x^2} \tag{7-9}$$

Note that to bend the plate to a cylindrical surface, not only M_x but also M_y must be involved. Without the action of M_y the plate will be bent to an anticlastic surface.

LONG RECTANGULAR PLATES WITH UNIFORM LOADING

The deflection surface of a long rectangular plate loaded with uniform loads can be assumed cylindrical. We can cut an elemental strip from the plate; the deflection of this strip is similar to the deflection of a bent beam. If the strip has a unit width, the equation for the deflection curve of the strip can be expressed as follows [1]:

$$M = \int_{-\frac{1}{2}}^{\frac{1}{2}} \sigma_x z \, dz = \frac{-Et^3}{12(1 - \nu^2)} \frac{d^2 w}{dx^2}$$

$$\text{Let } D = \frac{Et^3}{12(1 - \nu^2)}$$

$$M = -D \frac{d^2 w}{dx^2} \tag{7-10}$$

where   t = plate thickness
$\quad\quad\;\; \nu$ = Poisson's ratio
$\quad\quad\;$ w = deflection of the plate
$\quad\quad\;$ M = bending moment

Equation 7-10 is the same as Equation 7-8, which is a special case of pure bending when the plate is bent to a cylindrical surface. Plate deflection can be calculated by integrating Equation 7-7. The bending moment M can be easily derived if there is only a uniform lateral load acting on the plate.

Long Rectangular Plates with Simply Supported Edges

A unit strip of a long rectangular plate with simple supports is shown in Figure 7-2. It is a case of a uniformly loaded beam subjected to the action of an axial force P. If q is the intensity of the uniform load, the bending moment at any cross section is expressed by

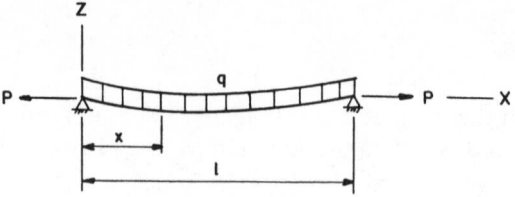

Figure 7-2.   Unit strip of a long rectangular plate.

$$M = \frac{q\ell}{2}x - \frac{qx^2}{2} - Pw$$

Substituting into Equation 7-7 we obtain

$$\frac{d^2w}{dx^2} - \frac{Pw}{D} = \frac{qx^2}{2D} - \frac{q\ell x}{2D} \qquad (7\text{-}11)$$

where

$$u = \frac{\ell}{2}\sqrt{\frac{P}{D}} \qquad (7\text{-}12)$$

The magnitude of the axial force P is such as to prevent the ends of the strip from moving along the axis. Equation 7-11 is solved in Reference 1. The deflections of the strip bar depend upon the quantity u, which is a function of the axial force P. The force P is determined from the condition that the longitudinal edges do not move in the plane of the plate. This means that the extension of the strip produced by the force P is equal to the difference between the arc of the deflected curve and the chord length. For small deflections this can be expressed by the following equation [2]:

$$\frac{P\ell(1 - \nu^2)}{Et} = \frac{1}{2}\int_0^\ell \left(\frac{dw}{dx}\right)^2 dx \qquad (7\text{-}13)$$

Exact expressions of Equation 7-10 are complicated. An approximate method of calculating the parameter u is provided in Reference 1. This method can be used with sufficient accuracy in practical application.

The deflection curve of the strip under consideration can be presented by

$$w = \frac{y_0}{1 + \alpha}\sin\frac{\pi x}{\ell} \qquad (7\text{-}14)$$

where α is the ratio of the axial force P to the Euler critical load for the strip, and y_0 is the deflection at the midspan of the strip produced by the lateral load alone.

$$y_0 = \frac{5q\ell^4}{384D}$$

$$\alpha = \frac{P}{P_{cr}} = \frac{P\ell^2}{\pi^2D} \tag{7-15}$$

From Equations 7-13 through 7-15 we obtain

$$\frac{y_0}{t} = \frac{(1 + \alpha)\sqrt{\alpha}}{\sqrt{3}}$$

$$u = 0.5\pi\sqrt{\alpha}$$

The y_0/t value for a uniformly loaded and simply supported bar can easily be calculated, and the equivalent u parameter can be found from Table 7-1. After the parameter u is obtained, the maximum moment and maximum deflection of the strip bar can be calculated from the following expressions:

$$M_{max} = M'g_u \tag{7-16}$$

$$w_{max} = y_0 f_u \tag{7-17}$$

where M' and y_0 are the maximum moment and deflection of the strip produced by the lateral loading alone, and g_u and f_u are functions of parameter u provided in Table 7-1. The maximum moment and deflection of the unit width strip can be calculated from the following expressions:

$$M_{max} = \frac{q\ell^2}{8}g_u \tag{7-18}$$

$$w_{max} = \frac{5q\ell^4}{384D}f_u \tag{7-19}$$

The axial force P is calculated from

$$P = \frac{4Du^2}{\ell^2} = \frac{Eu^2t^3}{3(1 - \nu^2)\ell^2}$$

Table 7-1
y_0/t and Parameter u for Simply Supported Bar

y_0/t	u	f_u	g_u	y_0/t	u	f_u	g_u
0.2008	0.4967	0.9088	0.9066	4.0000	2.7207	0.2481	0.2348
0.3098	0.7025	0.8328	0.8288	4.1678	2.7657	0.2420	0.2287
0.4111	0.8604	0.7685	0.7630	4.3377	2.8099	0.2362	0.2229
0.5112	0.9935	0.7134	0.7066	4.5099	2.8535	0.2307	0.2174
0.6124	1.1107	0.6656	0.6579	4.6842	2.8964	0.2254	0.2122
0.7155	1.2167	0.6238	0.6152	4.8606	2.9387	0.2203	0.2071
0.8212	1.3142	0.5869	0.5776	5.0390	2.9804	0.2155	0.2024
0.9295	1.4050	0.5542	0.5442	5.2196	3.0215	0.2109	0.1978
1.0407	1.4902	0.5249	0.5144	5.4022	3.0620	0.2065	0.1934
1.1547	1.5708	0.4985	0.4875	5.5869	3.1021	0.2022	0.1892
1.2716	1.6475	0.4746	0.4633	5.7734	3.1416	0.1981	0.1852
1.3914	1.7207	0.4529	0.4412	5.9621	3.1806	0.1942	0.1813
1.5140	1.7910	0.4331	0.4211	6.1527	3.2192	0.1905	0.1776
1.6395	1.8586	0.4150	0.4027	6.3453	3.2573	0.1868	0.1740
1.7678	1.9238	0.3983	0.3858	6.5397	3.2949	0.1834	0.1706
1.8988	1.9869	0.3828	0.3702	6.7361	3.3322	0.1800	0.1673
2.0325	2.0481	0.3686	0.3558	6.9344	3.3690	0.1768	0.1641
2.1689	2.1074	0.3553	0.3424	7.1345	3.4054	0.1736	0.1610
2.3079	2.1652	0.3430	0.3300	7.3365	3.4414	0.1706	0.1581
2.4495	2.2214	0.3315	0.3184	7.5403	3.4771	0.1677	0.1552
2.5936	2.2763	0.3207	0.3076	7.7460	3.5124	0.1649	0.1525
2.7403	2.3299	0.3106	0.2974	7.9534	3.5474	0.1622	0.1498
2.8895	2.3822	0.3012	0.2879	8.1627	3.5820	0.1595	0.1472
3.0411	2.4335	0.2922	0.2789	8.3737	3.6162	0.1570	0.1447
3.1950	2.4836	0.2838	0.2705	8.5865	3.6502	0.1545	0.1423
3.3514	2.5328	0.2759	0.2625	8.8010	3.6838	0.1521	0.1400
3.5101	2.5811	0.2684	0.2550	9.0173	3.7172	0.1498	0.1377
3.6711	2.6284	0.2613	0.2479	9.2353	3.7502	0.1475	0.1355
3.8344	2.6750	0.2545	0.2412	9.4550	3.7830	0.1453	0.1334
9.6764	3.8155	0.1432	0.1313	16.7844	4.6597	0.1005	0.0904
9.8995	3.8476	0.1411	0.1293	17.0518	4.6861	0.0995	0.0894
10.1242	3.8796	0.1391	0.1274	17.3205	4.7124	0.0985	0.0884
10.5507	3.9113	0.1372	0.1255	17.5906	4.7485	0.0975	0.0875
10.5787	3.9427	0.1353	0.1237	17.8621	4.7645	0.0966	0.0866
10.8084	3.9738	0.1335	0.1219	18.1350	4.7903	0.0956	0.0857
11.0397	4.0048	0.1317	0.1202	18.4093	4.8160	0.0947	0.0848
11.2736	4.0354	0.1299	0.1185	18.6849	4.8415	0.0938	0.0840
11.5071	4.0659	0.1282	0.1168	18.9619	4.8669	0.0929	0.0831
11.7433	4.0961	0.1266	0.1152	19.2402	4.8922	0.0920	0.0823
11.8909	4.1261	0.1249	0.1137	19.5198	4.9174	0.0912	0.0815
12.2202	4.1559	0.1234	0.1122	19.8008	4.9424	0.0903	0.0807
12.4610	4.1855	0.1218	0.1107	20.0832	4.9673	0.0895	0.0799
12.7034	4.2149	0.1203	0.1093	20.3668	4.9921	0.0887	0.0792
12.9473	4.2441	0.1189	0.1079	20.6518	5.0167	0.0879	0.0784
13.1927	4.2730	0.1174	0.1065	20.9381	5.0413	0.0871	0.0777
13.4397	4.3018	0.1161	0.1051	21.2256	5.0657	0.0863	0.0770
13.6881	4.3304	0.1147	0.1038	21.5145	5.0900	0.0856	0.0762
13.9381	4.3588	0.1134	0.1026	21.8047	5.1141	0.0848	0.0755
14.1896	4.3870	0.1121	0.1013	22.0962	5.1382	0.0841	0.0749
14.4425	4.4150	0.1108	0.1001	22.3889	5.1622	0.0834	0.0742
14.6969	4.4429	0.1096	0.0989	22.6830	5.1860	0.0827	0.0735
14.9528	4.4706	0.1083	0.0978	22.9793	5.2097	0.0820	0.0729
15.2102	4.4981	0.1072	0.0966	23.2748	5.2334	0.0813	0.0722
15.4690	4.5254	0.1060	0.0955	23.5726	5.2569	0.0806	0.0716
15.7292	4.5526	0.1049	0.0945	23.8717	5.2803	0.0800	0.0710
15.9909	4.5796	0.1037	0.0934	24.1720	5.3036	0.0793	0.0704
16.2540	4.6065	0.1027	0.0924	24.4736	5.3268	0.0787	0.0700
16.5185	4.6332	0.1016	0.0914	24.7764	5.3499	0.0780	0.0692

the axial stress $f_a = P/t$

$$f_a = \frac{E u^2}{3(1 - v^2)} \left(\frac{t}{\ell}\right)^2$$

and the bending stress $f_b = 6M/t^2$

$$f_b = = \frac{6M}{t^2} = \frac{3}{4} q \left(\frac{\ell}{t}\right)^2 g_u$$

$$f = f_a \pm f_b = \frac{E u^2}{3(1 - v^2)} \left(\frac{t}{\ell}\right)^2 \pm \frac{3}{4} q \left(\frac{\ell}{t}\right)^2 g_u \qquad (7\text{-}20)$$

Example 7-1

A simply supported long rectangular plate is loaded with 20 lb/in.^{2} uniform loading. The plate is ½ in. thick and 48 in. wide; calculate the maximum stress and deflection.

$$\frac{y_0}{t} = \frac{5q\ell^4}{384D} = \frac{5q(1 - v^2)}{32 E} \left(\frac{\ell}{t}\right)^4$$

$$= \frac{5 \times 20 \times 0.91}{32 \times 29} (96)^4 \times 10^{-6} = 8.3287$$

From Table 7-1,

$u = 3.6089$

$f_u = 0.1575$

$g_u = 0.1452$

$w_{max} = y_0 f_u = 4.16435 \times 0.1575 = 0.6559$ in.

$$M_{max} = \frac{q\ell^2}{8} g_u = \frac{20 \times 48^2}{8} \times 0.1452 = 836.352 \text{ in.-lb}$$

$$f_b = \frac{6M}{t^2} = \frac{6 \times 836.352}{0.25} = 20,072.45 \text{ lb/in.}^2$$

$f_a = P/t$

$$P = \frac{4Du^2}{\ell^2} = \frac{29 \times 10^6 \times 0.5^3 \times 3.6089^2}{3 \times 0.91 \times 48^2} = 7,506.07 \text{ lb}$$

$$f_a = 7,506.07/0.5 = 15,012.14 \text{ lb/in.}^2$$

$$f = f_a + f_b = 35,084.59 \text{ lb/in.}^2 = 35.085 \text{ kips/in.}^2$$

Long Rectangular Plate with Fixed Ends

A unit width strip of a fixed long rectangular plate with uniform load-ing is shown in Figure 7-3. The fixed end moment M_f and the maximum deflection w_{max} are solved in Reference 1 and can be expressed as fol-lows:

$$M_f = M' g_u'$$

$$w_{max} = y_0 f_u'$$

where M' and y_0 are the fixed end moments and the maximum deflection of the strip produced by the lateral loading alone, and g_u' and f_u' are func-tions of the parameter u. An approximate method for calculating u is ex-pressed as follows:
In the case of fixed-end condition the deflection curve of the strip can be represented by

$$w = \frac{0.5 \, y_0}{1 + 0.25\alpha} \left(1 - \cos\frac{2\pi x}{\ell}\right) \tag{7-21}$$

Substituting this expression in Equation 7-13 we obtain

$$\frac{y_0}{t} = \frac{(1 + 0.25\alpha)\sqrt{\alpha}}{\sqrt{3}}$$

$$u = 0.5\pi\sqrt{\alpha}$$

The y_0/t value of a fixed-end strip with uniform loading can be calculated by

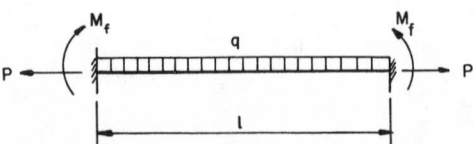

Figure 7-3. Unit strip of a fixed long rectangular plate.

$$\frac{y_0}{t} = \frac{q\,\ell^4}{384 Dt} = \frac{(1 - \nu^2)q}{32\,E}\left(\frac{\ell}{t}\right)^4$$

and the corresponding values for u, g_u', and f_u' can be found in Table 7-2. The maximum stress and deflection can be calculated from

$$w_{max} = \frac{(1 - \nu^2)q\ell^4}{32\,E\,t^3}\,f_u' \tag{7-22}$$

$$M_f = -\frac{q\ell^2}{12}\,g_u'$$

$$P = \frac{E\,u^2\,t^3}{3(1 - \nu^2)\ell^2}$$

$$f = f_a \pm f_b = \frac{E\,u^2}{3(1 - \nu^2)}\left(\frac{t}{\ell}\right)^2 \pm \frac{q}{2}\left(\frac{\ell}{t}\right)^2 g_u'$$

Table 7-2
y_0/t and Parameter u for Fixed Bar

y_0/t	u	f_u'	g_u'	y_0/t	u	f_u'	g_u'
0.1871	0.4967	0.9759	0.9839	1.7500	2.7207	0.5766	0.7070
0.2711	0.7025	0.9530	0.9686	1.8043	2.7657	0.5687	0.7011
0.3399	0.8604	0.9311	0.9539	1.8590	2.8099	0.5609	0.6955
0.4017	0.9935	0.9103	0.9398	1.9141	2.8535	0.5534	0.6899
0.4593	1.1107	0.8903	0.9264	1.9695	2.8964	0.5461	0.6845
0.5143	1.2167	0.8712	0.9134	2.0252	2.9387	0.5389	0.6792
0.5676	1.3142	0.8530	0.9010	2.0813	2.9804	0.5320	0.6741
0.6197	1.4050	0.8355	0.8891	2.1378	3.0215	0.5252	0.6690
0.6710	1.4902	0.8187	0.8776	2.1947	3.0620	0.5186	0.6641
0.7217	1.5708	0.8025	0.8665	2.2518	3.1021	0.5122	0.6593
0.7721	1.6475	0.7871	0.8559	2.3094	3.1416	0.5059	0.6545
0.8222	1.7207	0.7722	0.8456	2.3673	3.1806	0.4998	0.6499
0.8722	1.7910	0.7578	0.8357	2.4256	3.2192	0.4939	0.6454
0.9222	1.8586	0.7440	0.8261	2.4842	3.2573	0.4880	0.6410
0.9723	1.9238	0.7307	0.8168	2.5432	3.2949	0.4824	0.6367
1.0224	1.9869	0.7179	0.8078	2.6026	3.3322	0.4768	0.6324
1.0727	2.0481	0.7055	0.7992	2.6623	3.3690	0.4714	0.6283
1.1232	2.1074	0.6935	0.7907	2.7224	3.4054	0.4661	0.6242
1.1738	2.1652	0.6820	0.7826	2.7828	3.4414	0.4609	0.6202
1.2247	2.2214	0.6708	0.7747	2.8436	3.4771	0.4559	0.6163
1.2759	2.2763	0.6600	0.7670	2.9047	3.5124	0.4509	0.6125
1.3273	2.3299	0.6496	0.7596	2.9662	3.5474	0.4461	0.6087
1.3791	2.3822	0.6394	0.7524	3.0281	3.5820	0.4414	0.6050
1.4311	2.4335	0.6296	0.7453	3.0903	3.6162	0.4367	0.6014
1.4834	2.4836	0.6201	0.7385	3.1529	3.6502	0.4322	0.5978
1.5361	2.5328	0.6109	0.7319	3.2158	3.6838	0.4277	0.5943
1.5890	2.5811	0.6019	0.7254	3.2790	3.7172	0.4234	0.5909
1.6424	2.6284	0.5933	0.7191	3.3426	3.7502	0.4191	0.5875
1.6960	2.6750	0.5848	0.7129	3.4066	3.7830	0.4149	0.5842

Table 7-2 (continued)

y_0/t	u	f'_u	g'_u	y_0/t	u	f'_u	g'_u
3.4709	3.8155	0.4108	0.5810	5.4806	4.6597	0.3199	0.5058
3.5355	3.8476	0.4068	0.5778	5.5547	4.6861	0.3175	0.5037
3.6005	3.8796	0.4029	0.5746	5.6292	4.7124	0.3151	0.5016
3.6659	3.9113	0.3991	0.5715	5.7039	4.7385	0.3128	0.4996
3.7315	3.9427	0.3953	0.5685	5.7789	4.7645	0.3105	0.4976
3.7975	3.9738	0.3916	0.5665	5.8543	4.7903	0.3082	0.4956
3.8639	4.0048	0.3879	0.5626	5.9299	4.8160	0.3060	0.4937
3.9306	4.0354	0.3844	0.5597	6.0059	4.8415	0.3038	0.4917
3.9976	4.0659	0.3809	0.5568	6.0821	4.8669	0.3016	0.4898
4.0650	4.0961	0.3774	0.5540	6.1587	4.8922	0.2995	0.4879
4.1327	4.1261	0.3741	0.5512	6.2355	4.9174	0.2974	0.4861
4.2007	4.1559	0.3707	0.5485	6.3127	4.9424	0.2953	0.4842
4.2691	4.1855	0.3675	0.5456	6.3901	4.9673	0.2932	0.4824
4.3377	4.2149	0.3643	0.5432	6.4678	4.9921	0.2912	0.4806
4.4068	4.2441	0.3611	0.5406	6.5459	5.0167	0.2892	0.4789
4.4761	4.2730	0.3581	0.5380	6.6242	5.0413	0.2873	0.4771
4.5458	4.3018	0.3550	0.5355	6.7028	5.0657	0.2853	0.4754
4.6158	4.3304	0.3521	0.5330	6.7818	5.0900	0.2834	0.4736
4.6861	4.3588	0.3491	0.5306	6.8610	5.1141	0.2815	0.4719
4.7567	4.3870	0.3462	0.5282	6.9405	5.1382	0.2797	0.4703
4.8277	4.4150	0.3434	0.5258	7.0203	5.1622	0.2778	0.4686
4.8990	4.4429	0.3406	0.5234	7.1003	5.1860	0.2760	0.4670
4.9706	4.4706	0.3379	0.5211	7.1807	5.2097	0.2742	0.4653
5.0425	4.4981	0.3352	0.5188	7.2614	5.2334	0.2725	0.4637
5.1147	4.5254	0.3325	0.5166	7.3423	5.2569	0.2707	0.4622
5.1873	4.5526	0.3299	0.5144	7.4235	5.2803	0.2690	0.4606
5.2602	4.5796	0.3274	0.5122	7.5050	5.3036	0.2673	0.4590
5.3333	4.6065	0.3248	0.5100	7.5868	5.3268	0.2657	0.4575
5.4068	4.6332	0.3224	0.5079	7.6689	5.3499	0.2640	0.4560

DATA FOR RECTANGULAR PLATES [1,4]

The calculation of plate deflections and stresses is a very laborious process. When the plate is rectangular, the work of computation is so extensive that the tedious calculations become unpractical. In the course of time many cases have been worked out, the sources and details of which can be found in Timoshenko's work in *Theory of Plates and Shells* [1]. Tables and graphs have been developed and are provided here for practical application. These data are valid only for plates with small deflections. Poisson's ratio ν has been taken as 0.3, the usual value for steel and many other materials. In the tabulations, ν has been absorbed in the numerical coefficients. The provided coefficients B_x, B_y and A_u are dimensionless; they can be applied in both metric and Imperial units. Note that ordinates of some graphs have been factored for easier reading. In Case 1, for example, $B_x \times 10^{-3} = 4.872$. That is, $B_x = 4,872$.

The symbols appearing in the formulas are defined as follows:

M_x = bending moment per unit length acting on axes parallel to the y-axis

M_y = bending moment per unit length acting on axes parallel to the x-
    axis
f_{bx} = bending stress induced by M_x
f_{by} = bending stress induced by M_y
B_x, B_y = bending stress coefficients
A_u = deflection coefficient
E = modulus of elasticity (29,000 kips/in.^{2} for steel)
t = plate thickness
a, b = plate dimensions
q = uniform load

Example 7-2

A 15 ft by 3.5 ft simply supported rectangular plate is loaded with 368
lb/ft^2 uniform load; if the plate is $3/8$ in. thick calculate the maximum
stress and deflection.

$q = 368$ lb/ft^2 = 2.56 lb/in.^{2}

$t = 0.375$ in.

$b/a = 15/3.5 = 4.2857$

$a/t = 42/0.375 = 112$

From the Table or Graph provided in Case 1

$B_x \times 10^{-3} = 9.2$ $B_y \times 10^{-3} = 2.86$ $A_u \times 10^{-5} = 1.95$

max. stress $f_{bx} = B_x q = 9,200 \times 2.56 = 23,552$ lb/in.^{2}
                           = 23.55 kips/in.^{2}

      $f_{by} = 2,860 \times 2.56 = 7,322$ lb/in.^{2}

max. deflection $w_{max} = a\, A_u q/E$
                    = $42 \times 1.95 \times 10^5 \times 2.56/(29 \times 10^6)$
                    = 0.723 in.

Example 7-3

A simply supported rectangular plate is loaded with 1,800 kg/m^2 uniform load. If the plate dimensions are 4,572 mm by 1,067 mm, and the
maximum allowable stress and deflection are 165.47 MPa and 15 mm,
respectively, design the plate.

$b/a = 4,572/1,067 = 4.285$

$q = 1,800 \text{ kg/m}^2 = 0.18 \text{ kg/cm}^2$

max. allowable stress $= 165.74 \text{ MPa} = 1,686 \text{ kg/cm}^2$

$f_{bx} = q B_x = 1,687$

$B_x = 1,686/0.18 = 9,367 = 9.367 \times 10^{-3}$

From Case 1,

$a/t = 112$

$t_{\text{req'd}} = a/112 = 1,067/112 = 9.526 \text{ mm}$

Try $t = 10 \text{ mm}, a/t = 106.7$

From Case 1,

$A_u = 1.59 \times 10^5$

$w_{\max} = a A_u q/E = 1,067 \times 1.59 \times 10^5 \times 0.18/(20.37 \times 10^5)$
$= 14.99 \text{ mm}$

Use 10 mm plate.

CIRCULAR PLATES

The deflections and stresses of a circular plate with uniform load depend on one variable r only. This simplifies the general plate equations considerably. As shown in Figure 7-4, r is the radial distance of point A in the middle plane of the plate, and the distance AC is the tangential radius of curvature at point A, denoted by r_t.

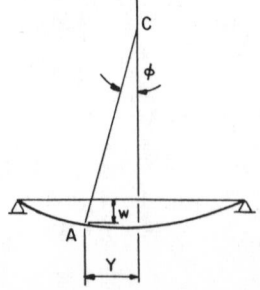

Figure 7-4. A deflected circular plate.

$$\frac{1}{r_t} = -\frac{1}{r}\frac{dw}{dr} = \frac{\phi}{r} \tag{7-24}$$

and the other radius of curvature r_n (the meridional) is given as

$$\frac{1}{r_n} = -\frac{d^2w}{dr^2} = \frac{d\phi}{dr} \tag{7-25}$$

Having Equations 7-24 and 7-25 for the principal curvatures, we can obtain the corresponding bending moments from Equations 7-1 and 7-2:

$$M_r = -D\left(\frac{d^2w}{dr^2} + \frac{\nu}{r}\frac{dw}{dr}\right) \tag{7-26}$$

$$M_t = -D\left(\frac{1}{r}\frac{dw}{dr} + \nu\frac{d^2w}{dr^2}\right) \tag{7-27}$$

where M_r = bending moments per unit length acting along circumferential sections of the plate

M_t = bending moments per unit length acting along the diametrical sections of the plate

Equations 7-26 and 7-27 contain only one variable w, which can be determined by considering the equilibrium of an element of the plate. As derived in Chapter 3 of *Theory of Plates and Shells* [1], the differential equation of a circular plate can be expressed as

$$\frac{d^3w}{dr^3} + \frac{1}{r}\frac{d^2w}{dr^2} - \frac{1}{r^2}\frac{dw}{dr} = \frac{Q}{D} \tag{7-28a}$$

where Q is the shearing force per unit length of a cylindrical section of radius r. Q can easily be calculated for the case of a symmetrically loaded circular plate. Equation 7-28a can be written in the following form for convenient integration:

$$\frac{d}{dr}\left[\frac{1}{r}\frac{d}{dr}\left(r\frac{dw}{dr}\right)\right] = \frac{Q}{D} \tag{7-28b}$$

Q can be expressed as a function of r, and for a circular plate uniformly loaded with q, we obtain

$$2\pi rQ = 2\pi \int_0^r qr\,dr$$

Using this relation we finally obtain

$$\frac{1}{r}\frac{d}{dr}\left\{r\frac{d}{dr}\left[r\frac{d}{dr}\left(r\frac{dw}{dr}\right)\right]\right\} = \frac{q}{D} \tag{7-29}$$

Equation 7-29 can easily be integrated if the uniform load q is expressed as a function of r. For example, if a circular plate of radius R carries uniform load q, Q at a distance r from the center can be calculated from

$$2\pi\, rQ = \pi r^2 q \text{ or } Q = qr/2$$

Substituting in Equation 7-28b and integrating, the following solution is obtained:

$$\frac{dw}{dr} = \frac{qr^3}{16D} + \frac{C_1 r}{2} + \frac{C_2}{r} \tag{7-30}$$

$$w = \frac{qr^4}{64D} + \frac{C_1 r^2}{4} + C_2 \ln\frac{r}{R} + C_3 \tag{7-31}$$

C_1, C_2, and C_3 are constants of integration, which may be found from boundary conditions of the plate. Given a circular plate with clamped edges, for example, the slope at the center and edge must be zero, and the deflection at the edge must be zero. This can be expressed as follows:

$$\frac{dw}{dr} = 0 \text{ at } r = 0 \text{ and } r = R$$

$$w = 0 \text{ at } r = R$$

Substituting these boundary conditions in Equations 7-30 and 7-31, the following expressions for slope and deflection are obtained:

$$\frac{dw}{dr} = -\frac{qr}{16D}(R^2 - r^2) \tag{7-32}$$

$$w = \frac{q}{64D}(R^2 - r^2)^2 \tag{7-33}$$

Having Equation 7-32, the slope equation, we can obtain the bending moments

$$M_r = \frac{q}{16} [R^2(1 + \nu) - r^2(3 + \nu)] \tag{7-34}$$

$$M_t = \frac{q}{16} [R^2(1 + \nu) - r^2(1 + 3\nu)] \tag{7-35}$$

From Equation 7-33, the maximum deflection is at the center of the plate and is given by

$$w_{max} = \frac{qR^4}{64D} \tag{7-36}$$

and the bending moments at the edge are

$$M_r = - \frac{q R^2}{8} \text{ at } r = R \tag{7-37}$$

$$M_t = - \nu \frac{q R^2}{8} \text{ at } r = R \tag{7-38}$$

At the center of the plate where r = 0,

$$M_r = M_t = \frac{q R^2}{8} (1 + \nu) \tag{7-39}$$

DATA FOR CIRCULAR PLATES

Many cases of circular plate have been worked out, the sources and results of which can be found in References 1, 3, and 4.

Tables and graphs have been developed and are presented here for practical application. Poisson's ratio ν has been taken as 0.3 and has been absorbed in the numerical coefficients. The bending coefficient B_r and deflection coefficient A_r are dimensionless; they can be applied in both metric and Imperial units.

Example 7-4

A circular plate of 4 ft radius is clamped along the edge and loaded with 350 lb/ft^2 uniform load; design the plate and calculate the maximum deflection (A36 steel).

$$q = 350 \text{ lb/ft}^2 = 2.431 \text{ lb/in.}^2$$

The maximum allowable stress $= 24$ kips/in.^{2}

$B_r = f_{br}/q = 24,000/2.431 = 9,872$

$B_r \times 10^{-3} = 9.872$

From Case C1

$R/t = 115$

$t_{req'd} = R/115 = 48/115 = 0.4174$ in.

Use 7/16 plate

$R/t = 48 \times 16/7 = 109.7$ (say 110)

$A_r \times 10^{-4} = 23$

$w_{max} = RA_r q/E = 48 \times 230,000 \times 2.431/(29 \times 10^6) = 0.925$ in.

Using Equation 7-37, $M_r = 2.41 \times 48^2/8 = 700$ in.-lb/in.

$t^2 = 6M_r/24,000 = 6 \times 700/24,000 = 0.175 \qquad t = 0.418$ in.

Use 7/16 plate

From Equation 7-36

$$W_{max} = qR^4/(64D)$$
$$= 10.92 \times 2.431 \times 48^4/(64 \times 29 \times 10^6 \times 0.4375^3)$$
$$= 0.92 \text{ in.}$$

REFERENCES

1. Timoshenko, S. and Woinowsky-Krieger, S., *Theory of Plates and Shells,* McGraw-Hill Book Company, New York, 1959.
2. Timoshenko, S., *Strength of Materials* Parts I & II, D. Van Norstrand Company, 1956.
3. Hartog, J. P., *Advanced Strength of Materials,* McGraw-Hill Book Company, 1952.
4. Roark, R. J. and Young, W. C., *Formulas for Stress and Strain,* Fifth Edition, McGraw-Hill Book Company, 1975.

Cases for Chapter 7

Case 1

Simply supported rectangular plate with uniform loads

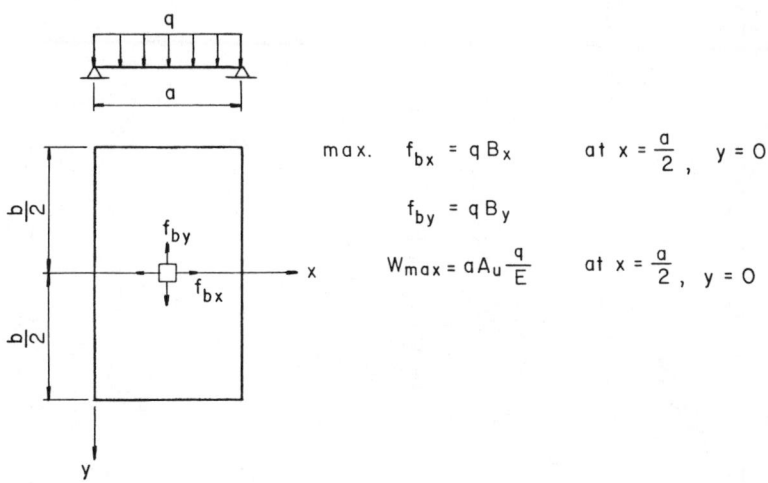

$$\text{max.} \quad f_{bx} = q\,B_x \quad\quad \text{at } x = \frac{a}{2}, \; y = 0$$

$$f_{by} = q\,B_y$$

$$W_{max} = a\,A_u\,\frac{q}{E} \quad\quad \text{at } x = \frac{a}{2}, \; y = 0$$

B_x

a/t b/a	80	100	110	120	130	140	160	180	200
1.0	1840	2874	3478	4139	4857	5633	7357	9312	11496
1.1	2128	3324	4022	4787	5618	6515	8510	10770	13296
1.2	2408	3700	4552	5417	6358	7374	9631	12189	15048
1.3	2665	4164	5038	5996	7037	8161	10660	13492	16656
1.4	2899	4530	5481	6523	7656	8879	11597	14677	18120
1.5	3119	4872	5895	7016	8234	9549	12472	15785	19488
1.6	3310	5172	6258	7448	8741	10137	13240	16757	20688
1.7	3487	5448	6592	7845	9207	10678	13947	17652	21792
1.8	3640	5688	6882	8191	9613	11148	14561	18429	22752
1.9	3783	5910	7151	8510	9988	11584	15130	19148	23640
2.0	3905	6102	7383	8787	10312	11960	15621	19770	24408
2.5	4335	6774	8197	9755	11448	13277	17341	21948	27096
3.0	4566	7134	8632	10273	12056	13983	18263	23114	28536
4.0	4742	7410	8966	10670	12523	14524	18970	24008	29640
5.0	4785	7476	9046	10766	12634	14653	19139	24222	29904
∞	4800	7500	9075	10800	12675	14700	19200	24300	30000

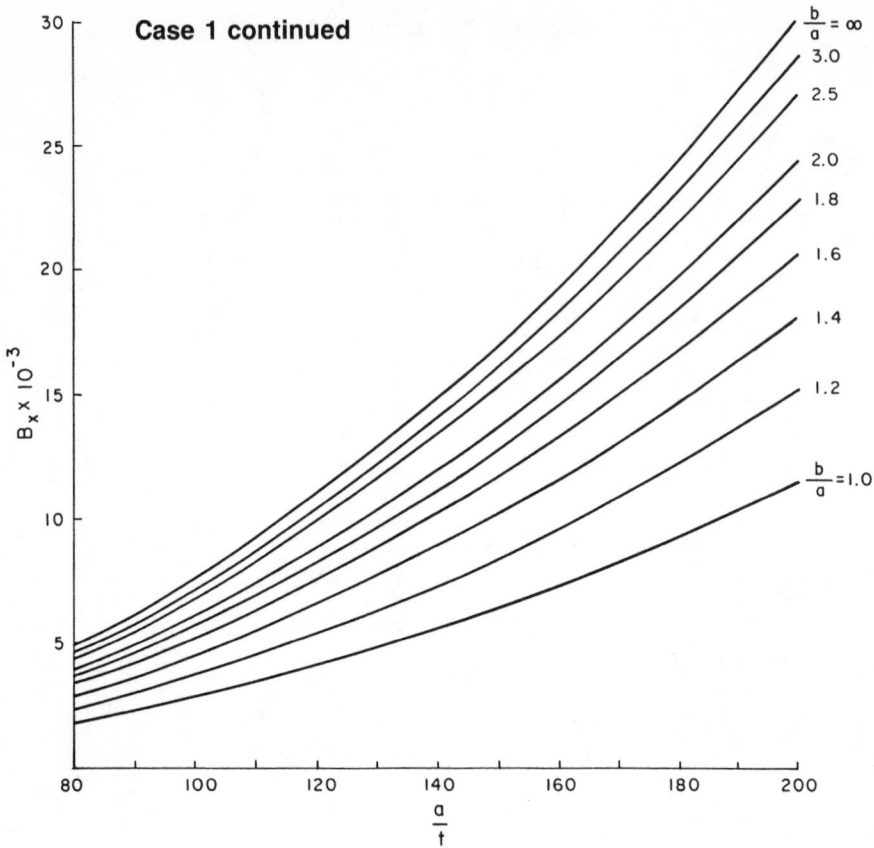

Case 1 continued

$\frac{b}{a} = \infty$
3.0
2.5
2.0
1.8
1.6
1.4
1.2
$\frac{b}{a} = 1.0$

$B_x \times 10^{-3}$

$\frac{a}{t}$

$$\mathbf{B_y}$$

a/t b/a	80	100	110	120	130	140	160	180	200
1.0	1840	2874	3478	4139	4857	5633	7357	9312	11496
1.1	1893	2958	3579	4260	4999	5798	7573	9584	11832
1.2	1924	3006	3637	4329	5080	5892	7695	9740	12024
1.3	1932	3018	3652	4346	5100	5915	7726	9778	12072
1.4	1928	3012	3645	4337	5090	5904	7711	9759	12048
1.5	1912	2988	3616	4303	5050	5857	7649	9681	11952
1.6	1889	2952	3572	4251	4989	5786	7757	9565	11808
1.7	1866	2916	3528	4199	4928	5715	7465	9448	11664
1.8	1840	2874	3478	4139	4857	5633	7357	9312	11496
1.9	1809	2826	3420	4069	4776	5539	7235	9156	11304
2.0	1782	2784	3369	4009	4705	5457	7127	9020	11136
2.5	1651	2580	3122	3715	4360	5057	6605	8359	10320
3.0	1559	2436	2948	3508	4117	4775	6236	7893	9744
4.0	1475	2304	2788	3318	3894	4516	5898	7465	9216
5.0	1440	2250	2723	3240	3803	4410	5760	7290	9000
∞	1440	2250	2723	3240	3803	4410	5760	7290	9000

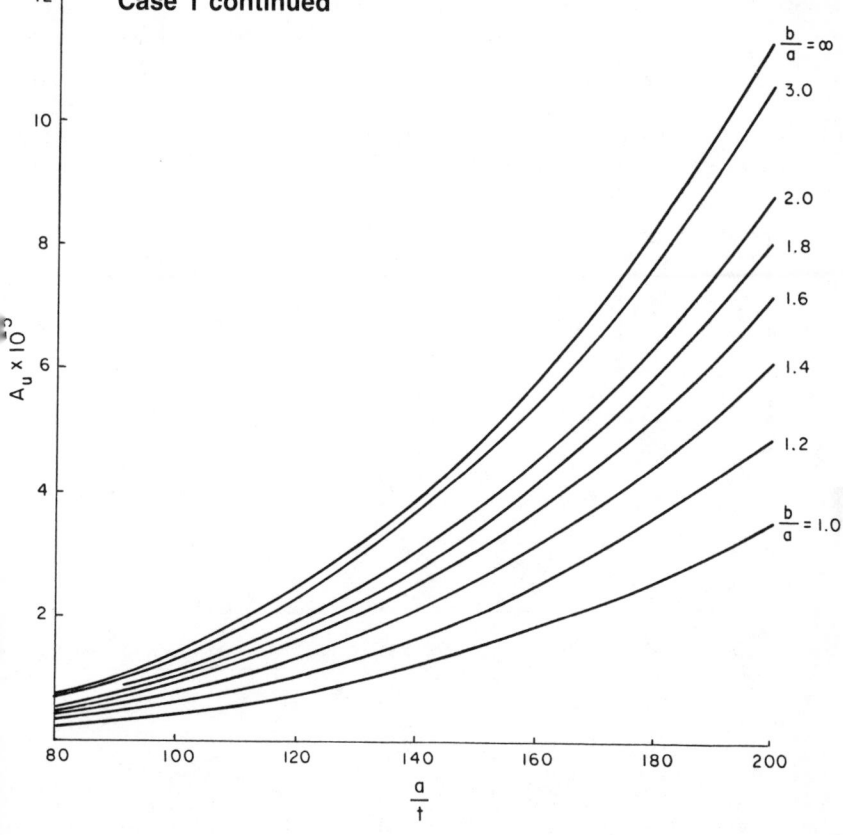

$A_u \times 10^{-3}$

a/t b/a	80	100	110	120	130	140	160	180	200
1.0	22.700	44.335	59.010	76.611	97.404	121.656	181.597	258.563	354.468
1.1	27.117	52.962	70.492	91.518	116.358	145.328	216.932	308.874	423.696
1.2	31.533	61.589	81.975	106.425	135.310	167.000	252.268	359.186	492.710
1.3	35.671	69.670	92.730	120.389	153.064	191.173	285.367	406.313	557.357
1.4	39.417	76.860	102.468	133.032	169.138	211.250	315.335	448.982	615.888
1.5	43.163	84.302	112.206	145.675	185.212	231.326	345.303	491.652	674.419
1.6	46.406	90.636	120.637	156.619	199.127	248.705	371.245	528.589	725.088
1.7	49.369	96.424	128.340	166.620	211.843	264.586	394.951	562.342	771.389
1.8	52.053	101.665	135.316	175.677	223.358	278.969	416.421	592.911	813.322
1.9	54.457	106.361	141.566	183.791	233.675	291.854	435.654	620.296	850.886
2.0	56.637	110.620	147.235	191.151	243.031	303.540	453.098	645.134	884.957
3.0	68.378	133.552	177.757	230.777	293.413	366.466	547.027	778.873	1068.413
4.0	71.677	139.994	186.333	241.910	307.568	384.145	573.417	816.447	1119.955
5.0	72.516	141.632	188.512	244.741	311.166	388.639	580.126	826.000	1133.059
∞	72.795	142.178	189.239	245.684	312.366	390.138	582.363	829.184	1137.427

Case 2

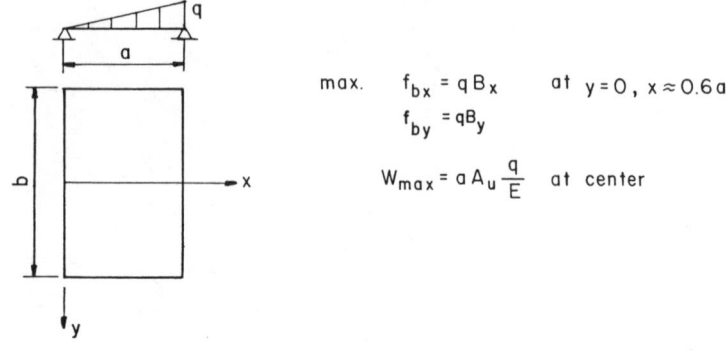

Simply supported rectangular plate with hydrostatic load

$$\text{max.} \quad f_{bx} = q\,B_x \qquad \text{at} \quad y = 0\,, \; x \approx 0.6\,a$$

$$f_{by} = qB_y$$

$$W_{max} = a\,A_u\,\frac{q}{E} \quad \text{at center}$$

B_x

a/t b/a	80	100	110	120	130	140	160	180	200
1.0	1014	1584	1917	2281	2677	3105	4055	5132	6336
1.1	1160	1812	2193	2609	3062	3552	4639	5871	7248
1.2	1298	2028	2454	2920	3427	3975	5192	6571	8112
1.3	1425	2226	2694	3205	3762	4363	5699	7212	8904
1.4	1544	2412	2919	3473	4076	4728	6175	7815	9648
1.5	1647	2574	3115	3707	4350	5045	6589	8340	10296
1.6	1743	2724	3296	3923	4604	5339	6973	8826	10896
1.7	1828	2856	3456	4113	4827	5598	7311	9253	11424
1.8	1905	2976	3600	4285	5030	5833	7619	9642	11904
1.9	1970	3078	3724	4432	5202	6033	7880	9973	12312
2.0	2031	3174	3841	4571	5364	6221	8125	10284	12696
3.0	2346	3666	4436	5279	6196	7185	9385	11878	14664
4.0	2427	3792	4588	5460	6408	7432	9708	12286	15168
5.0	2450	3828	4632	5512	6469	7503	9800	12403	15312
∞	2458	3840	4646	5530	6490	7526	9830	12442	15360

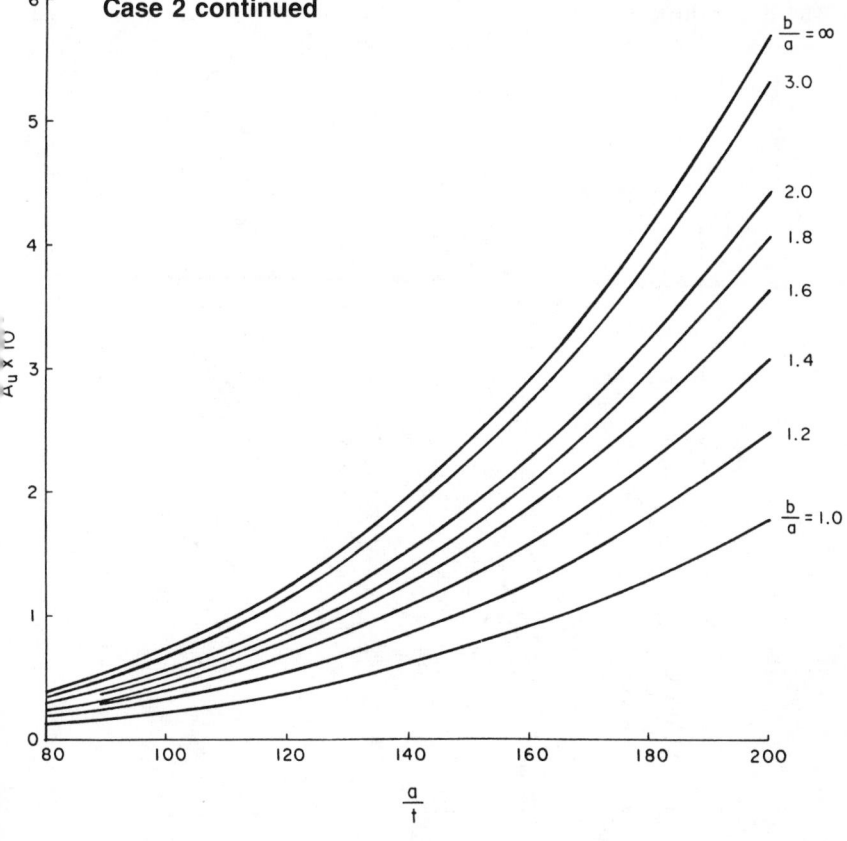

Case 2 continued

$\frac{b}{a} = \infty$

3.0

2.0

1.8

1.6

1.4

1.2

$\frac{b}{a} = 1.0$

$A_u \times 10$

$\frac{a}{t}$

B_y

a/t b/a	80	100	110	120	130	140	160	180	200
1.0	941	1470	1778	2117	2484	2881	3763	4763	5880
1.1	964	1506	1822	2169	2545	2952	3855	4880	6024
1.2	976	1524	1844	2195	2576	2987	3901	4938	6096
1.3	979	1530	1851	2203	2586	2999	3917	4957	6120
1.4	976	1524	1844	2195	2576	2987	3901	4938	6096
1.5	968	1512	1830	2177	2555	2964	3870	4899	6048
1.6	956	1494	1808	2151	2525	2928	3825	4841	5976
1.7	945	1476	1786	2125	2495	2893	3779	4782	5904
1.8	930	1452	1757	2091	2454	2846	3717	4705	5808
1.9	914	1428	1728	2056	2413	2799	3656	4627	5712
2.0	899	1404	1699	2022	2373	2752	3594	4549	5616
3.0	795	1242	1503	1788	2099	2434	3180	4024	4968
4.0	753	1176	1423	1693	1987	2305	3011	3810	4704
5.0	741	1158	1401	1668	1957	2270	2965	3752	4632
∞	737	1152	1394	1659	1947	2258	2949	3733	4608

Case 2 continued

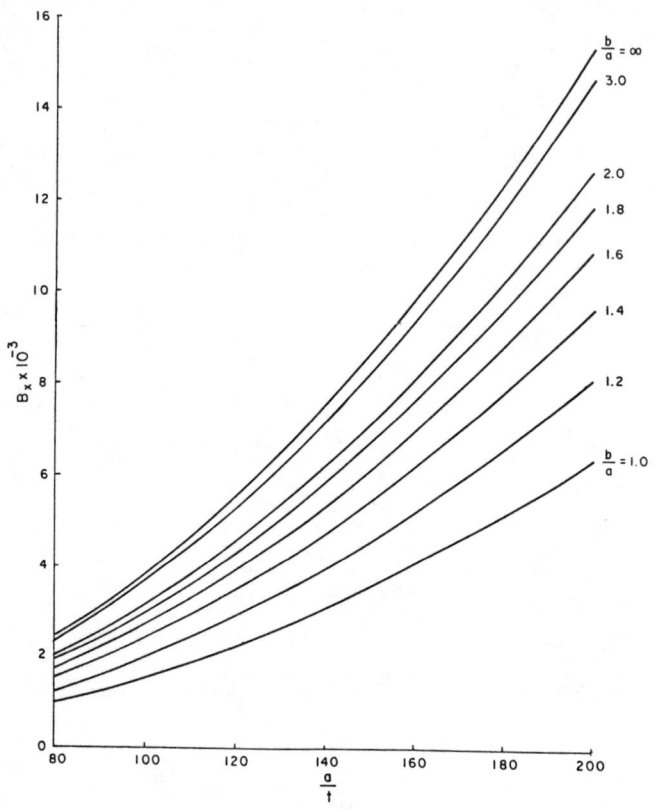

$A_u \times 10^{-3}$

a/t b/a	80	100	110	120	130	140	160	180	200
1.0	11.350	22.168	29.505	38.306	48.702	60.828	90.798	129.281	177.341
1.1	13.586	26.536	35.319	45.854	58.299	72.814	108.690	154.726	212.285
1.2	15.767	30.794	40.987	53.213	67.655	84.499	126.134	179.593	246.355
1.3	17.835	34.835	46.365	60.195	76.532	95.587	142.683	203.157	278.678
1.4	19.736	38.548	51.307	66.610	84.689	105.775	157.891	224.810	308.381
1.5	21.581	42.151	56.103	72.837	92.606	115.663	172.651	245.826	337.210
1.6	23.203	45.318	60.318	78.310	99.564	124.353	185.625	264.295	362.544
1.7	24.656	48.157	64.097	83.216	105.801	132.143	197.252	280.853	385.258
1.8	25.998	50.778	67.586	87.744	111.559	139.335	207.987	296.137	406.224
1.9	27.228	53.180	70.783	91.896	116.837	145.927	217.827	310.148	425.443
2.0	28.291	55.255	73.545	95.481	121.396	151.620	226.325	322.248	442.042
3.0	34.217	66.830	88.951	115.483	146.826	183.383	273.737	389.755	534.643
4.0	35.839	69.997	93.166	120.955	153.784	192.072	286.709	408.224	559.978
5.0	36.230	70.762	94.184	122.276	155.463	194.170	289.840	412.682	566.093
∞	36.398	71.089	94.620	122.842	156.183	195.069	291.181	414.592	568.714

Case 3

Simply supported rectangular plate with hydrostatic load (a > b)

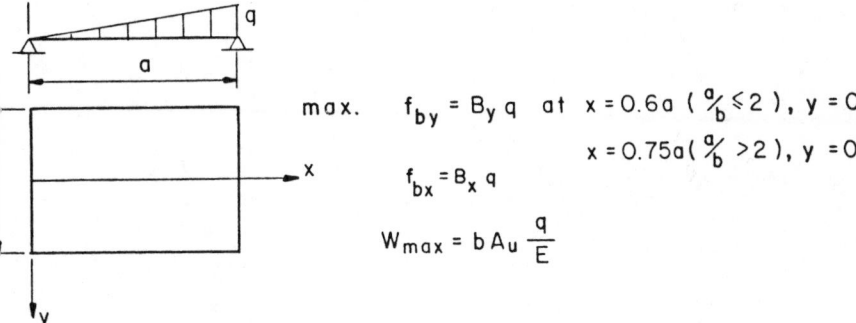

$$\text{max.} \quad f_{by} = B_y\, q \quad \text{at} \quad x = 0.6a \ (\tfrac{a}{b} \leqslant 2), \ y = 0$$
$$x = 0.75a \,(\tfrac{a}{b} > 2), \ y = 0$$
$$f_{bx} = B_x\, q$$
$$W_{max} = b\, A_u\, \frac{q}{E}$$

B_y

b/t a/b	80	100	110	120	130	140	160	180	200
1.1	1094	1710	2069	2462	2890	3352	4378	5540	6840
1.2	1240	1938	2345	2791	3275	3798	4961	6279	7752
1.3	1382	2160	2614	3110	3650	4234	5530	6998	8640
1.4	1521	2376	2875	3421	4015	4657	6082	7698	9504
1.5	1644	2568	3107	3698	4340	5033	6574	8320	10272
1.6	1755	2742	3318	3948	4634	5374	7020	8884	10968
1.7	1862	2910	3521	4190	4918	5704	7450	9428	11640
1.8	1955	3054	3695	4398	5161	5986	7818	9895	12216
1.9	2047	3198	3870	4605	5405	6268	8187	10362	12792
2.0	2127	3324	4022	4787	5618	6515	8509	10770	13296
3.0	2604	4068	4922	5858	6875	7973	10414	13180	16272
4.0	2792	4362	5278	6281	7372	8550	11167	14133	17448

Case 3 continued

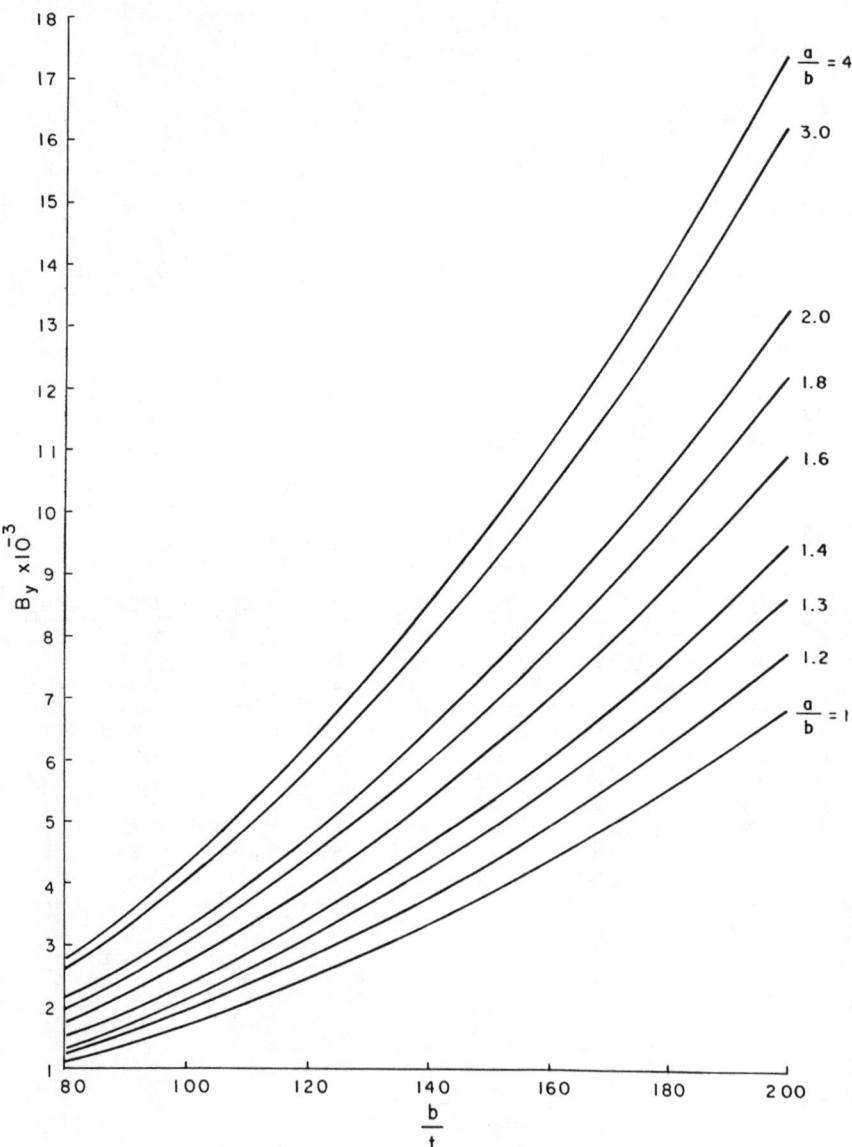

B_x

b/t a/b	80	100	110	120	130	140	160	180	200
1.1	1060	1656	2004	2385	2799	3246	4240	5366	6624
1.2	1117	1746	2113	2514	2951	3422	4470	5657	6984
1.3	1167	1824	2207	2626	3083	3575	4669	5910	7296
1.4	1210	1890	2287	2722	3194	3704	4838	6124	7560
1.5	1244	1944	2352	2799	3285	3810	4977	6299	7776
1.6	1271	1986	2403	2860	3356	3893	5084	6435	7944
1.7	1294	2022	2447	2912	3417	3963	5176	6551	8088
1.8	1309	2046	2476	2946	3458	4010	5238	6629	8184
1.9	1325	2070	2505	2981	3498	4057	5299	6707	8280
2.0	1336	2088	2526	3007	3529	4092	5345	6765	8352
3.0	1325	2070	2505	2981	3498	4057	5299	6707	8280
4.0	1252	1956	2367	2817	3306	3834	5007	6337	7824

$A_u \times 10^{-3}$

b/t a/b	80	100	110	120	130	140	160	180	200
1.1	13.698	26.754	35.610	46.231	58.779	73.413	109.584	156.029	214.032
1.2	15.990	31.231	41.569	53.968	68.615	85.698	127.923	182.140	249.850
1.3	18.171	35.490	47.237	61.327	77.972	97.385	145.367	206.978	283.920
1.4	20.295	39.640	52.760	68.497	87.088	108.771	162.364	231.178	317.117
1.5	22.308	43.571	57.993	75.290	95.725	119.558	178.466	254.105	348.566
1.6	24.153	47.174	62.789	81.517	103.642	129.447	193.226	275.121	377.395
1.7	25.887	50.560	67.295	87.367	111.079	138.736	207.092	294.864	404.477
1.8	27.452	53.617	71.364	92.651	117.797	147.126	219.616	312.696	428.938
1.9	28.962	56.566	75.289	97.745	124.275	155.216	231.693	329.891	452.525
2.0	30.303	59.186	78.777	102.274	130.033	162.407	242.427	345.175	473.491
3.0	38.690	75.566	100.579	130.579	166.019	207.354	309.520	440.703	604.531
4.0	41.988	82.009	109.154	141.712	180.174	225.033	335.910	478.278	656.074

Case 4

Simply supported rectangular plate with concentric
load over small area of radius r

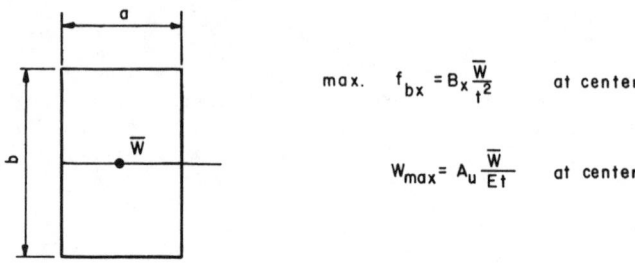

$$\text{max.} \quad f_{bx} = B_x \times \frac{\overline{W}}{t^2} \quad \text{at center}$$

$$W_{max} = A_u \frac{\overline{W}}{Et} \quad \text{at center}$$

B_x

a/r	b/a 1.0	1.2	1.4	1.6	1.8	2.0	∞
4	0.7879	0.8905	0.9569	0.9980	1.0228	1.0376	1.0576
5	0.9264	1.0290	1.0954	1.1365	1.1613	1.1761	1.1962
6	1.0396	1.1422	1.2086	1.2496	1.2745	1.2893	1.3093
7	1.1352	1.2379	1.3043	1.3453	1.3701	1.3849	1.4050
8	1.2181	1.3208	1.3871	1.4282	1.4530	1.4678	1.4879
9	1.2912	1.3939	1.4602	1.5013	1.5261	1.5409	1.5610
10	1.3566	1.4593	1.5256	1.5667	1.5915	1.6063	1.6264
12	1.4698	1.5724	1.6388	1.6799	1.7047	1.7195	1.7396
14	1.5655	1.6681	1.7345	1.7756	1.8004	1.8152	1.8352
16	1.6484	1.7510	1.8174	1.8584	1.8833	1.8981	1.9181
18	1.7215	1.8241	1.8905	1.9315	1.9564	1.9712	1.9912
20	1.7869	1.8896	1.9559	1.9969	2.0218	2.0366	2.0566

$A_u \times 10^{-3}$

b/a	a/t 80	100	110	120	130	140	160	180	200
1.0	0.8107	1.2667	1.5327	1.8241	2.1408	2.4828	3.2428	4.1042	5.0669
1.1	0.8841	1.3814	1.6715	1.9892	2.3345	2.7075	3.5363	4.4756	5.5255
1.2	0.9456	1.4775	1.7877	2.1276	2.4969	2.8959	3.7823	4.7870	5.9099
1.4	1.0371	1.6205	1.9608	2.3336	2.7387	3.1762	4.1486	5.2505	6.4821
1.6	1.0972	1.7144	2.0745	2.4688	2.8974	3.3603	4.3890	5.5548	6.8578
1.8	1.1322	1.7690	2.1405	2.5474	2.9897	3.4673	4.5287	5.7317	7.0762
2.0	1.1539	1.8029	2.1815	2.5962	3.0469	3.5337	4.6154	5.8414	7.2116
3.0	1.1811	1.8455	2.2330	2.6575	3.1189	3.6171	4.7244	5.9794	7.3819
∞	1.1846	1.8509	2.2396	2.6635	3.1281	3.6278	4.7384	5.9970	7.4038

Case 4 continued

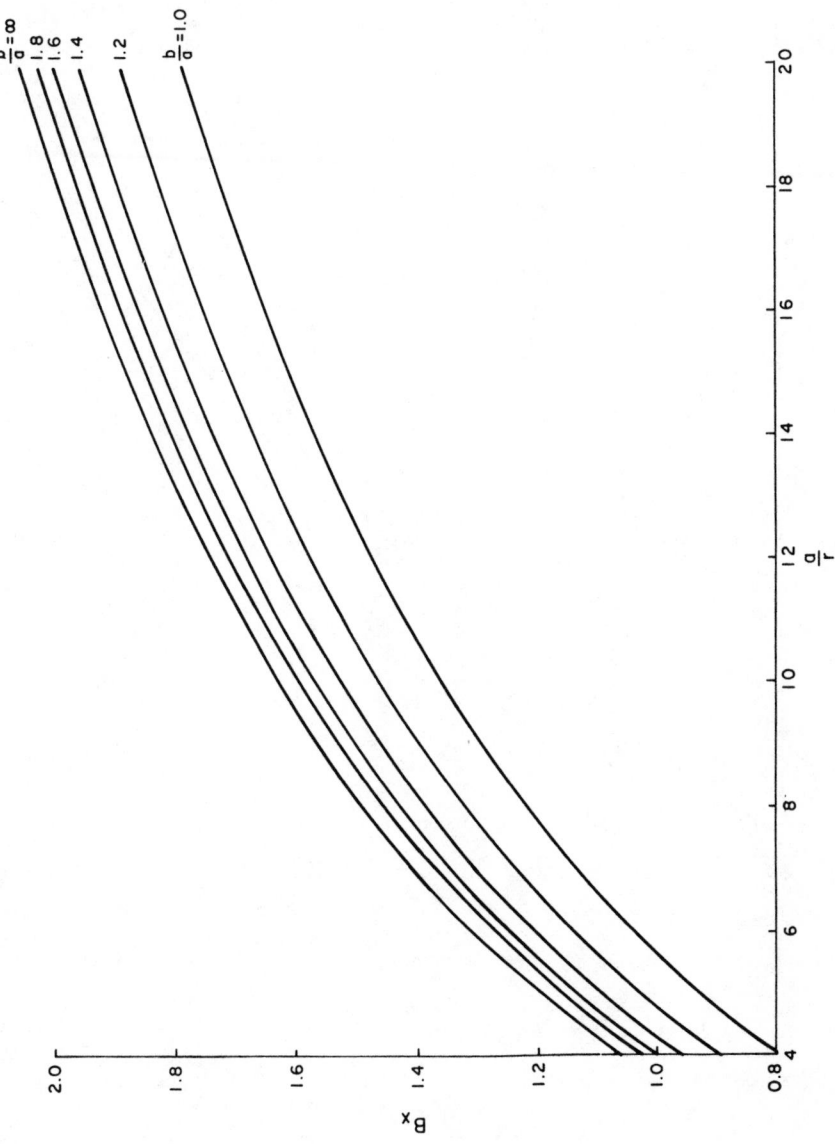

Case 4 continued

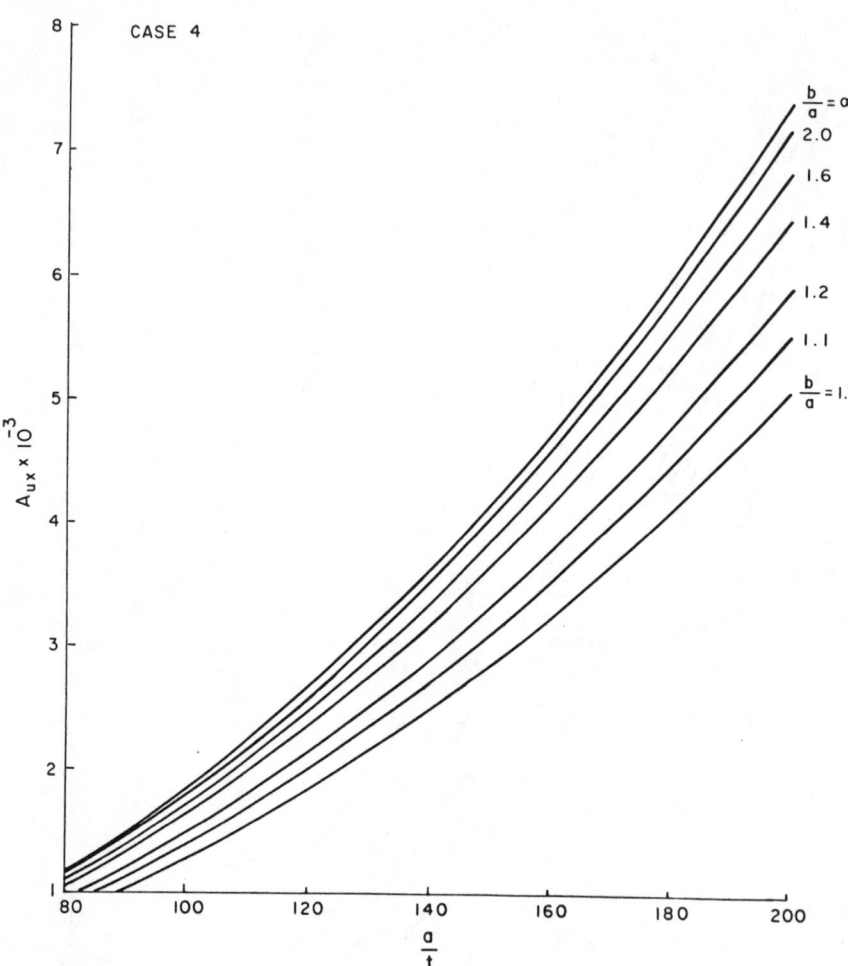

Case 5

Simply supported rectangular plate with partial load

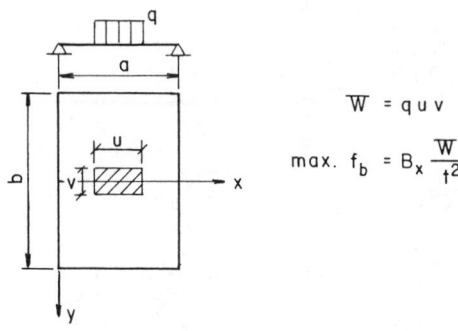

$$\overline{W} = q\,u\,v$$

$$\text{max. } f_b = B_x \frac{\overline{W}}{t^2}$$

B_x

	u/a						
	v/a	0	0.2	0.4	0.6	0.8	1.0
$\frac{b}{a} = 1.0$	0	—	1.506	1.080	0.846	0.672	0.552
	0.2	1.848	1.284	0.966	0.762	0.618	0.504
	0.4	1.392	1.074	0.846	0.678	0.552	0.456
	0.6	1.128	0.900	0.726	0.594	0.486	0.402
	0.8	0.930	0.756	0.618	0.510	0.420	0.342
	1.0	0.762	0.630	0.516	0.426	0.348	0.288
$\frac{b}{a} = 1.4$	0	—	1.794	1.380	1.098	0.906	0.750
	0.2	1.992	1.434	1.116	0.912	0.750	0.618
	0.4	1.566	1.242	1.008	0.828	0.690	0.570
	0.6	1.314	1.086	0.906	0.756	0.630	0.516
	0.8	1.122	0.948	0.804	0.672	0.564	0.468
	1.0	0.972	0.834	0.708	0.600	0.504	0.420
	1.2	0.846	0.732	0.624	0.534	0.450	0.372
	1.4	0.738	0.636	0.546	0.462	0.390	0.324
$\frac{b}{a} = 2.0$	0	—	1.764	1.350	1.074	0.888	0.732
	0.2	2.082	1.512	1.194	0.978	0.810	0.666
	0.4	1.650	1.326	1.086	0.900	0.750	0.618
	0.6	1.398	1.170	0.984	0.828	0.690	0.570
	0.8	1.218	1.044	0.888	0.756	0.636	0.528
	1.0	1.074	0.930	0.804	0.690	0.582	0.480
	1.2	0.966	0.846	0.732	0.630	0.534	0.444
	1.4	0.864	0.762	0.666	0.576	0.486	0.408
	1.6	0.780	0.690	0.606	0.522	0.444	0.372
	1.8	0.708	0.624	0.546	0.474	0.402	0.336
	2.0	0.642	0.564	0.498	0.432	0.366	0.306

Case 5 continued

$$B_y \ (f_{by} = B_y \ W/t^2 \text{ at center})$$

	v/a	u/a 0	0.2	0.4	0.6	0.8	1.0
$\frac{b}{a} = 1.0$	0	—	1.831	1.398	1.129	0.928	0.763
	0.2	1.500	1.275	1.073	0.904	0.758	0.627
	0.4	1.074	0.964	0.845	0.730	0.620	0.515
	0.6	0.835	0.763	0.679	0.594	0.509	0.424
	0.8	0.672	0.617	0.554	0.487	0.419	0.349
	1.0	0.549	0.505	0.454	0.400	0.344	0.287
$\frac{b}{a} = 1.4$	0	—	1.808	1.376	1.109	0.910	0.748
	0.2	1.455	1.252	1.051	0.884	0.741	0.613
	0.4	1.048	0.944	0.826	0.712	0.605	0.502
	0.6	0.817	0.748	0.665	0.581	0.498	0.415
	0.8	0.663	0.611	0.548	0.482	0.415	0.346
	1.0	0.552	0.510	0.460	0.406	0.349	0.292
	1.2	0.469	0.434	0.392	0.346	0.298	0.249
	1.4	0.404	0.374	0.338	0.299	0.258	0.215
$\frac{b}{a} = 2.0$	0	—	1.779	1.349	1.084	0.888	0.730
	0.2	1.417	1.223	1.024	0.859	0.719	0.594
	0.4	1.014	0.915	0.799	0.687	0.583	0.484
	0.6	0.785	0.719	0.638	0.556	0.475	0.395
	0.8	0.632	0.582	0.520	0.457	0.392	0.327
	1.0	0.522	0.482	0.433	0.381	0.328	0.274
	1.2	0.440	0.407	0.366	0.323	0.278	0.232
	1.4	0.378	0.350	0.314	0.278	0.239	0.200
	1.6	0.331	0.305	0.275	0.242	0.209	0.175
	1.8	0.293	0.271	0.244	0.215	0.185	0.155
	2.0	0.263	0.244	0.219	0.193	0.167	0.139

Case 6

Fixed rectangular plate with uniform loads

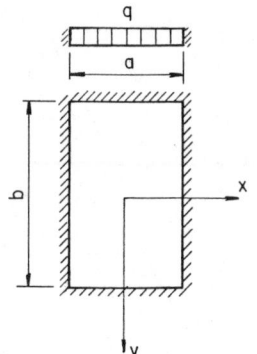

max. $f_{bx} = q B_x$ at $x = \dfrac{a}{2}$, $y = 0$

$W_{max} = a A_u \dfrac{q}{E}$ at $x = 0$, $y = 0$

B_x

a/t b/a	80	100	110	120	130	140	160	180	200
1.0	1970	3078	3724	4432	5202	6033	7880	9973	12312
1.1	2231	3486	4218	5020	5891	6833	8924	11295	13944
1.2	2454	3834	4639	5521	6480	7515	9815	12422	15336
1.3	2638	4122	4988	5936	6966	8079	10552	13355	16488
1.4	2788	4356	5271	6273	7362	8538	11151	14113	17424
1.5	2907	4542	5496	6541	7676	8902	11628	14716	18168
1.6	2995	4680	5663	6739	7909	9173	11981	15163	18720
1.7	3068	4794	5800	6903	8102	9396	12273	15533	19176
1.8	3118	4872	5895	7016	8234	9549	12472	15785	19488
1.9	3156	4932	5968	7102	8335	9667	12626	15980	19728
2.0	3183	4974	6019	7163	8406	9749	12733	16116	19896
∞	3199	4998	6048	7197	8447	9796	12795	16194	19992

$A_u \times 10^{-3}$

a/t b/a	80	100	110	120	130	140	160	180	200
1.0	7.045	13.759	18.314	23.776	30.229	37.755	56.358	80.244	110.074
1.1	8.387	16.380	21.802	28.305	35.987	44.947	67.093	95.528	131.040
1.2	9.617	18.782	24.999	32.456	41.265	51.539	76.933	109.539	150.259
1.3	10.679	20.857	27.761	36.041	45.823	57.232	85.431	121.639	166.858
1.4	11.574	22.604	30.087	39.060	49.662	62.027	92.588	131.829	180.835
1.5	12.300	24.024	31.976	41.514	52.781	65.922	98.402	140.108	192.192
1.6	12.859	25.116	33.429	43.400	55.180	68.918	102.875	146.477	200.928
1.7	13.307	25.990	34.592	44.910	57.099	71.316	106.453	151.571	207.917
1.8	13.698	26.754	35.610	46.231	58.779	73.413	109.584	156.029	214.032
1.9	13.922	27.191	36.191	46.986	59.738	74.612	111.374	158.577	217.526
2.0	14.201	27.737	36.918	47.929	60.938	76.110	113.610	161.761	221.894
∞	14.537	28.392	37.790	49.061	62.377	77.908	116.294	165.582	227.136

Case 6 continued

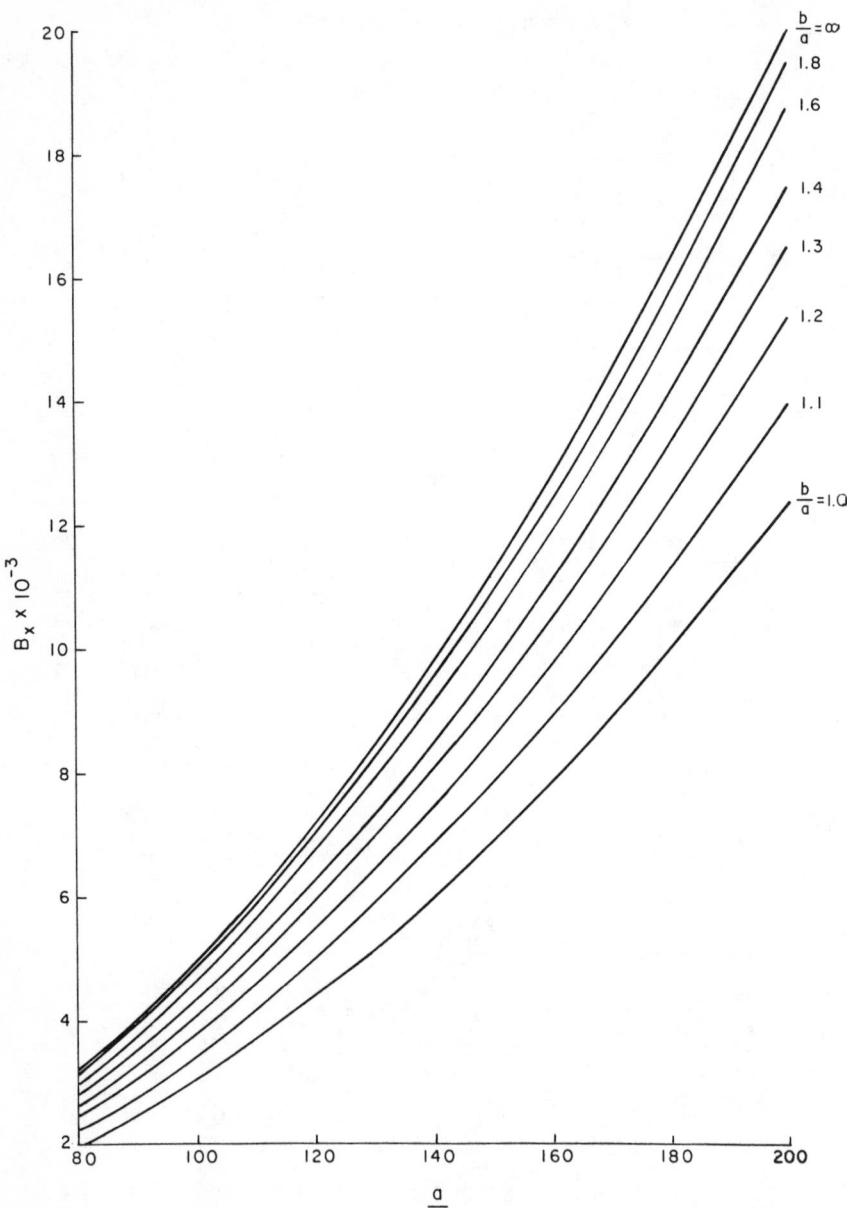

Case 6 continued

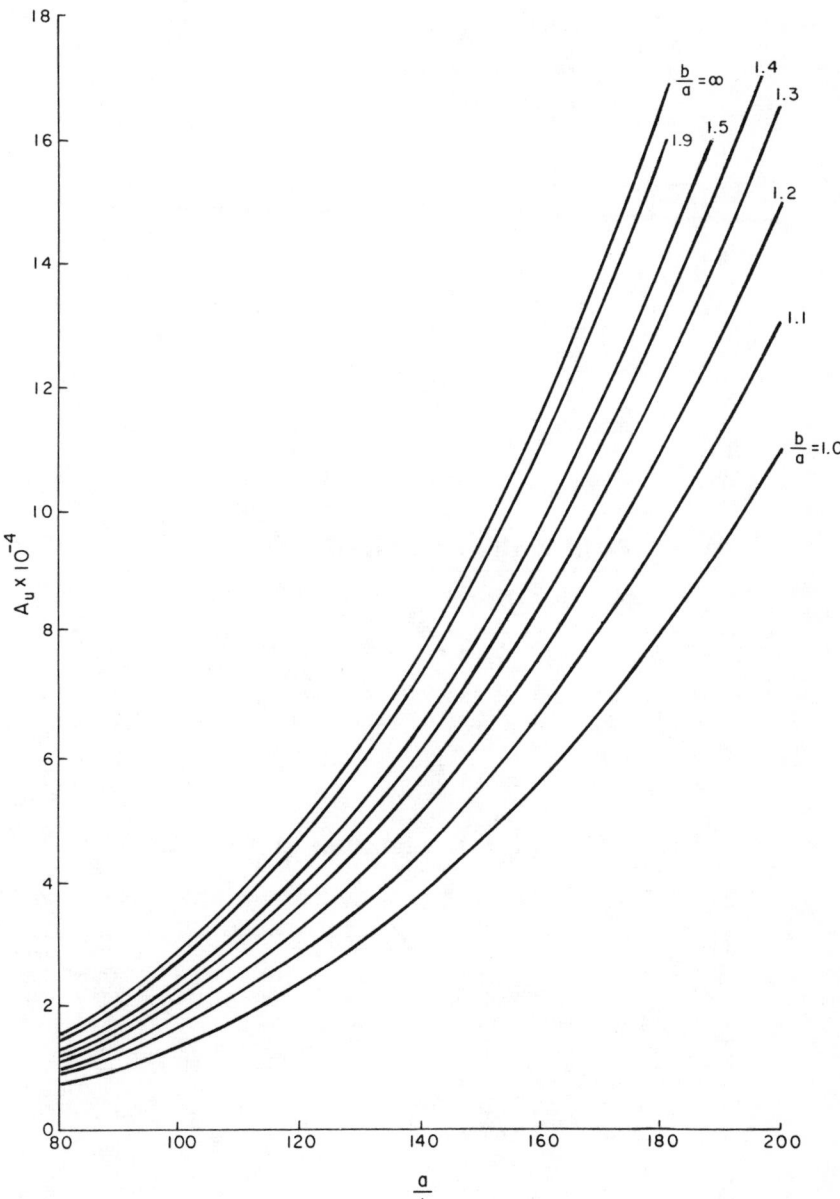

Case 7

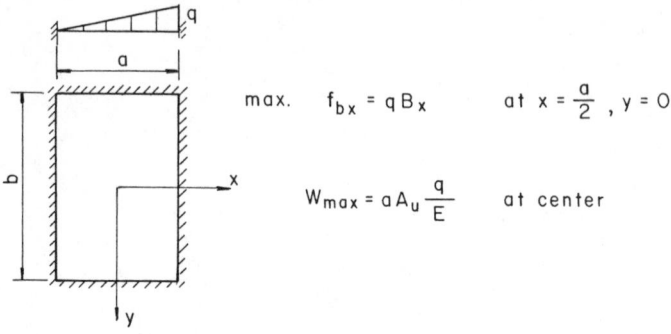

Fixed rectangular plate with hydrostatic load

max. $f_{bx} = q B_x$ at $x = \dfrac{a}{2}$, $y = 0$

$W_{max} = a A_u \dfrac{q}{E}$ at center

B_x (d = the smaller of a and b)

d/t b/a	80	100	110	120	130	140	160	180	200
0.5	442	690	835	994	1166	1352	1766	2236	2760
0.667	718	1122	1358	1616	1896	2199	2872	3635	4488
1.0	1283	2004	2425	2886	3387	3928	5130	6493	8016
1.5	1774	2772	3354	3992	4685	5433	7096	8981	11088
∞	1920	3000	3630	4320	5070	5880	7680	9720	12000

$A_u \times 10^{-3}$ (d = the smaller of a and b)

d/t b/a	80	100	110	120	130	140	160	180	200
0.5	0.447	0.874	1.163	1.510	1.919	2.397	3.578	5.095	6.989
0.667	1.213	2.369	3.154	4.095	5.206	6.502	9.706	13.820	18.957
1.0	3.522	6.880	9.157	11.888	11.115	18.878	28.179	40.122	55.037
1.5	6.150	12.012	15.988	20.757	26.390	32.961	49.201	70.054	96.096
∞	7.268	14.196	18.895	24.531	31.189	58.147	58.147	82.791	113.568

Case 7 continued

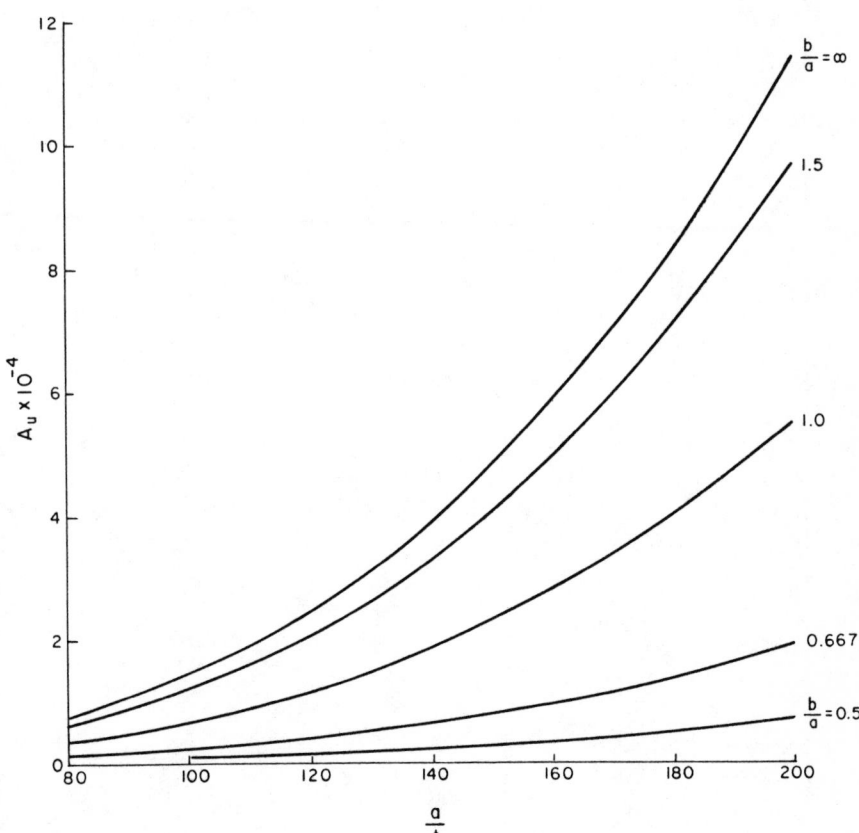

Case 7 continued

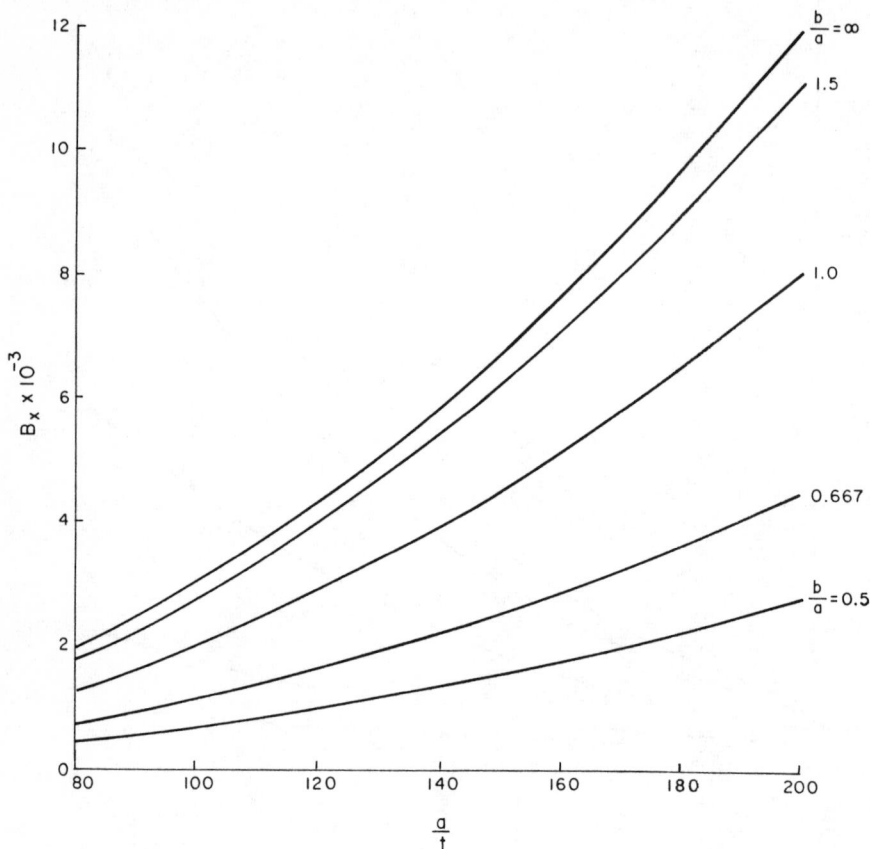

Case 8

Fixed rectangular plate under central load

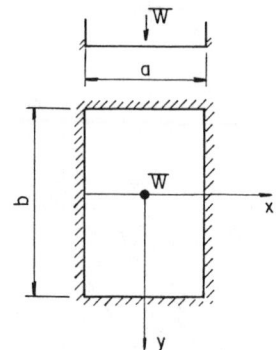

max. $f_b = B_y \dfrac{W}{a^2}$ at $x = 0$, $y = \dfrac{b}{2}$

$W_{max} = A_u \dfrac{W}{E\,t}$ at center

$B_y \times 10^{-3}$

a/t b/a	80	100	110	120	130	140	160	180	200
1.0	4.827	7.542	9.126	10.860	12.746	14.782	19.308	24.436	30.168
1.2	5.722	8.940	10.817	12.874	15.109	17.522	22.886	28.966	35.760
1.4	6.159	9.624	11.645	13.859	16.265	18.863	24.637	31.182	38.496
1.6	6.340	9.906	11.986	14.265	16.741	19.416	25.360	32.095	39.624
1.8	6.401	10.002	12.102	14.403	16.903	19.604	25.605	32.406	40.008
2.0	6.428	10.044	12.153	14.463	16.974	19.686	25.713	32.543	40.176
∞	6.451	10.080	12.197	14.515	17.035	19.757	25.805	32.659	40.320

$A_u \times 10^{-3}$

a/t b/a	80	100	110	120	130	140	160	180	200
1.0	0.3914	0.6115	0.7399	0.8806	1.0335	1.1986	1.5654	1.9813	2.4461
1.2	0.4522	0.7065	0.8549	1.0174	1.1940	1.3848	1.8087	2.2891	2.8261
1.4	0.4829	0.7546	0.9130	1.0866	1.2752	1.4790	1.9317	2.4448	3.0183
1.6	0.4976	0.7775	0.9408	1.1196	1.3140	1.5239	1.9904	2.5191	3.1100
1.8	0.5032	0.7862	0.9514	1.1322	1.3287	1.5410	2.0128	2.5474	3.1450
2.0	0.5046	0.7884	0.9540	1.1353	1.3324	1.5453	2.0184	2.5545	3.1537
∞	0.5067	0.7917	0.9580	1.1400	1.3380	1.5517	2.0268	2.5661	3.1668

Case 8 continued

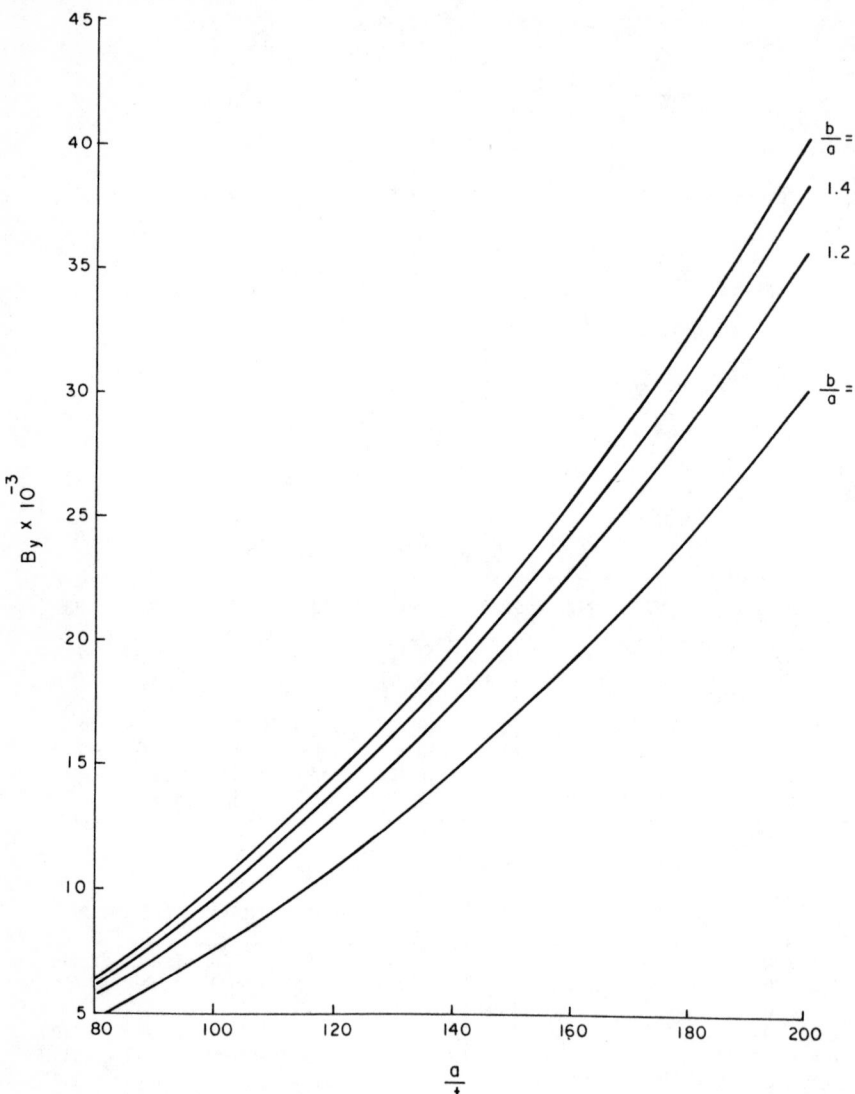

Case 9

Short edges fixed and long edges simply supported
rectangular plate with uniform load

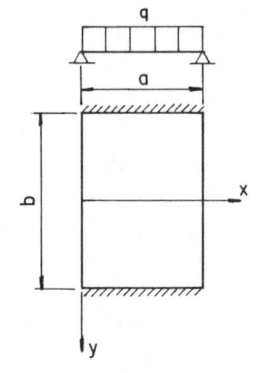

max. $f_{by} = B_y\, q$ at $x = \dfrac{a}{2}$, $y = \dfrac{b}{2}$

$W_{max} = a\, A_u \dfrac{q}{E}$ at $x = \dfrac{a}{2}$, $y = 0$

B_y

a/t b/a	80	100	110	120	130	140	160	180	200
1.0	2676	4182	5060	6022	7068	8197	10706	13550	16728
1.1	3022	4722	5714	6800	7980	9255	12088	15299	18888
1.2	3333	5208	6302	7500	8802	10208	13332	16874	20832
1.3	3602	5628	6810	8104	9511	11031	14408	18235	22512
1.4	3832	5988	7245	8623	10120	11736	15329	19401	23952
1.5	4028	6294	7616	9063	10637	12336	16113	20393	25176
1.6	4186	6540	7913	9418	11053	12818	16742	21190	26160
1.7	4308	6732	8146	9694	11377	13195	17234	21812	26928
1.8	4424	6912	8364	9953	11681	13548	17695	22395	27648
1.9	4508	7044	8523	10143	11904	13806	18033	22823	28176
2.0	4573	7146	8647	10290	12077	14006	18294	23153	28584
3.0	4785	7476	9046	10765	12634	14653	19139	24222	29904
∞	4800	7500	9075	10800	12675	14700	19200	24300	30000

$A_u \times 10^{-3}$

a/t b/a	80	100	110	120	130	140	160	180	200
1.0	10.735	20.966	27.906	36.230	46.063	57.532	85.878	122.276	167.731
1.1	14.034	27.409	36.482	47.363	60.218	75.211	112.268	159.850	219.274
1.2	17.835	34.835	46.365	60.195	76.532	95.587	142.683	203.157	278.678
1.3	21.693	42.370	56.394	73.215	93.086	116.262	173.546	247.100	338.957
1.4	25.719	50.232	66.859	86.801	110.360	137.837	205.750	292.953	401.856
1.5	29.688	57.985	77.178	100.198	127.393	159.111	237.507	338.170	463.882
1.6	33.714	65.848	87.643	113.785	144.667	180.686	269.712	384.023	526.781
1.7	37.348	72.946	97.091	126.050	160.261	200.163	298.785	425.419	583.565
1.8	40.926	79.934	106.393	138.127	175.616	219.340	327.411	466.177	639.475
1.9	44.169	86.268	114.823	149.071	189.531	236.719	353.354	503.115	690.144
2.0	47.188	92.165	122.671	159.261	202.486	252.900	377.507	537.505	737.318
3.0	65.303	127.546	169.763	220.399	280.218	349.985	522.427	743.846	1020.365
∞	72.795	142.178	189.240	245.684	312.366	390.138	582.363	829.184	1137.427

Case 9 continued

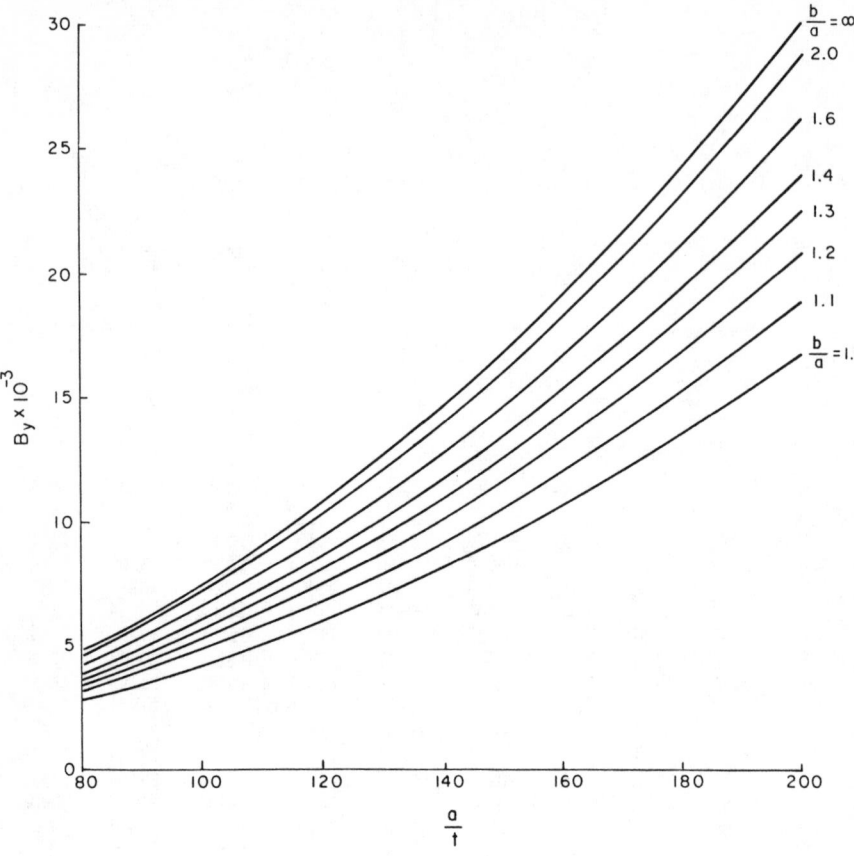

Case 10

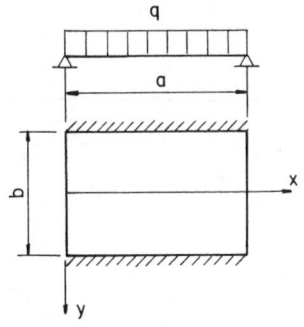

Long edges fixed and short edges simply supported
rectangular plate with uniform load

$$\text{max.} \quad f_{by} = B_y\, q \quad \text{at } x = \frac{a}{2},\ y = \frac{b}{2}$$

$$W_{max} = a A_u\, \frac{q}{E} \quad \text{at } x = \frac{a}{2},\ y = 0$$

B_y

a/t a/b	80	100	110	120	130	140	160	180	200
1.1	2838	4434	5365	6385	7493	8691	11351	14366	17736
1.2	2961	4626	5597	6661	7818	9067	11843	14988	18504
1.3	3049	4764	5764	6860	8051	9337	12196	15435	19056
1.4	3110	4860	5881	6998	8213	9526	12442	15746	19440
1.5	3156	4932	5968	7102	8335	9667	12626	15980	19728
2.0	3233	5052	6113	7275	8538	9902	12933	16368	20208
∞	3199	4998	6048	7197	8447	9796	12795	16194	19992

$A_u \times 10^{-3}$

a/t a/b	80	100	110	120	130	140	160	180	200
1.1	11.685	22.823	30.477	39.438	50.142	62.626	93.482	133.103	182.582
1.2	12.468	24.352	32.412	42.080	53.500	66.821	99.744	142.019	194.813
1.3	13.083	25.553	34.011	44.155	56.140	70.117	104.664	149.024	204.422
1.4	13.418	26.208	34.883	45.287	57.579	71.915	107.348	152.845	209.664
1.5	13.810	26.972	35.900	46.608	59.258	74.012	110.479	157.303	215.779
2.0	14.537	28.392	37.790	49.061	62.377	77.908	116.294	165.582	227.136
∞	14.537	28.392	37.790	49.061	62.377	77.908	116.294	165.582	227.136

Case 10 continued

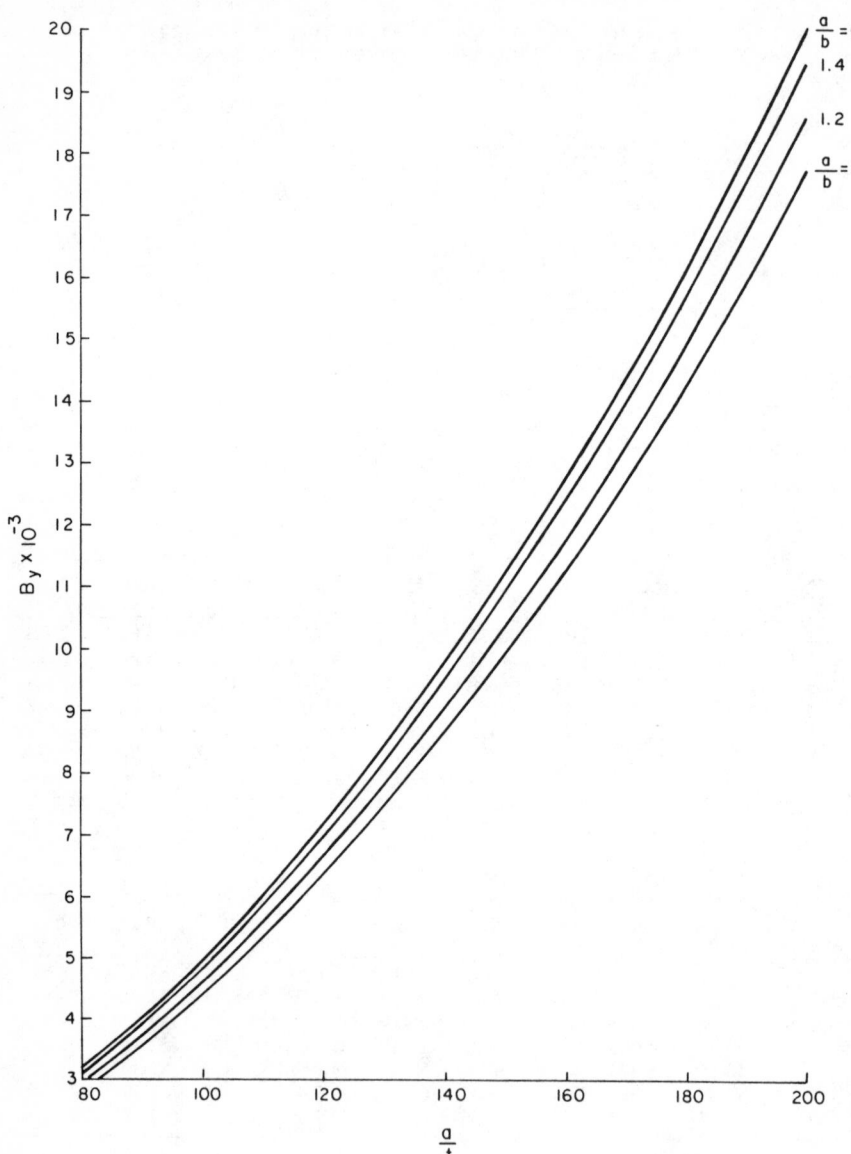

Case 11

Two edges fixed and two edges simply supported
rectangular plate with hydrostatic load

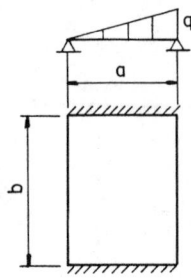

max. $f_{by} = B_y\, q$ at $x = \dfrac{a}{2}$, $y = \dfrac{b}{2}$ ($b \geqslant a$)

$x = \dfrac{3}{4}a$, $y = \dfrac{b}{2}$ ($b < a$)

B_y (d = the smaller of a and b)

d/t b/a	80	100	110	120	130	140	160	180	200
0.5	2381	3720	4501	5357	6287	7291	9523	12053	14880
0.75	1728	2700	3268	3888	4563	5292	6912	8748	10800
1.00	1344	2100	2541	3024	3549	4116	5376	6804	8400
1.25	1728	2700	3268	3888	4563	5292	6912	8748	10800
1.50	1958	3060	3703	4406	5171	5998	7834	9914	12240
2.0	2304	3600	4356	5184	6084	7056	9216	11664	14400
∞	2419	3780	4574	5443	6388	7409	9677	12247	15120

Case 11 continued

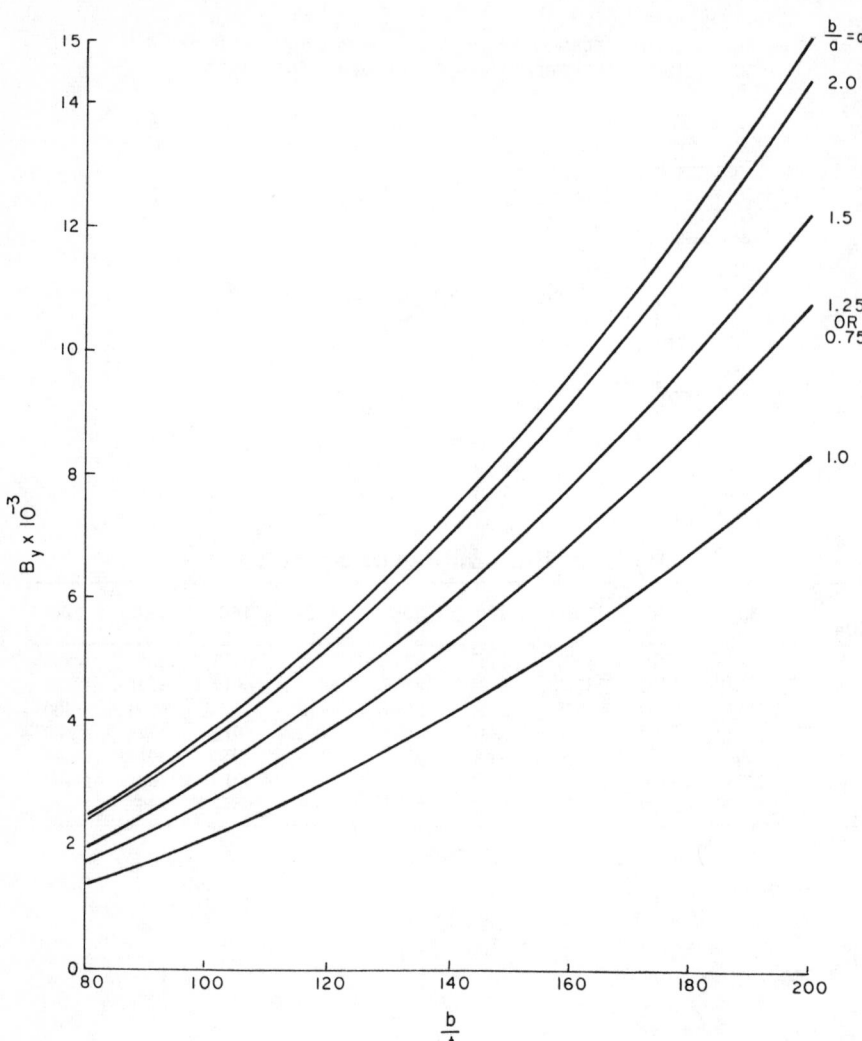

Case 12

Three edges simply supported and one edge fixed
rectangular plate with uniform load

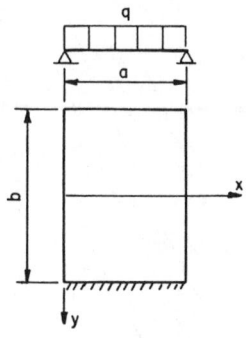

$$\text{max.} \quad f_{by} = B_y\, q \qquad \text{at} \quad x = \frac{a}{2}\ ,\ y = \frac{b}{2}$$

$$W_{max} = a\, A_u\, \frac{q}{E} \qquad \text{at} \quad x = \frac{a}{2}\ ,\ y = 0$$

B_y

a/t b/a	80	100	110	120	130	140	160	180	200
1.0	3226	5040	6098	7258	8518	9878	12902	16330	20160
1.1	3533	5520	6679	7949	9329	10819	14131	17885	22080
1.2	3763	5880	7115	8467	9937	11525	15053	19051	23520
1.3	3994	6240	7550	8986	10546	12230	15974	20218	24960
1.4	4186	6540	7913	9418	11053	12818	16742	21190	26160
1.5	4301	6720	8131	9677	11357	13171	17203	21773	26880
2.0	4685	7320	8857	10541	12371	14347	18739	23717	29280
∞	4800	7500	9075	10800	12675	14700	19200	24300	30000

$A_u \times 10^{-3}$

a/t b/a	80	100	110	120	130	140	160	180	200
1.0	15.655	30.576	40.697	52.835	67.175	83.900	125.239	178.319	244.608
1.1	19.569	38.220	50.870	66.044	83.969	104.875	156.549	222.899	305.760
1.2	24.041	46.956	62.498	81.140	103.162	128.847	192.332	273.847	375.648
1.3	27.955	54.600	72.673	94.349	119.956	149.822	223.642	318.427	436.800
1.4	32.428	63.336	84.300	109.445	139.149	173.794	259.424	369.376	506.688
1.5	35.783	69.888	93.021	120.766	153.544	191.773	286.261	407.587	559.104
2.0	51.997	101.556	135.171	175.489	223.119	278.670	415.973	592.275	812.448
∞	72.684	141.960	188.949	245.307	311.886	389.538	581.468	827.911	1135.680

Case 12 continued

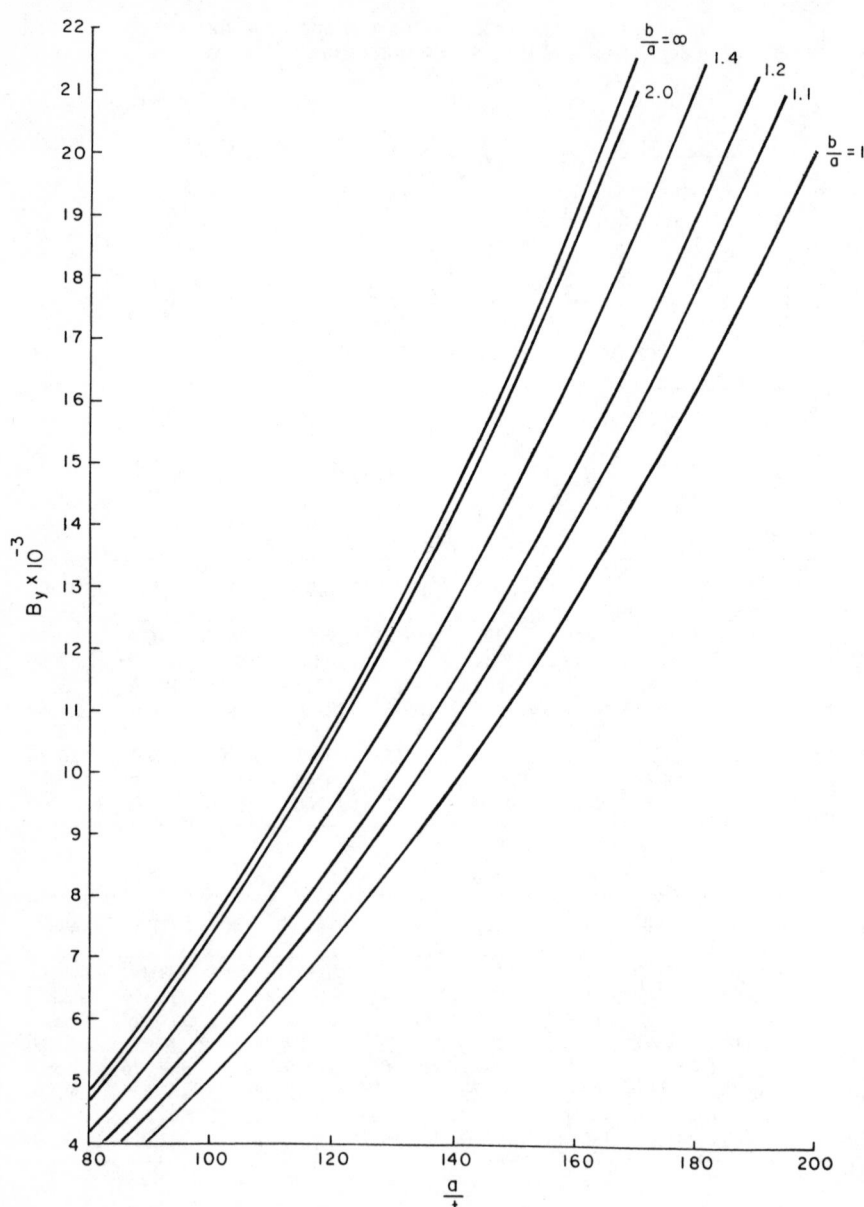

Case 12 continued

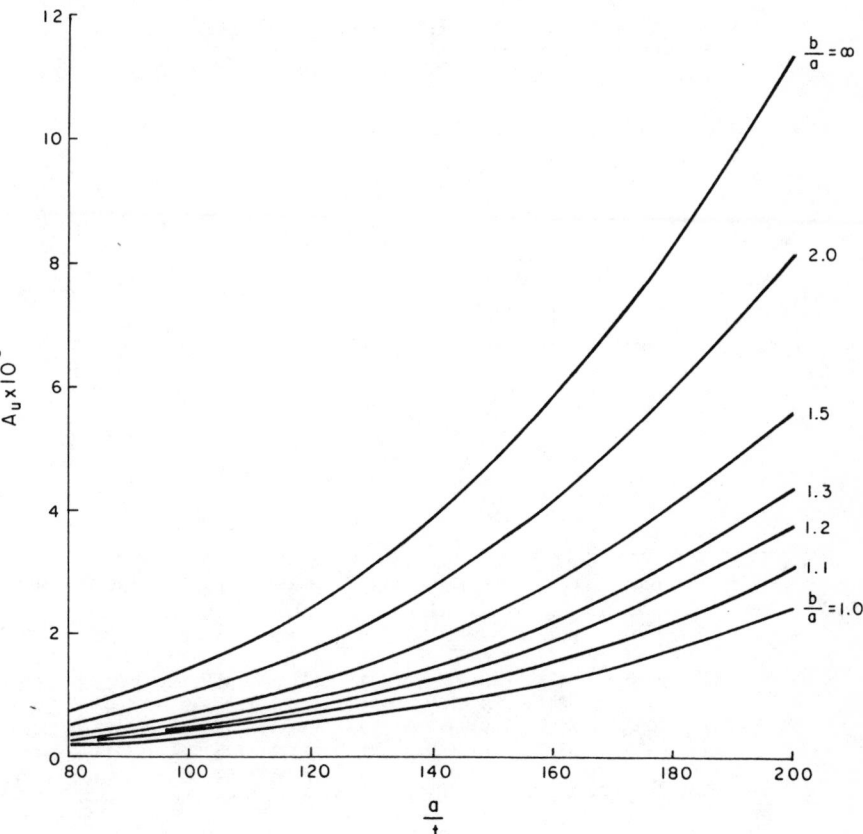

Case 13

One long edge fixed and three other edges simply supported
rectangular plate with uniform load

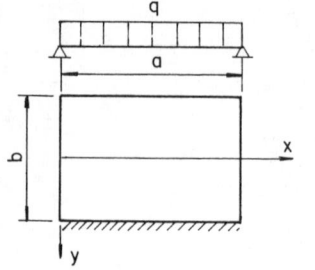

$$\text{max.} \quad f_{by} = B_y q \quad \text{at} \quad x = \frac{a}{2}, \quad y = \frac{b}{2}$$

$$W_{max} = b A_u \frac{q}{E} \quad \text{at} \quad x = \frac{a}{2}, \quad y = 0$$

B_y

b/t a/b	80	100	110	120	130	140	160	180	200
1.1	3533	5520	6679	7949	9329	10819	14131	17885	22080
1.2	3763	5880	7115	8467	9937	11525	15053	19051	23520
1.3	3955	6180	7478	8899	10444	12113	15821	20023	24720
1.4	4147	6480	7841	9331	10951	12701	16589	20995	25920
1.5	4262	6660	8059	9590	11255	13054	17050	21578	26640
2.0	4685	7320	8857	10541	12371	14347	18739	23717	29280
∞	4800	7500	9075	10800	12675	14700	19200	24300	30000

$A_u \times 10^{-3}$

b/t a/b	80	100	110	120	130	140	160	180	200
1.1	17.891	34.944	46.511	60.383	76.772	95.886	143.130	203.793	279.552
1.2	19.569	38.220	50.871	66.044	83.969	104.876	156.549	222.899	305.760
1.3	21.246	41.496	55.231	71.705	91.167	113.865	169.968	242.005	331.968
1.4	22.364	43.680	58.138	75.479	95.965	119.858	178.913	254.742	349.440
1.5	23.482	45.864	61.045	79.253	100.763	125.851	187.859	267.479	366.912
2.0	27.396	53.508	71.219	92.462	117.557	146.826	219.169	312.059	428.064
∞	29.073	56.784	75.219	98.123	124.754	155.815	232.587	331.164	454.272

Case 14

Three edges simply supported and one edge fixed
rectangular plate with hydrostatic load

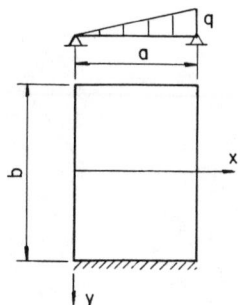

$$\max. \quad f_{by} = B_y q \quad at \quad x = \frac{a}{2}, \; y = \frac{b}{2}$$

					$\mathbf{B_y}$				
a/t	80	100	110	120	130	140	160	180	200
b/a									
1.0	1613	2520	3049	3629	4259	4939	6451	8165	10080
1.5	2150	3360	4066	4838	5678	6586	8602	10886	13440
2.0	2342	3660	4429	5270	6185	7174	9370	11858	14640
∞	2381	3720	4501	5357	6287	7291	9523	12053	14880

Case 14 continued

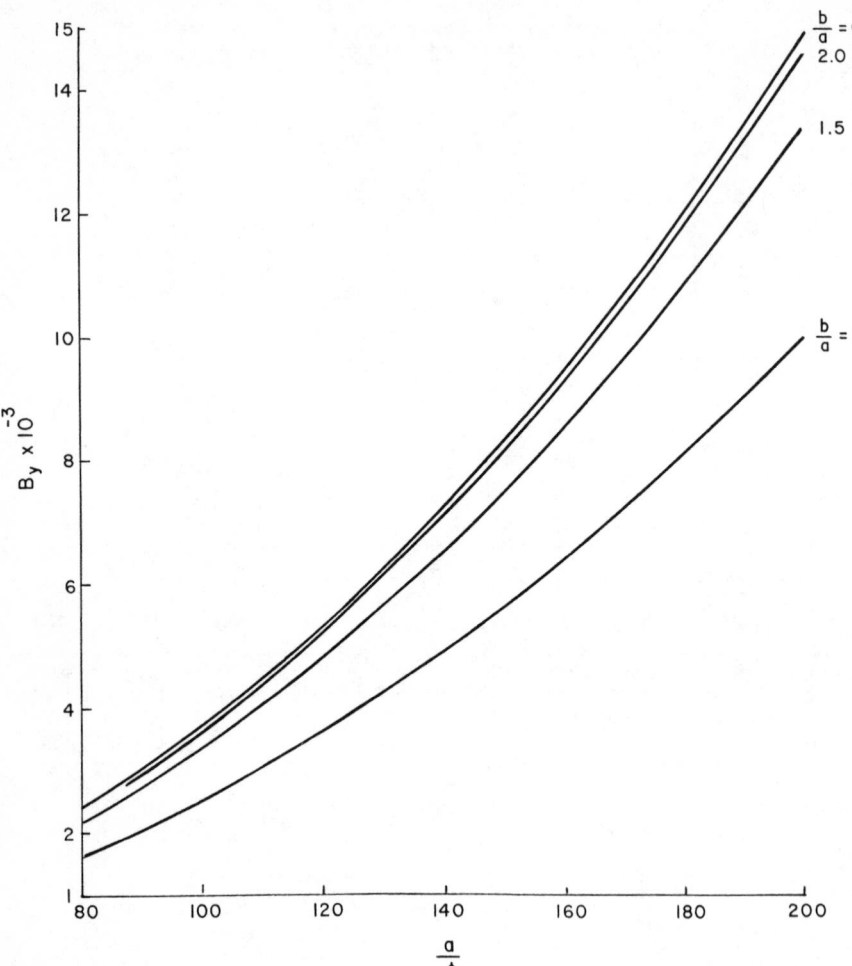

Case 15

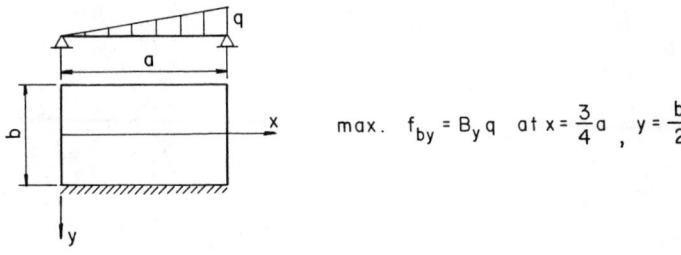

Same as Case 14 but a>b

max. $f_{by} = B_y q$ at $x = \frac{3}{4}a$, $y = \frac{b}{2}$

B_y

b/t a/b	80	100	110	120	130	140	160	180	200
1.0	1613	2520	3049	3629	4259	4939	6451	8165	10080
1.5	2304	3600	4356	5184	6084	7056	9216	11664	14400
2.0	2803	4380	5300	6307	7402	8585	11213	14191	17520
∞	3610	5640	6824	8122	9532	11054	14438	18274	22560

Case 15 continued

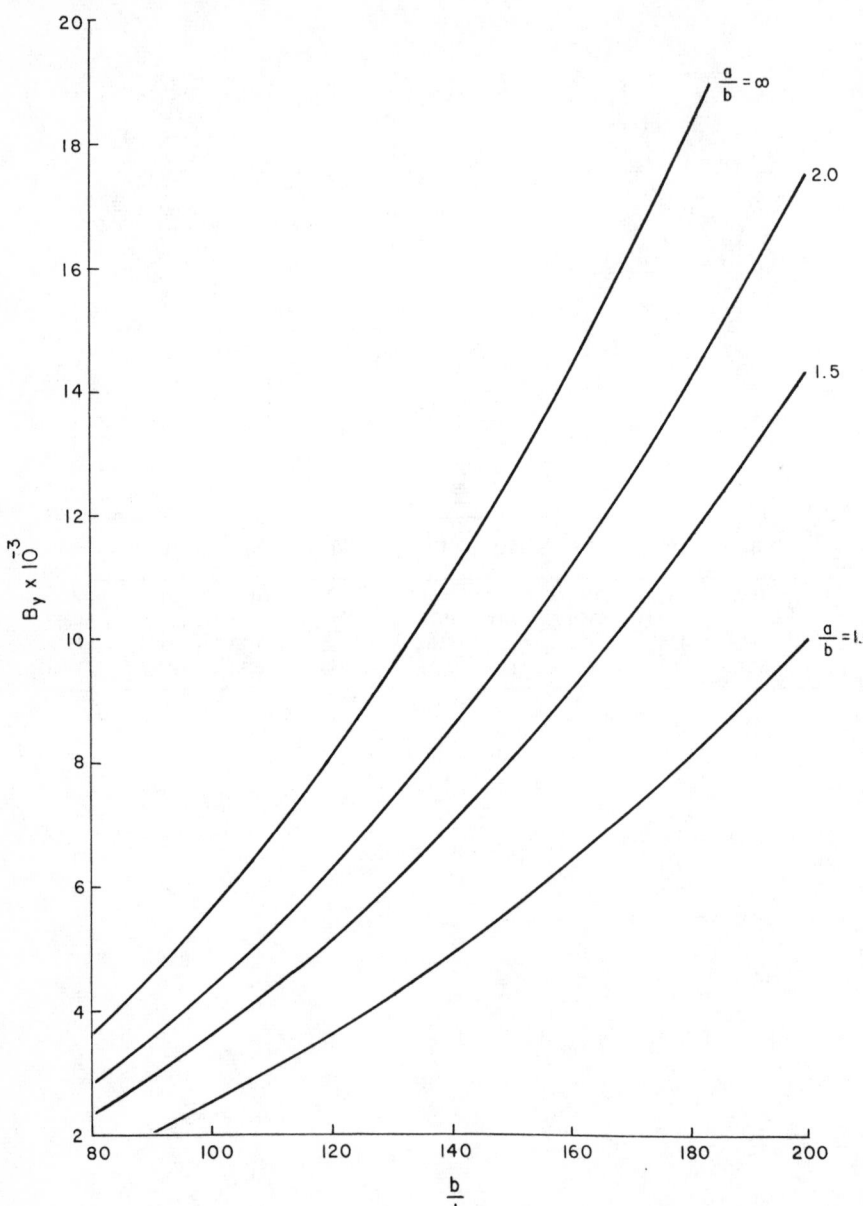

Case 16

One long edge fixed and other three edges simply supported
rectangular plate with hydrostatic load

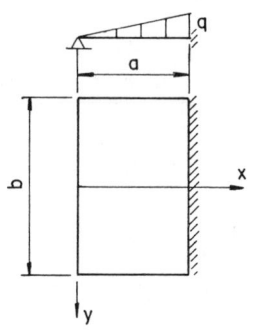

max. $f_{bx} = B_x q$ at $x = a$, $y = 0$

$W_{max} = a A_u \dfrac{q}{E}$ at $x = \dfrac{a}{2}$, $y = 0$

B_x

a/t b/a	80	100	110	120	130	140	160	180	200
1.0	1843	2880	3485	4147	4867	5645	7373	9331	11520
1.5	2342	3660	4429	5270	6185	7174	9370	11858	14640
2.0	2419	3780	4574	5443	6388	7409	9677	12247	15120
∞	2573	4020	4864	5789	6794	7879	10291	13025	16080

$A_u \times 10^{-3}$

a/t b/a	80	100	110	120	130	140	160	180	200
1.0	7.268	14.196	18.895	24.531	31.189	38.954	58.147	82.791	113.568
1.5	16.773	32.760	43.604	56.609	71.974	89.893	134.185	191.056	262.080
2.0	25.160	49.140	65.405	84.914	107.961	134.840	201.277	286.584	393.120
∞	36.342	70.980	94.474	122.653	155.943	194.769	290.734	413.955	567.840

Case 16 continued

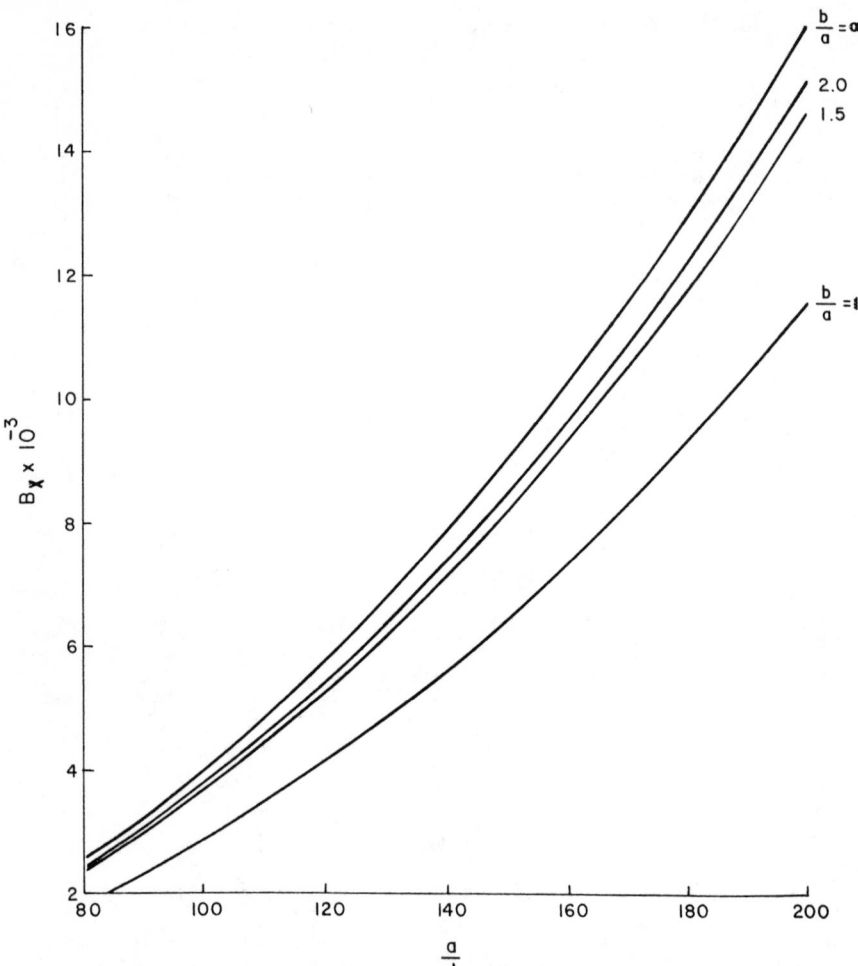

Case 17

One short edge fixed and other three edges simply supported
rectangular plate with hydrostatic load

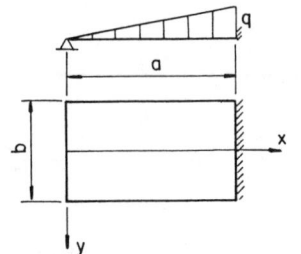

max. $f_{bx} = B_x q$ at $x = a$, $y = 0$

$W_{max} = b A_u \dfrac{q}{E}$ at $x = \dfrac{a}{2}$, $y = 0$

B_x

b/t a/b	80	100	110	120	130	140	160	180	200
1.0	1843	2880	3485	4147	4867	5645	7373	9331	11520
1.5	2726	4260	5155	6134	7199	8350	10906	13802	17040
2.0	3226	5040	6098	7258	8518	9878	12902	16330	20160
∞	4800	7500	9075	10800	12675	14700	19200	24300	30000

$A_u \times 10^{-3}$

b/t a/b	80	100	110	120	130	140	160	180	200
1.0	7.268	14.196	18.890	24.531	31.189	38.954	58.147	82.791	113.568
1.5	16.773	32.760	43.604	56.609	71.974	89.893	134.185	191.056	262.080
2.0	25.160	49.140	65.405	84.914	107.961	134.840	201.277	286.584	393.120
∞	36.342	70.980	94.474	122.653	155.943	194.769	290.734	413.955	567.840

Case 17 continued

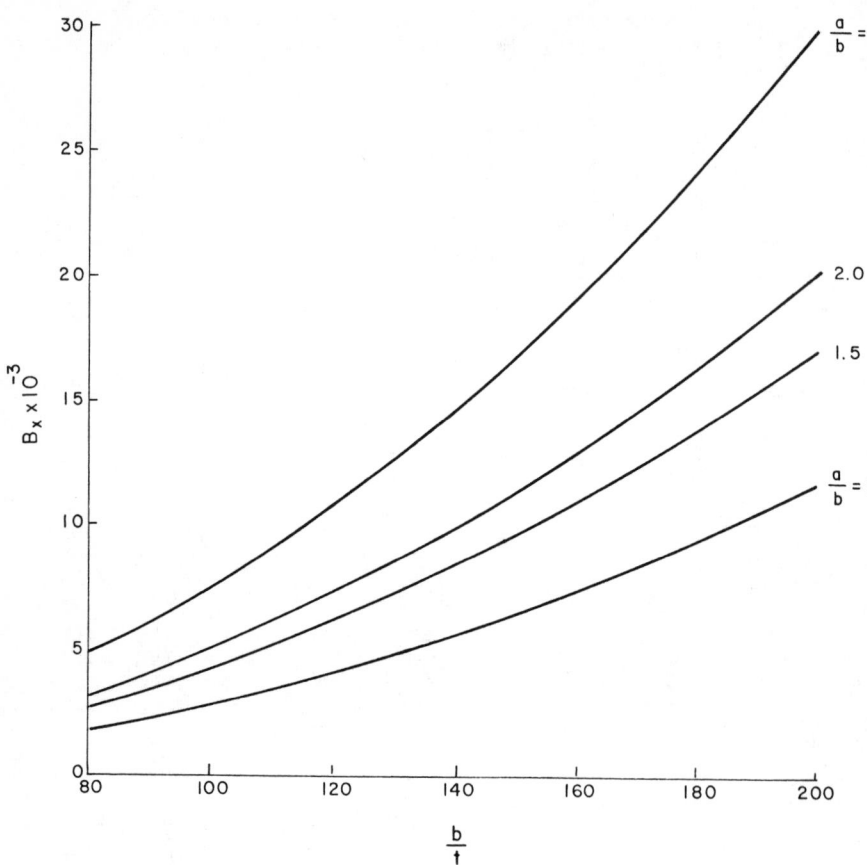

Case 18

Long edges simply supported one short edge fixed and the
other edge free rectangular plate with uniform load

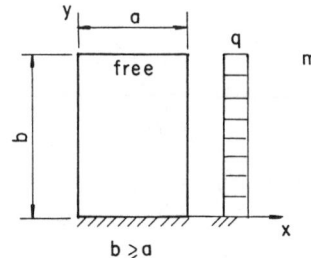

max. $f_{by} = B_y q$ at $x = \dfrac{a}{2}$, $y = 0$

$W_{max} = a A_u \dfrac{q}{E}$ at $x = \dfrac{a}{2}$, $y = b$

B_y

a/t b/a	80	100	110	120	130	140	160	180	200
1.0	4570	7140	8639	10282	12067	13994	18278	23134	28560
1.5	4762	7440	9002	10714	12574	14582	19046	24106	29760
2.0	4800	7500	9075	10800	12675	14700	19200	24300	30000
∞	4800	7500	9075	10800	12675	14700	19200	24300	30000

A_u × 10⁻³

a/t b/a	80	100	110	120	130	140	160	180	200
1.0	63.179	123.396	164.240	213.228	271.101	338.599	505.430	719.645	987.168
1.5	78.834	153.972	204.937	266.064	338.276	422.499	630.669	897.964	1231.776
2.0	83.866	163.800	218.018	283.046	359.869	449.467	670.925	955.282	1310.40
∞	84.984	165.984	220.925	286.820	364.667	455.460	679.870	968.019	1327.872

Case 18 continued

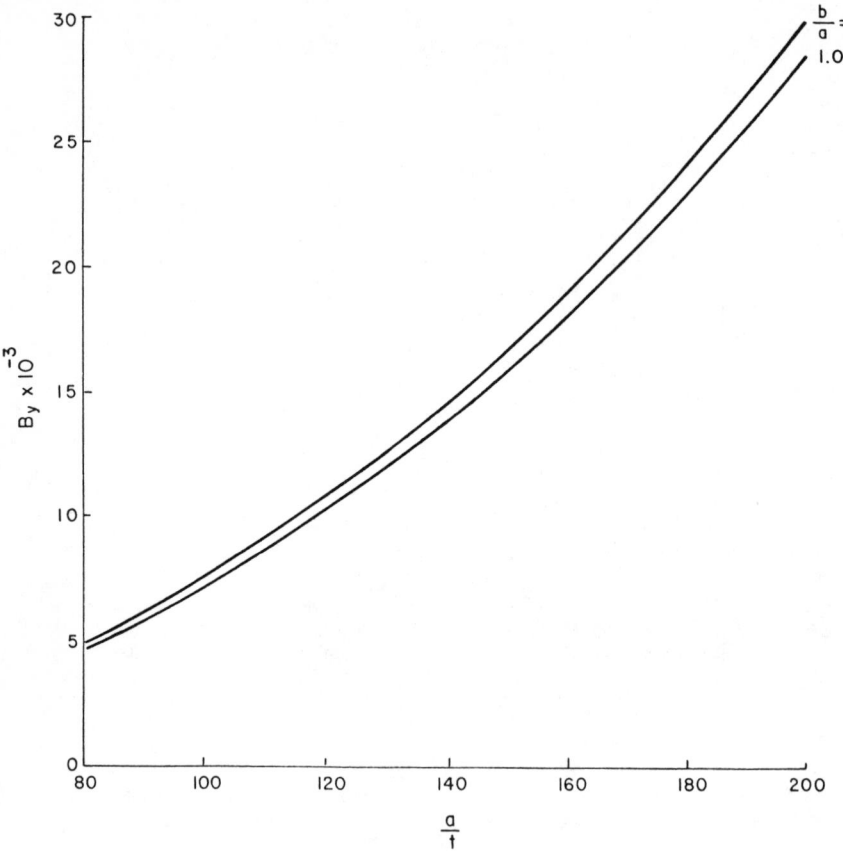

Case 19

Short edges simply supported one long edge fixed and
the other edge free rectangular plate with uniform
load

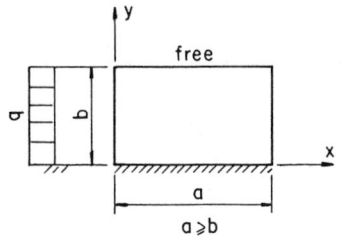

$$\text{max.} \quad f_{by} = B_y q \quad \text{at } x = \frac{a}{2}, \quad y = 0$$

$$W_{max} = bA_u \frac{q}{E} \quad \text{at } x = \frac{a}{2}, \quad y = b$$

B_y

a/t a/b	80	100	110	120	130	140	160	180	200
1.0	4570	7140	8639	10282	12067	13994	18278	23134	28560
1.5	8717	13620	16480	19613	23018	26695	34867	44129	54480
2.0	12250	19140	23159	27562	32347	37514	48998	62014	76560
3.0	16435	25680	31073	36979	43399	50333	65741	83203	102720
∞	19200	30000	36300	43200	50700	58800	76800	97200	120000

A_u × 10⁻³

a/t a/b	80	100	110	120	130	140	160	180	200
1.0	63.179	123.396	164.240	213.228	271.101	338.599	505.43	719.645	987.168
1.5	187.300	365.820	489.906	632.137	803.707	1003.81	1498.40	2133.46	2926.56
2.0	325.399	635.544	845.909	1098.22	1396.29	1743.93	2603.19	3706.49	5084.35
3.0	525.558	1026.48	1366.25	1773.76	2255.18	2816.66	4204.46	5986.43	8211.84
∞	698.880	1365.00	1816.82	2358.72	2998.91	3745.56	5591.04	7960.68	10920.00

Case 19 continued

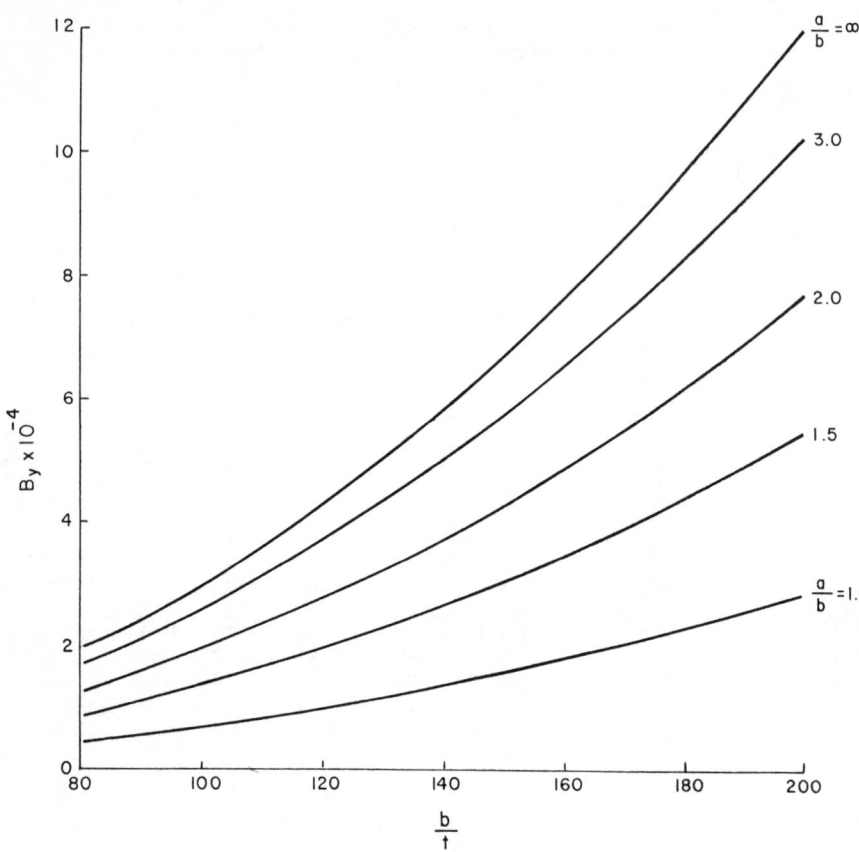

Case 20

Three edges simply supported and one edge free
rectangular plate with uniform load

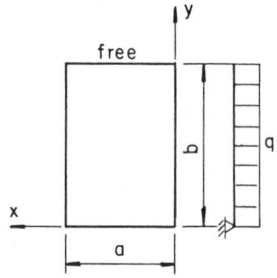

max. $f_{bx} = B_x q$ at $x = \dfrac{a}{2}$, $y = b$

$W_{max} = a A_u \dfrac{q}{E}$ at $x = \dfrac{a}{2}$, $y = b$

B_x

a/t a/b	80	100	110	120	130	140	160	180	200
0.5	2304	3600	4356	5184	6084	7056	9216	11664	14400
0.667	3187	4980	6026	7171	8416	9761	12749	16135	19920
0.714	3379	5280	6389	7603	8923	10349	13517	17107	21120
0.769	3610	5640	6824	8122	9532	11054	14438	18274	22560
0.833	3840	6000	7260	8640	10140	11760	15360	19440	24000
0.909	4109	6420	7768	9245	10850	12583	16435	20800	25680
1.0	4301	6720	8131	9677	11357	13171	17203	21773	26880
1.1	4493	7020	8494	10109	11864	13759	17971	22748	28080
1.2	4646	7260	8785	10454	12269	14230	18586	23522	29040
1.3	4762	7440	9002	10714	12574	14582	19046	24106	29760
1.4	4838	7560	9148	10886	12776	14818	19354	24494	30240
1.5	4915	7680	9293	11059	12979	15053	19661	24883	30720
2.0	5069	7920	9583	11405	13385	15523	20275	25661	31600
∞	5107	7980	9650	11491	13486	15641	20429	25855	31920

$A_u \times 10^{-3}$

a/t a/b	80	100	110	120	130	140	160	180	200
0.5	39.696	77.532	103.195	133.975	170.338	212.748	317.571	452.167	620.256
0.667	54.121	105.706	140.694	182.659	232.235	290.056	432.970	616.475	845.645
0.714	57.196	111.712	148.688	193.038	245.430	306.537	457.571	651.502	893.693
0.769	61.054	119.246	158.717	206.058	261.984	327.212	488.433	695.445	953.971
0.833	64.744	126.454	168.310	218.512	277.819	346.989	517.954	737.477	1011.63
0.909	68.882	134.534	179.065	232.475	295.572	369.162	551.053	784.605	1076.28
1.0	71.901	140.431	186.914	242.665	308.527	385.343	575.206	818.995	1123.45
1.1	74.976	146.437	194.908	253.043	321.722	401.824	599.807	854.022	1171.50
1.2	77.380	151.133	201.158	261.157	332.039	414.708	619.040	881.406	1209.06
1.3	79.225	154.736	205.954	267.384	339.956	424.597	633.800	902.423	1237.89
1.4	80.623	157.466	209.588	272.102	345.954	432.088	644.982	918.344	1259.73
1.5	81.741	159.650	212.495	275.876	350.752	438.081	653.928	931.081	1277.20
2.0	84.257	164.564	219.035	284.367	361.548	451.565	674.056	959.740	1316.52
∞	85.096	166.202	221.215	287.198	365.147	456.057	680.765	969.292	1329.62

Case 20 continued

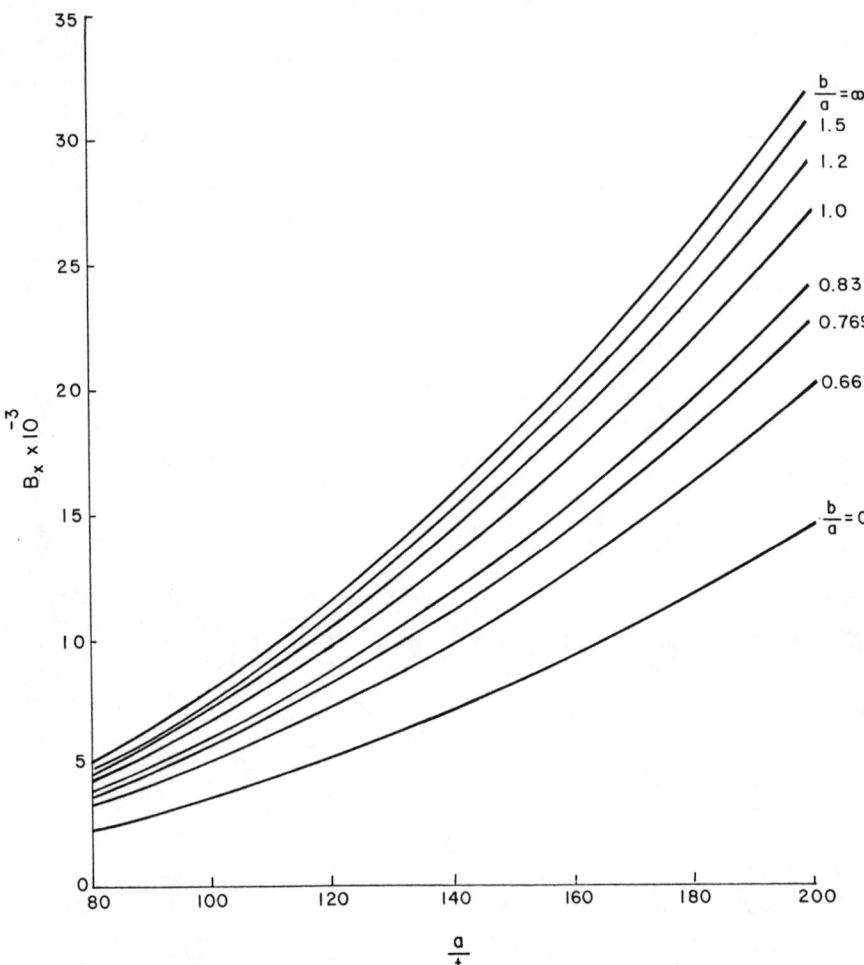

Case 20 continued

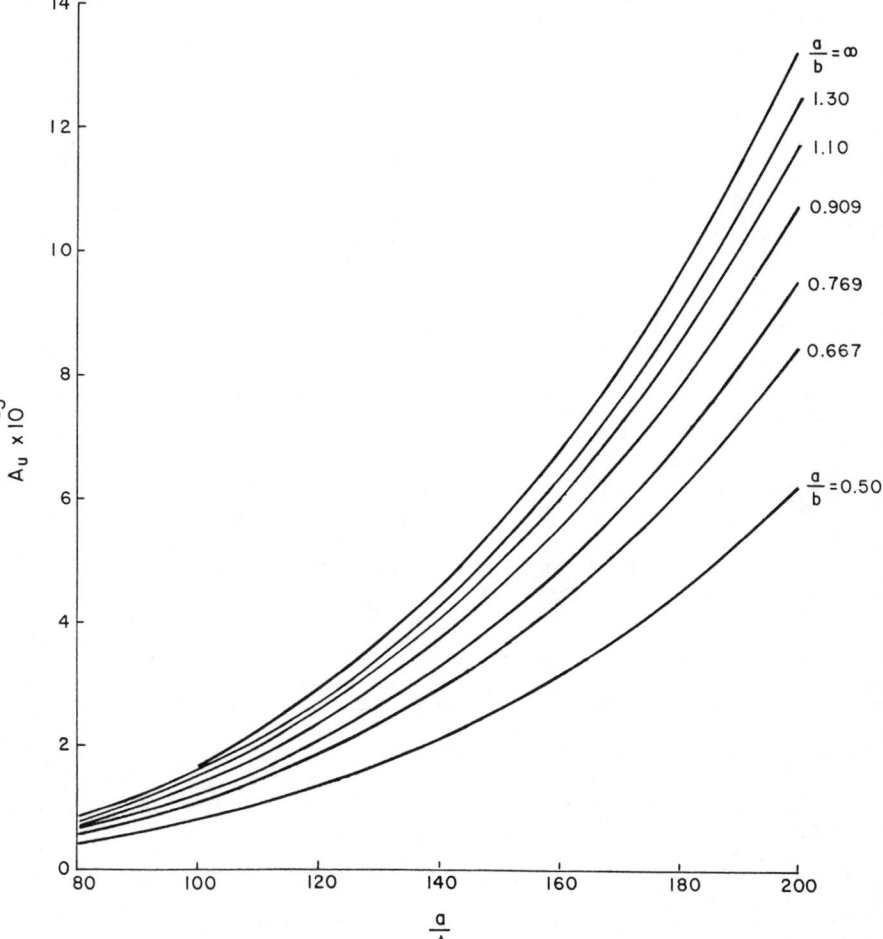

Case 21

Same as Case 20 but with hydrostatic load

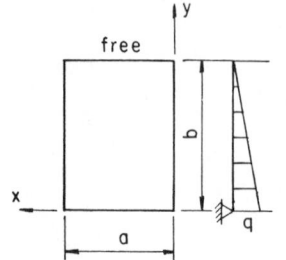

max. $f_{bx} = B_x\,q$ at $x = \dfrac{a}{2}$, $y = b$ $(a > b)$

$\qquad\qquad\qquad\qquad x = \dfrac{a}{2}$, $y = \dfrac{b}{2}$ $(b \lessgtr a)$

$f_{by} = B_y\,q$

$W_{max} = a\,A_u\,\dfrac{q}{E}$

B_x

a/t b/a	80	100	110	120	130	140	160	180	200
0.5	756	1182	1430	1702	1998	2317	3026	3830	4728
0.667	1018	1590	1924	2290	2687	3116	4070	5152	6360
1.0	1271	1986	2403	2860	3356	3893	5084	6435	7944
1.5	1740	2718	3289	3914	4593	5327	6958	8806	10872
2.0	2031	3174	3841	4571	5364	6221	8125	10284	12696
∞	2400	3750	4538	5400	6338	7350	9600	12150	15000

$A_u \times 10^{-3}$

a/t b/a	80	100	110	120	130	140	160	180	200
0.5	12.859	25.116	33.429	43.400	55.180	68.918	102.875	146.477	200.928
0.667	16.997	33.197	44.185	57.364	72.933	91.092	135.974	193.604	265.574
1.0	20.575	40.186	53.487	69.441	88.288	110.269	164.600	234.362	321.485
1.5	24.880	48.594	64.679	83.970	106.761	133.342	199.041	283.400	388.752
2.0	29.800	58.204	77.469	100.576	127.873	159.711	238.402	339.443	465.629
∞	36.398	71.089	94.620	122.842	156.182	195.069	291.181	414.592	568.714

B_y

a/t b/a	80	100	110	120	130	140	160	180	200
0.50	461	720	871	1037	1217	1411	1843	2333	2880
0.667	599	936	1133	1348	1582	1835	2396	3033	3744
1.00	822	1284	1554	1849	2170	2517	3287	4160	5136
1.50	887	1386	1677	1996	2342	2717	3548	4491	5544
2.00	852	1332	1612	1918	2251	2611	3410	4316	5328
∞	718	1122	1358	1616	1896	2199	2872	3635	4488

Case 21 continued

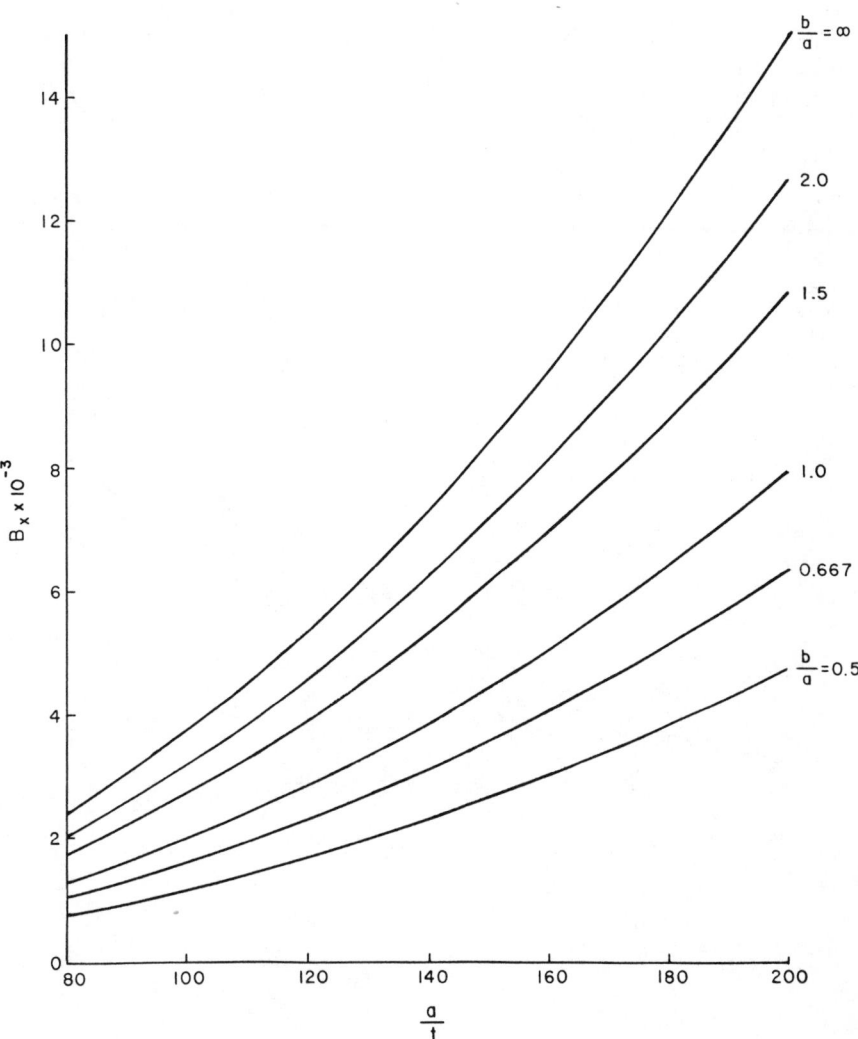

Case 21 continued

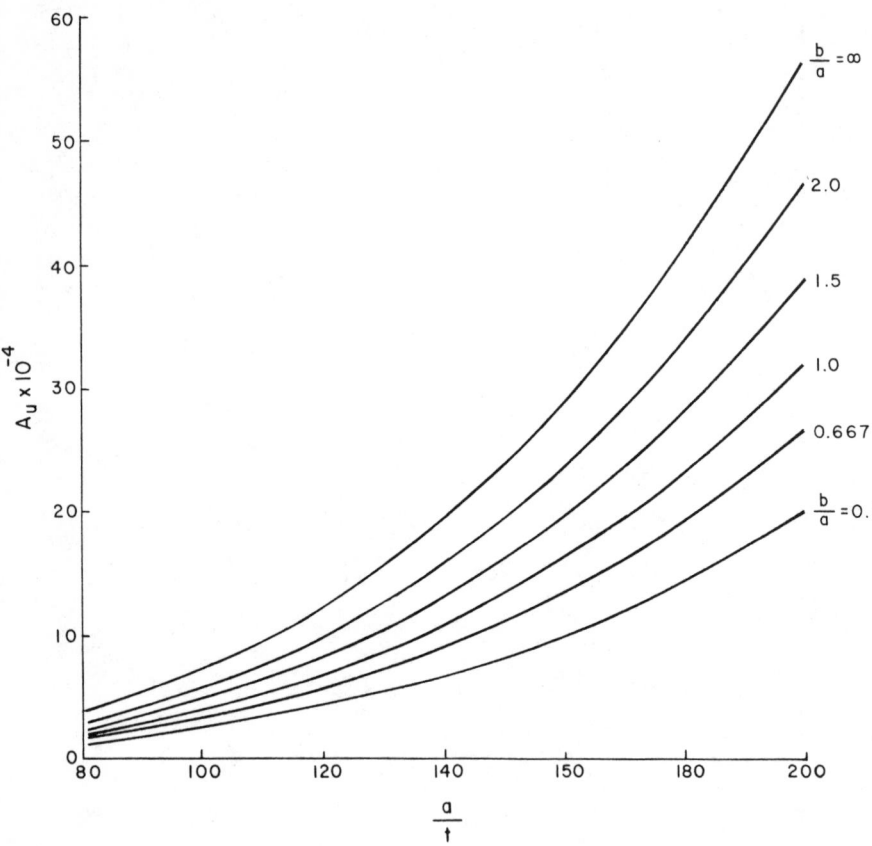

Case C1

Circular Plate with fixed edge under uniform load

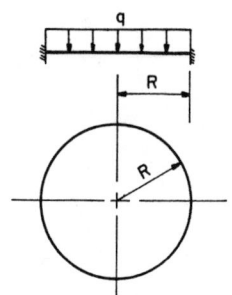

Max. Stress $f_{br} = \dfrac{6M_r}{t^2}$ at the edge

$$f_{br} = B_r q$$

$$W_{max} = R\, A_r\, \dfrac{q}{E} \quad \text{at center of the plate}$$

R/t	B$_r$	A$_r \times 10^{-3}$
55	2269	28.388
60	2700	36.855
65	3169	46.858
70	3675	58.524
75	4219	71.982
80	4800	87.360
85	5419	104.785
90	6075	124.386
95	6769	146.290
100	7500	170.625
105	8269	197.520
110	9075	227.102
115	9919	259.500
120	10800	294.840
125	11719	333.252
130	12675	374.863
135	13669	419.801
140	14700	468.195
145	15769	520.172
150	16875	575.860
155	18019	635.386
160	19200	698.880
165	20419	766.469
170	21675	838.281
175	22969	914.444
180	24300	995.085

Case C1 continued

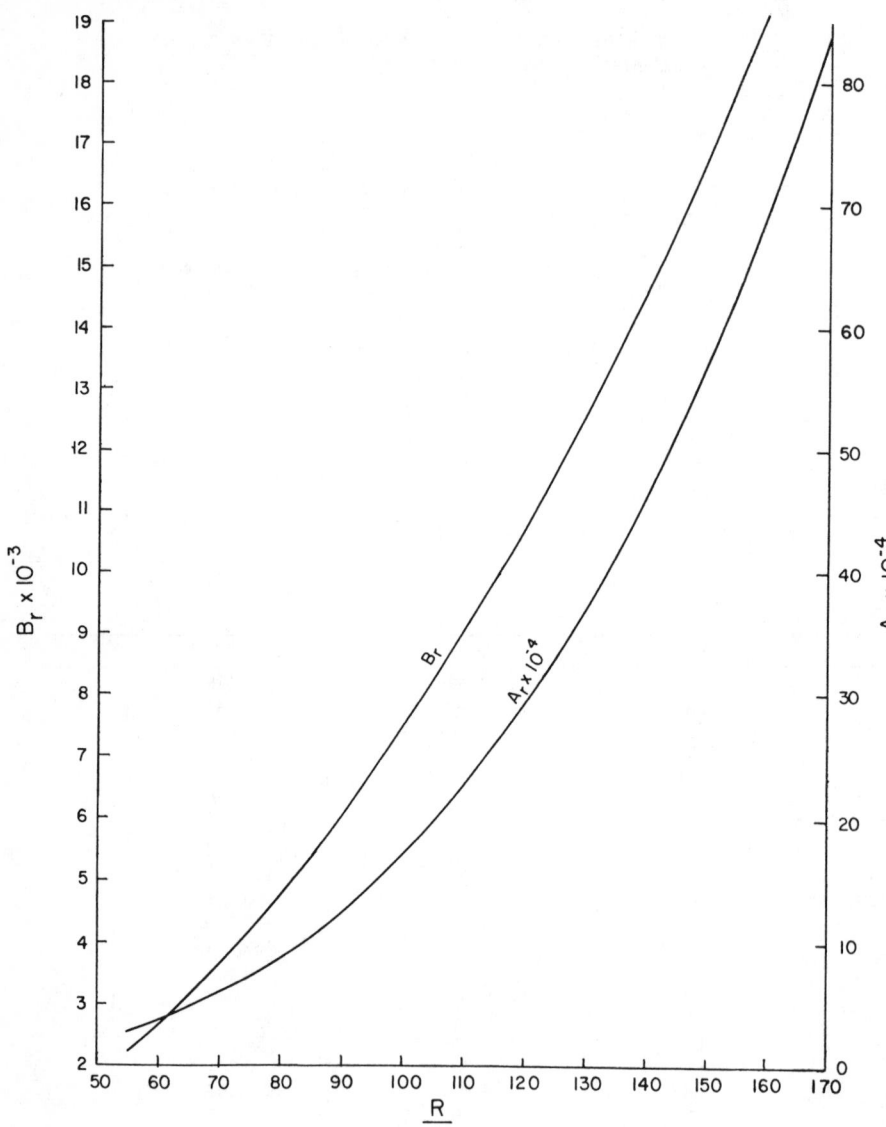

Case C2

Simply supported Circular Plate with uniform load

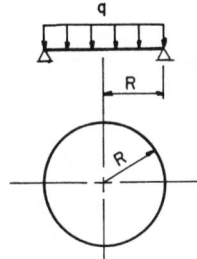

Max. moment $M_r = M_t$ at center of the Plate

Max. Stress $f_{br} = B_r\, q$

$W_{max} = RA_r\, \dfrac{q}{E}$ at center

R/t	B$_r$	A$_r$ × 10^{-3}
55	3743	115.740
60	4455	150.260
65	5228	191.040
70	6064	238.600
75	6961	293.470
80	7920	356.160
85	8941	427.200
90	10024	507.110
95	11168	595.230
100	12375	693.080
105	13643	805.270
110	14974	925.880
115	16370	1057.960
120	17820	1202.040
125	19336	1358.640
130	20914	1528.290
135	22553	1711.500
140	24255	1908.800
145	26018	2120.700
150	27844	2347.740
155	29731	2590.420
160	31680	2849.280
165	33691	3124.840
170	35764	3417.610
175	37898	3728.115
180	40095	4056.885

Case C2 continued

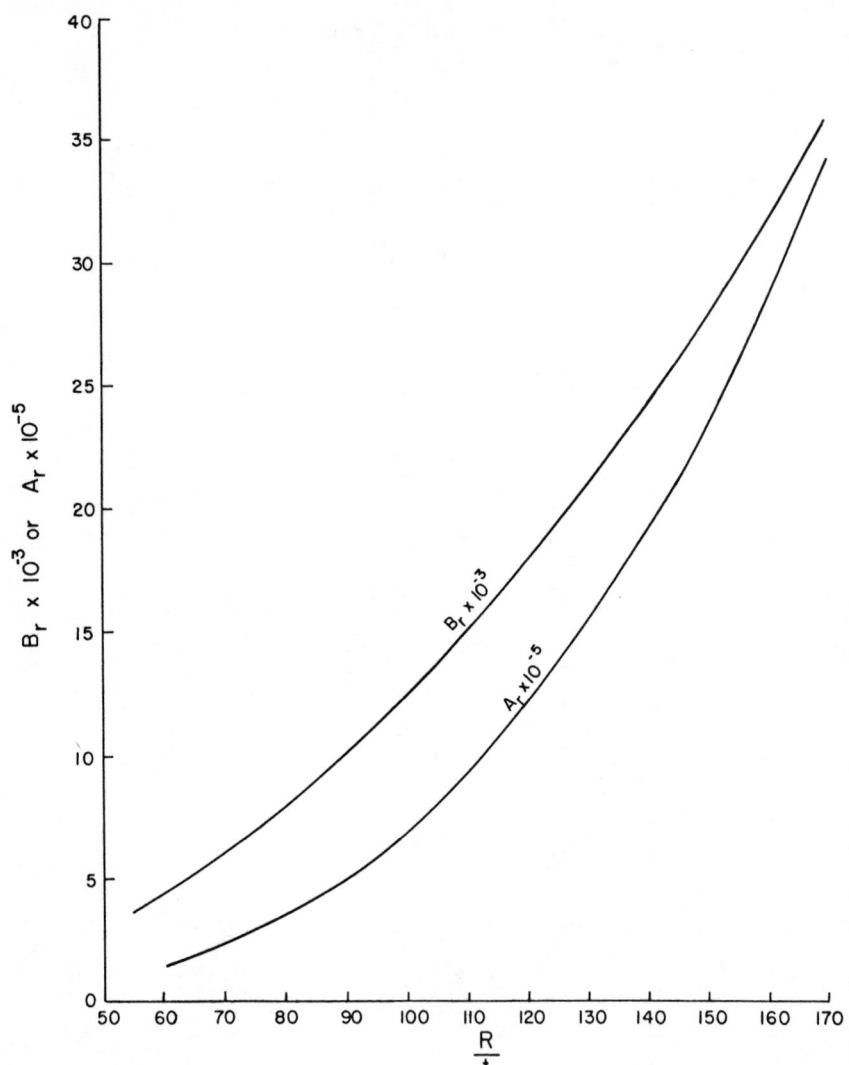

Case C3

Fixed Circular Plate with concentic load $\overline{W}$

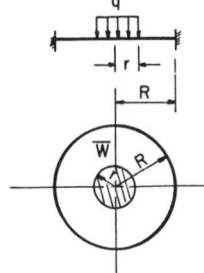

$$\overline{W} = \pi r^2 q$$

Max. Stress $f_b = B_r \dfrac{\overline{W}}{t^2}$

$$W_{max} = A_r \dfrac{\overline{W}}{Et}$$

r/R	B_r
0.05	1.8595
0.10	1.4311
0.20	1.0114
0.30	0.7752
0.40	0.6184
0.50	0.5078
0.57	0.4497
0.58	0.3972
0.60	0.3915
0.70	0.3605
0.80	0.3247
0.90	0.2841
1.00	0.2387

A_r

r/R R/t	0.05	0.20	0.40	0.60	0.80	1.00
55	651.02	595.15	481.97	358.88	361.44	164.29
60	774.76	708.26	573.58	427.10	348.61	195.52
65	909.27	831.22	673.15	501.25	409.13	229.47
70	1054.54	964.02	780.70	581.33	474.49	266.13
75	1210.57	1106.65	896.21	667.34	544.70	305.50
80	1377.36	1259.12	1019.69	759.29	619.74	347.59
85	1554.91	1421.44	1151.13	857.17	699.63	392.40
90	1743.22	1593.58	1290.54	960.98	784.36	439.92
95	1942.29	1775.56	1437.92	1070.72	873.94	490.16
100	2152.12	1967.38	1593.26	1186.39	968.35	543.12
105	2372.71	2169.04	1756.57	1307.99	1067.61	598.79
110	2604.07	2380.53	1927.85	1435.53	1171.70	657.17
115	2846.18	2601.87	2107.10	1569.00	1280.64	718.27
120	3099.06	2833.03	2294.30	1708.40	1394.42	782.09
125	3362.69	3074.04	2489.47	1853.73	1513.05	848.62
130	3637.09	3324.88	2692.62	2005.00	1636.51	917.87
135	3922.24	3585.56	2903.72	2163.20	1764.82	989.83
140	4218.16	3856.07	3122.80	2325.32	1897.97	1064.51
150	4842.27	4426.61	3584.84	2669.38	2178.79	1222.01
160	5509.43	5036.50	4078.76	3037.16	2478.98	1390.38
170	6219.63	5685.74	4604.53	3428.67	2798.53	1569.61

Case C3 continued

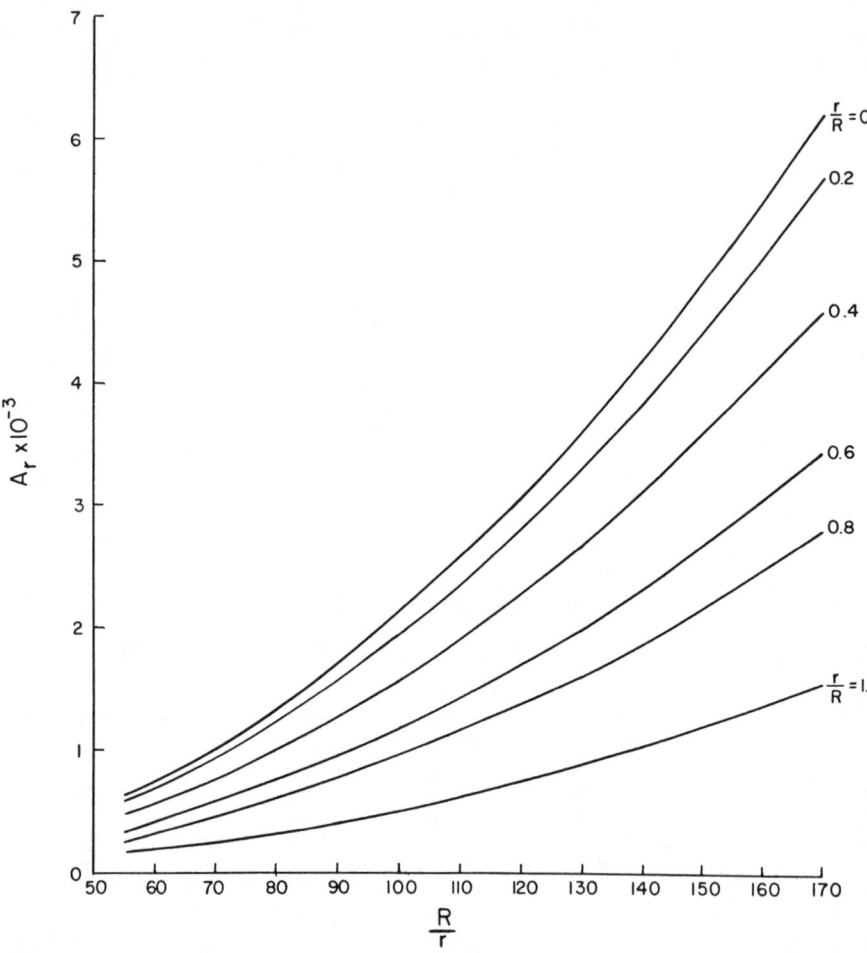

Case C3 continued

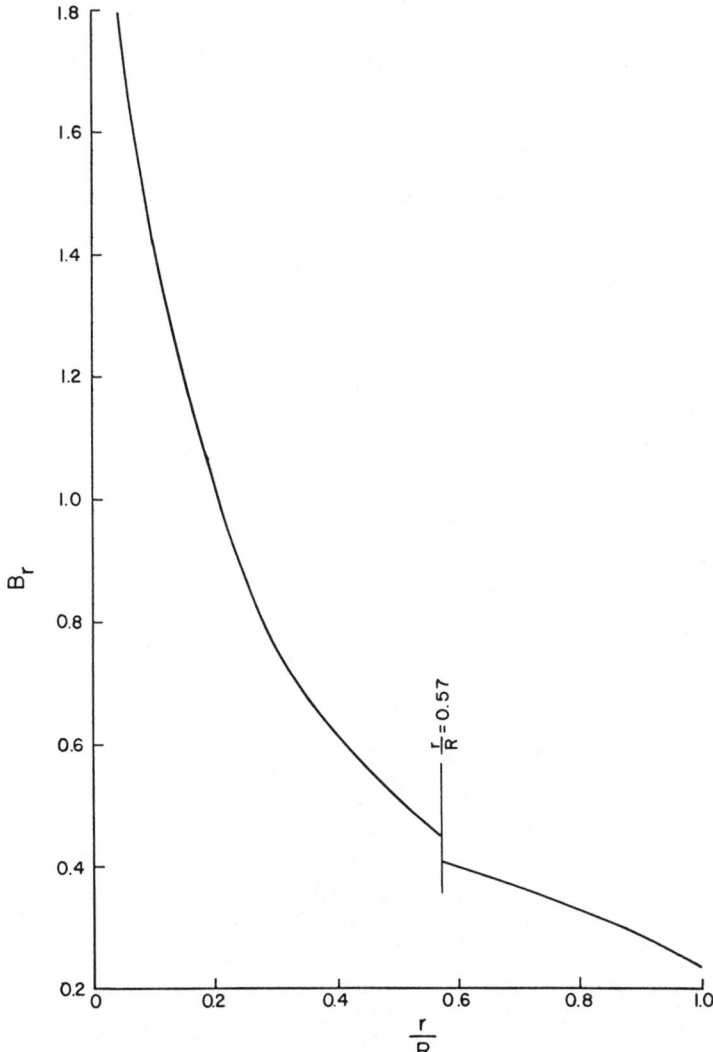

Case C4

Simply supported Circular Plate with concentric load $\overline{W}$

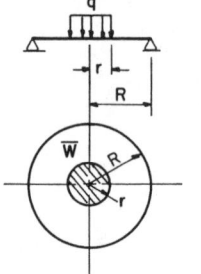

$$\overline{W} = \pi r^2 q$$

Max. Stress $f_b = B_r \dfrac{\overline{W}}{t^2}$

$$W_{max} = A_r \dfrac{\overline{W}}{Et}$$

r/R	B_r
0.05	2.337
0.10	1.907
0.20	1.473
0.30	1.217
0.40	1.033
0.50	0.887
0.60	0.765
0.70	0.658
0.80	0.563
0.90	0.475
1.00	0.394

A_r

r/R R/t	0.05	0.2	0.4	0.6	0.8	1.0
55	1660.79	1585.96	1412.11	1188.14	935.38	669.81
60	1976.47	1887.42	1680.53	1413.98	1113.18	797.13
65	2319.61	2215.10	1972.29	1659.47	1306.44	935.52
70	2690.20	2568.99	2287.39	1924.59	1515.16	1084.98
75	3088.24	2949.10	2625.83	2209.35	1739.34	1245.51
80	3513.73	3359.42	2987.61	2513.75	1978.98	1417.12
85	3966.67	3787.95	3372.74	2837.79	2234.09	1599.79
90	4447.06	4246.70	3781.20	3181.46	2504.65	1793.54
95	4954.90	4731.67	4213.00	3544.78	2790.68	1998.35
100	5490.20	5242.84	4668.15	3927.73	3092.16	2214.24
110	6643.14	6343.84	5648.46	4752.56	3741.51	2679.23
120	7905.88	7549.69	6722.13	5655.94	4452.71	3188.51
130	9278.43	8860.40	7889.17	6637.87	5225.75	3742.07
140	10760.79	10275.97	9149.57	7698.36	6060.64	4339.92
150	12352.94	11796.39	10503.33	8837.40	6957.36	4982.05
160	14054.91	13421.68	11950.45	10055.00	7915.93	5668.46
170	15866.67	15151.81	13490.94	11351.15	8936.34	6399.00

Case C4 continued

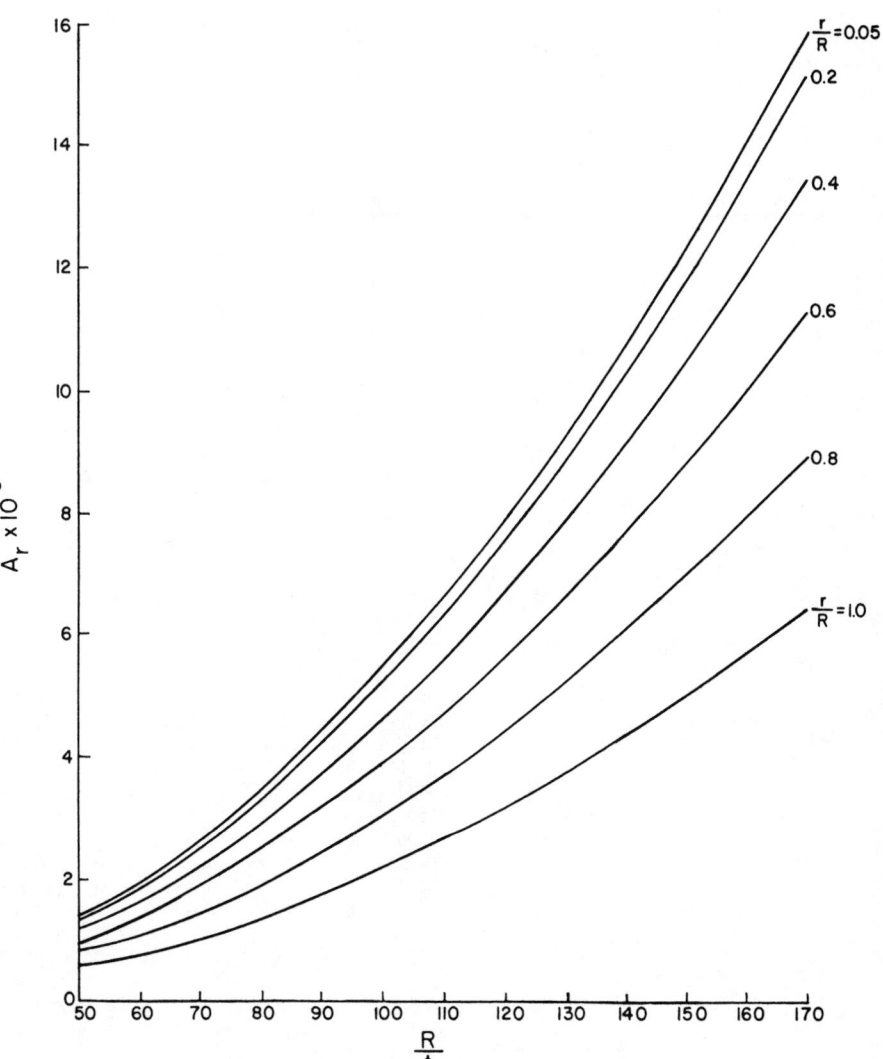

Case C5

Simply supported Circular Plate with hole under uniform load

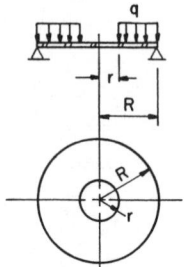

Max. Stress $f_b = B_r\, q$

$W_{max} = R\, A_r\, \dfrac{q}{E}$

B_r

R/r R/t	1.25	1.50	2.0	3.0	4.0	5.0
55	1791	2952	4356	5687	6292	6625
60	2131	3514	5184	6788	7488	7884
65	2501	4124	6084	7943	8788	9253
70	2901	4782	7056	9212	10192	10731
75	3330	5490	8100	10575	11700	12319
80	3789	6246	9216	12032	13312	14016
85	4277	7052	10404	13583	15028	15823
90	4795	7906	11664	15228	16848	17739
95	5343	8808	12996	16967	18772	19765
100	5920	9760	14400	18800	20800	21900
110	7163	11810	17424	22748	25168	26499
120	8525	14054	20736	27072	29952	31536
130	10005	16494	24336	31772	35152	37011
140	11603	19130	28224	36848	40768	42924
150	13320	21960	32400	42300	46800	49275
160	15155	24986	36864	48128	53248	56064
170	17109	28206	41616	54332	60112	63291

Case C5 continued

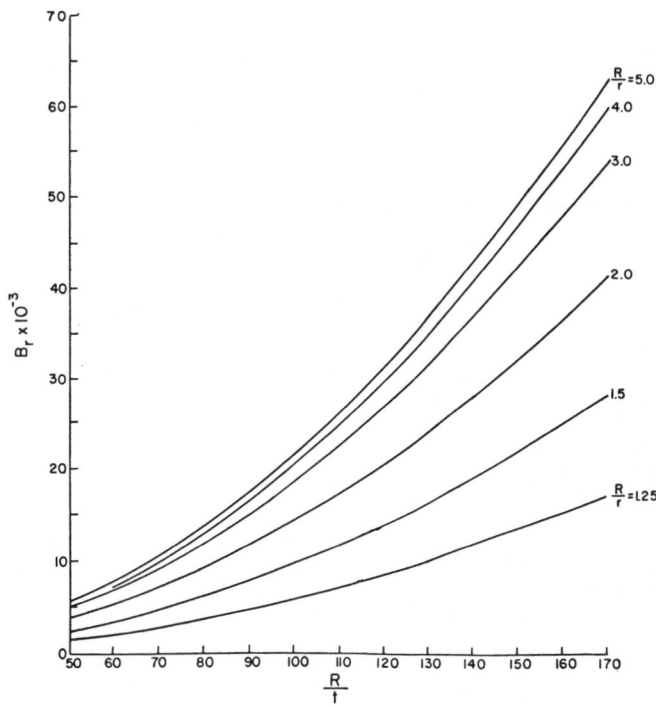

$A_r \times 10^{-3}$

R/r R/t	1.25	1.50	2.0	3.0	4.0	5.0
55	30.613	68.879	110.473	137.093	138.091	135.263
60	39.744	89.424	143.424	177.984	179.280	175.608
65	50.531	113.695	182.351	226.291	227.939	223.270
70	63.112	142.002	227.752	282.632	284.690	278.859
75	77.625	174.656	280.125	347.625	350.156	342.984
80	94.208	211.968	339.968	421.888	424.960	416.256
85	112.999	254.248	407.779	506.039	509.724	499.284
90	134.136	301.806	484.056	600.696	605.070	592.677
95	157.757	354.953	569.297	706.477	711.621	697.046
100	184.000	414.000	664.000	824.000	830.000	813.000
110	244.904	551.034	883.784	1096.744	1104.730	1082.103
120	317.952	715.392	1147.392	1423.872	1434.240	1404.864
130	404.248	909.558	1458.808	1810.328	1823.510	1786.161
140	504.896	1136.016	1822.016	2261.056	2277.520	2230.872
150	621.000	1397.250	2241.000	2781.000	2801.250	2743.875
160	753.664	1695.744	2719.744	3375.104	3399.680	3330.048
170	903.992	2033.983	3262.232	4048.312	4077.790	3994.269

Case C5 continued

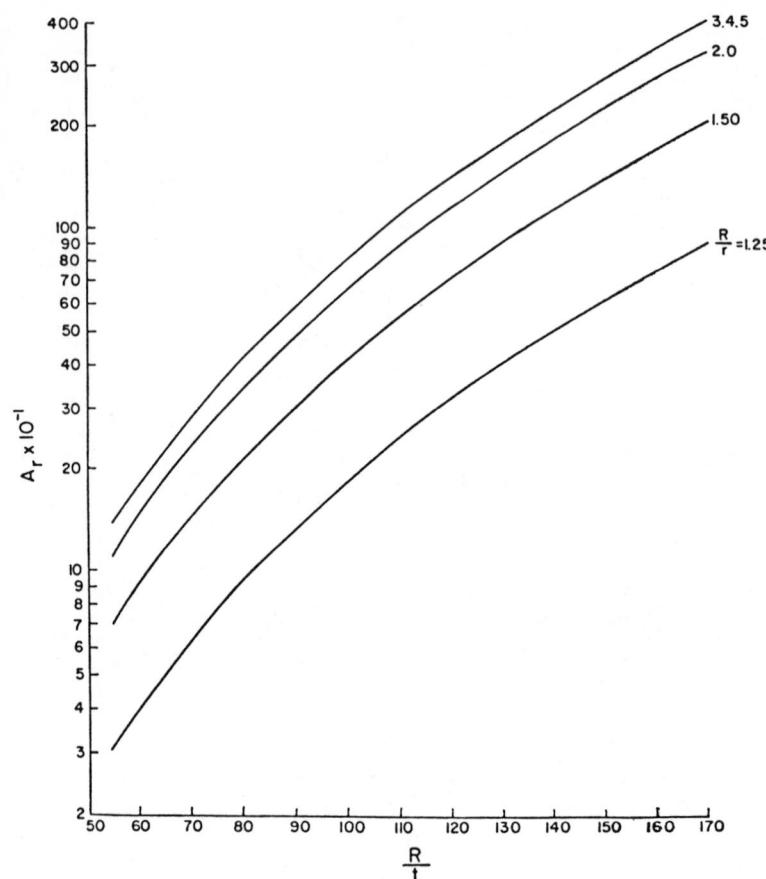

Case C6

Fixed Circular Plate with hole under
uniform load

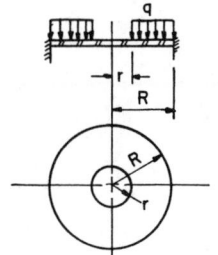

Max. Stress $f_b = B_r q$

$$W_{max} = R A_r \frac{q}{E}$$

B_r

R/r R/t	1.25	1.5	2.0	3.0	4.0	5.0
55	318	784	1452	1987	2148	2208
60	378	932	1728	2365	2556	2628
65	444	1094	2028	2776	3000	3084
70	515	1269	2352	3219	3479	3577
75	591	1457	2700	3696	3994	4106
80	672	1658	3072	4205	4544	4672
85	759	1871	3468	4747	5130	5274
90	851	2098	3888	5322	5751	5913
95	948	2338	4332	5930	6408	6588
100	1050	2590	4800	6570	7100	7300
110	1270	3134	5808	7950	8591	8833
120	1512	3730	6912	9461	10224	10512
130	1775	4377	8112	11103	11999	12337
140	2058	5076	9408	12877	13916	14300
150	2363	5828	10800	14783	15975	16425
160	2688	6630	12288	16819	18176	18688
170	3035	7485	13872	18987	20519	21097

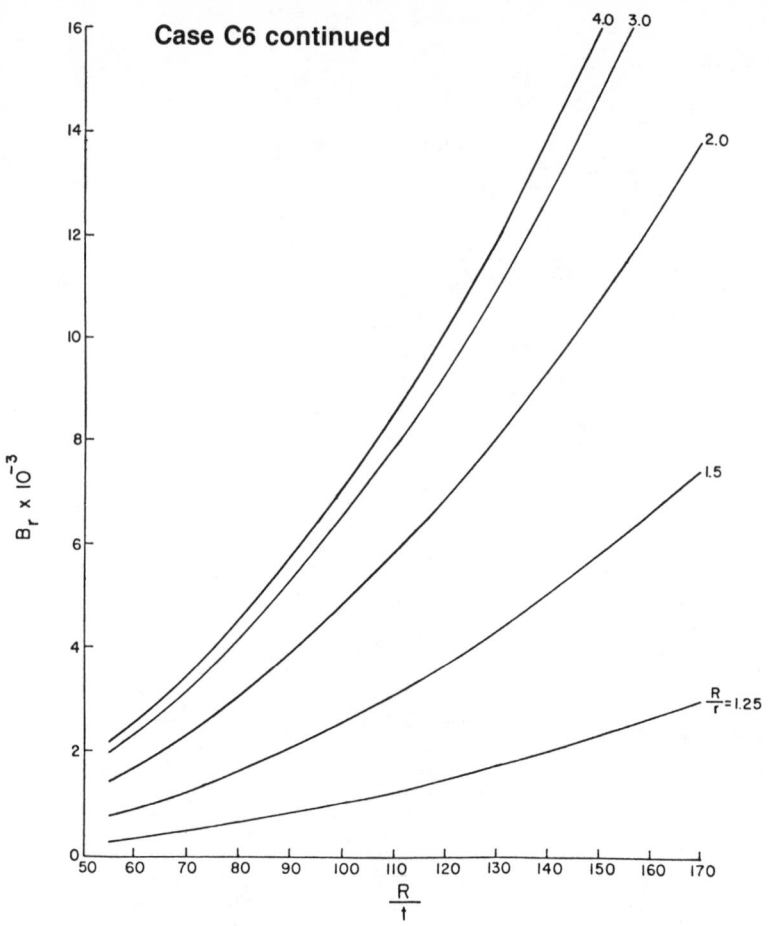

Case C6 continued

$A_r \times 10^{-3}$

R/r R/t	1.25	1.50	2.0	3.0	4.0	5.0
55	0.331	2.313	9.567	21.629	26.953	29.116
60	0.430	3.002	12.420	28.080	34.992	37.800
65	0.547	3.817	15.791	35.701	44.489	74.060
70	0.683	4.767	19.723	44.590	55.566	60.025
75	0.839	5.864	24.258	54.844	68.344	73.829
80	1.019	7.117	29.440	66.560	82.944	89.600
85	1.222	8.537	35.312	79.836	99.488	107.472
90	1.450	10.133	41.918	94.770	118.100	127.575
95	1.706	11.918	49.299	111.459	138.895	150.040
100	1.990	13.900	57.500	130.000	162.000	175.000
110	2.694	18.501	76.533	173.030	215.620	232.925
120	3.439	24.019	99.360	224.640	279.936	302.400
130	4.372	30.538	126.328	285.610	355.914	384.475
140	5.460	38.142	157.780	356.720	444.528	480.200
150	6.717	46.913	194.063	438.750	546.750	590.625
160	8.150	56.934	235.520	532.480	662.552	716.800
170	9.777	68.291	282.498	638.690	795.906	859.775

Case C6 continued

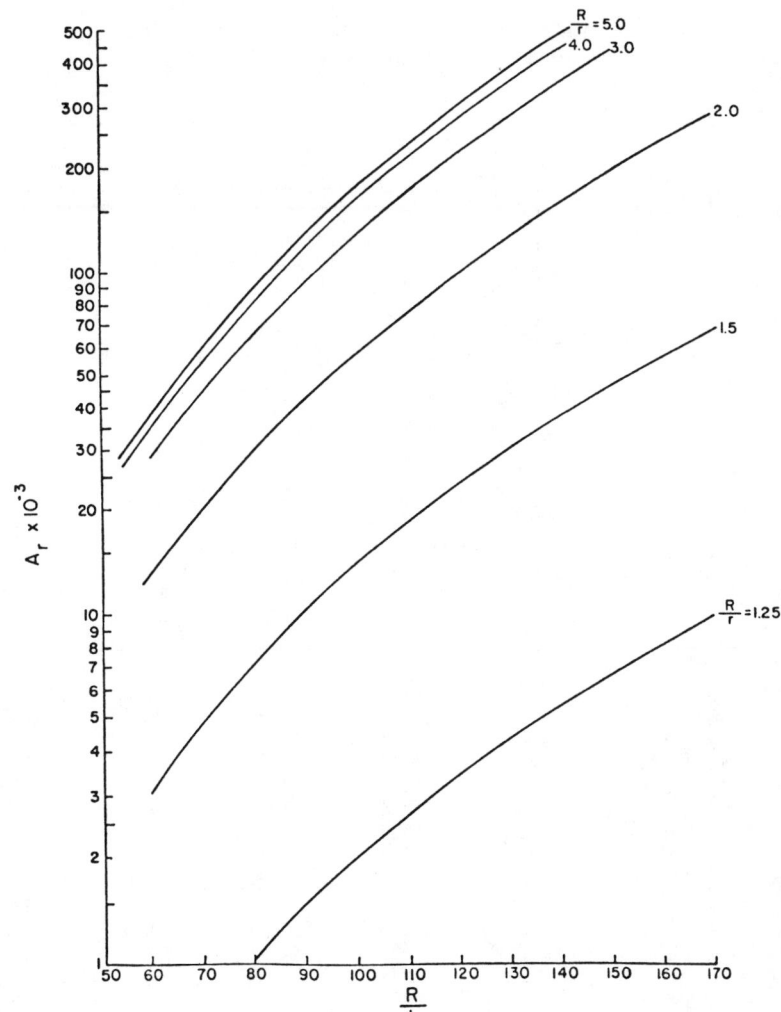

Case C7

Circular Plate with hole built in at the inside edge and uniformly loaded

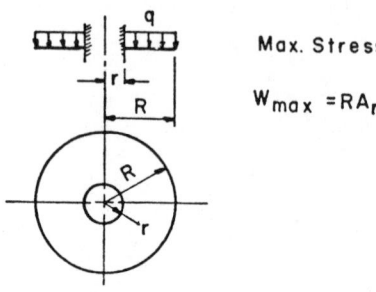

Max. Stress $f_b = B_r \, q$

$$W_{max} = RA_r \frac{q}{E}$$

B_r

R/r R/t	1.25	1.50	2.00	3.00	4.00	5.00
55	408	1240	3146	6504	9045	11162
60	486	1476	3744	7740	10764	13284
65	570	1732	4394	9084	12633	15590
70	662	2009	5096	10535	14651	18081
75	760	2306	5850	12094	16819	20756
80	864	2624	6656	13760	19136	23616
85	975	2962	7514	15534	21603	26660
90	1094	3321	8424	17415	24219	29889
95	1218	3700	9386	19404	26985	33302
100	1350	4100	10400	21500	29900	36900
110	1634	4961	12584	26015	36179	44649
120	1944	5904	14976	30960	43056	53136
130	2282	6929	17576	36335	50531	62361
140	2646	8036	20384	42140	58604	72324
150	3038	9225	23400	48375	67275	83025
160	3456	10496	26624	55040	76544	94464
170	3902	11849	30056	62135	86411	106641

Case C7 continued

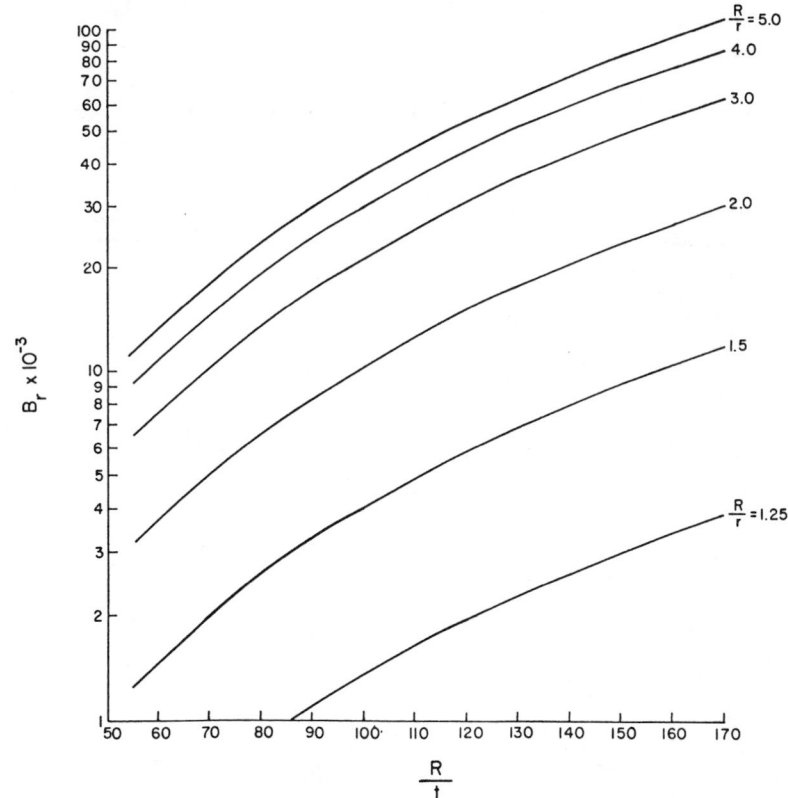

$A_r \times 10^{-3}$

R/r R/t	1.25	1.50	2.0	3.0	4.0	5.0
55	0.383	2.995	15.639	48.748	74.536	93.836
60	0.497	3.888	20.304	63.288	96.768	121.824
65	0.632	4.943	25.815	80.465	123.032	154.889
70	0.789	6.174	32.242	100.499	153.664	193.452
75	0.970	7.594	39.656	123.609	189.000	237.938
80	1.178	9.216	48.128	150.020	229.397	288.768
85	1.413	11.054	57.728	179.939	275.128	346.367
90	1.677	13.122	68.526	213.597	326.592	411.156
95	1.972	15.433	80.593	251.211	384.104	483.560
100	2.300	18.000	94.000	293.000	448.000	564.000
110	3.061	23.958	125.114	389.983	596.288	750.684
120	3.974	31.104	162.432	506.304	774.144	974.592
130	5.053	39.546	206.518	643.721	984.256	1239.11
140	6.311	49.392	257.936	803.992	1229.31	1547.62
150	7.763	60.750	317.250	988.875	1512.00	1903.50
160	9.421	73.728	385.024	1200.13	1835.00	2310.14
170	11.300	88.434	461.822	1439.51	2201.03	2770.93

Case C7 continued

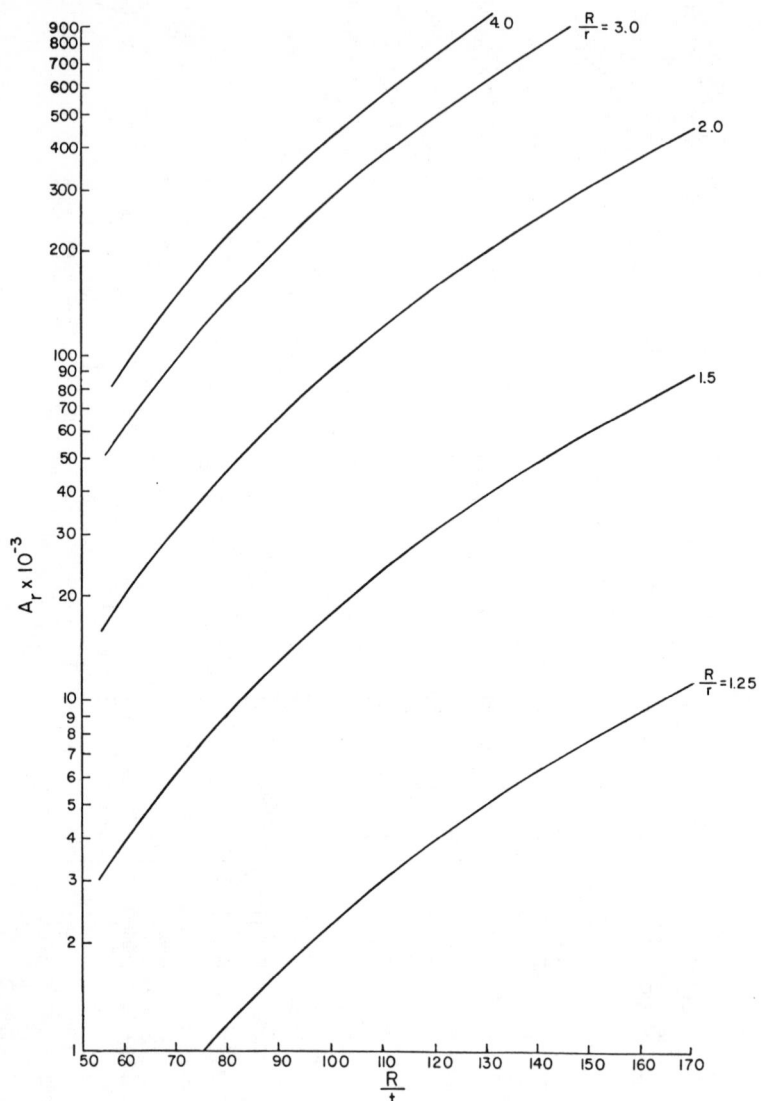

Case C8

Circular Plate with hole simply supported at inside edge and uniformly loaded

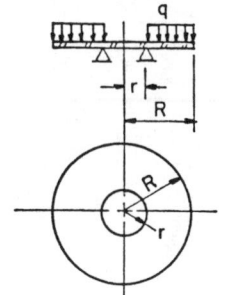

Max. Stress $f_b = B_r\, q$

$W_{max} = R\,A_r\,\dfrac{q}{E}$

B_r

R/r R/t	1.25	1.50	2.0	3.0	4.0	5.0
55	1997	3600	6171	10104	13008	15428
60	2376	4284	7344	12024	15480	18360
65	2789	5028	8619	14112	18168	21548
70	3234	5831	9996	16366	21070	24990
75	3713	6694	11475	18788	24188	28688
80	4224	7616	13056	21376	27520	32640
85	4769	8598	14739	24132	31068	36848
90	5346	9639	16524	27054	34830	41310
95	5957	10740	18411	30144	38808	46028
100	6600	11900	20400	33400	43000	51000
110	7986	14399	24684	40414	52030	61710
120	9504	17136	29376	48096	61920	73440
130	11154	20111	34476	56446	72670	86190
140	12936	23324	39984	65464	84280	99960
150	14850	26775	45900	75150	96750	114750
160	16896	30464	52224	85504	110080	130560
170	19074	34391	58956	96526	124270	147390

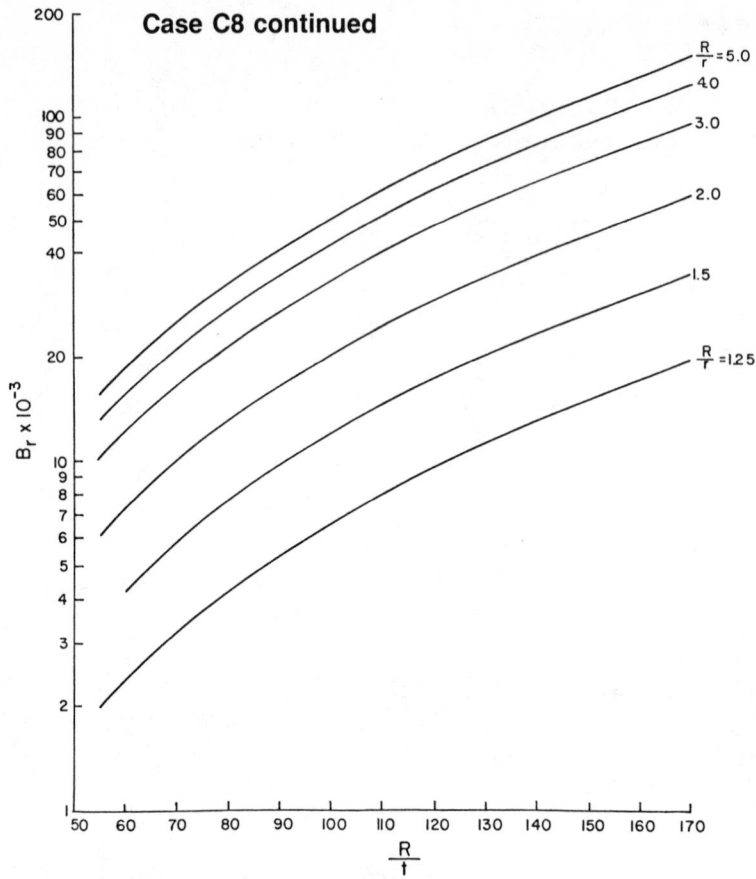

$A_r \times 10^{-3}$

R/r R/t	1.25	1.50	2.0	3.0	4.0	5.0
55	33.61	81.69	150.07	202.98	216.29	217.95
60	43.63	106.06	194.83	263.52	280.80	282.96
65	55.47	134.84	247.71	335.04	357.01	359.76
70	69.29	168.41	309.39	418.46	445.90	449.33
75	85.22	207.14	380.53	514.69	548.44	552.66
80	103.42	251.39	461.82	624.64	665.60	670.72
85	124.05	301.54	553.94	749.23	798.36	804.50
90	147.26	357.94	657.56	889.38	947.70	954.99
95	173.19	420.97	773.35	1046.00	1114.59	1123.16
100	202.00	491.00	902.00	1220.00	1300.00	1310.00
110	268.86	653.52	1200.56	1623.82	1730.30	1743.61
120	349.06	848.45	1558.66	2108.16	2246.40	2263.68
130	443.79	1078.73	1981.69	2680.34	2856.10	2878.07
140	554.29	1347.30	2475.09	3347.68	3567.20	3594.64
150	681.75	1657.13	3044.25	4117.50	4387.50	4421.25
160	827.39	2011.14	3694.59	4997.80	5324.80	5365.76
170	992.43	2412.28	4431.53	5993.60	6386.90	6436.30

Case C8 continued

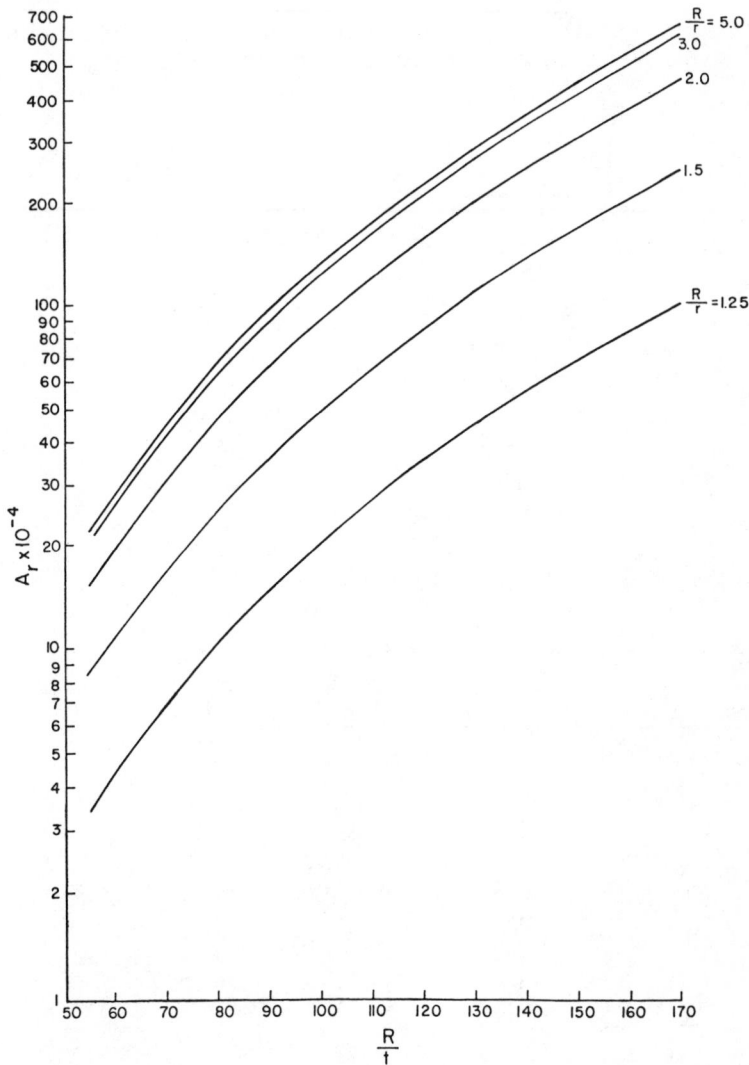

Case C9

Circular Plate   fixed at the inner edge and loaded by $\overline{W}$ linearly distributed along the outer edge

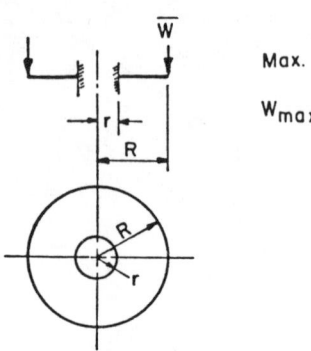

Max. Stress $f_b = B_r \dfrac{\overline{W}}{t^2}$

$W_{max} = A_r \dfrac{\overline{W}}{E t}$

R/r	1.25	1.50	2.0	3.0	4.0	5.0
B_r	0.227	0.428	0.753	1.210	1.510	1.750

A_r

R/r R/t	1.25	1.50	2.0	3.0	4.0	5.0
55	15.43	75.32	265.29	632.23	886.33	1058.75
60	18.36	89.64	315.72	752.40	1054.80	1260.00
65	21.55	105.20	370.53	833.03	1237.93	1478.75
70	24.99	122.01	429.73	1024.10	1435.70	1715.00
75	28.69	140.06	493.31	1175.63	1648.13	1968.75
80	32.64	159.36	561.28	1337.60	1875.20	2240.00
85	36.85	179.90	633.63	1510.03	2116.93	2528.75
90	41.31	201.69	710.37	1692.90	2373.30	2835.00
95	46.03	224.72	791.49	1886.23	2644.33	3158.75
100	51.00	249.00	877.00	2090.00	2930.00	3500.00
110	61.71	301.29	1061.17	2528.90	3545.30	4235.00
120	73.44	358.56	1262.88	3009.60	4219.20	5040.00
130	86.19	420.81	1482.13	3532.10	4951.70	5915.00
140	99.96	488.04	1718.92	4096.40	5742.80	6860.00
150	114.75	560.25	1973.25	4702.50	6592.50	7875.00

Case C9 continued

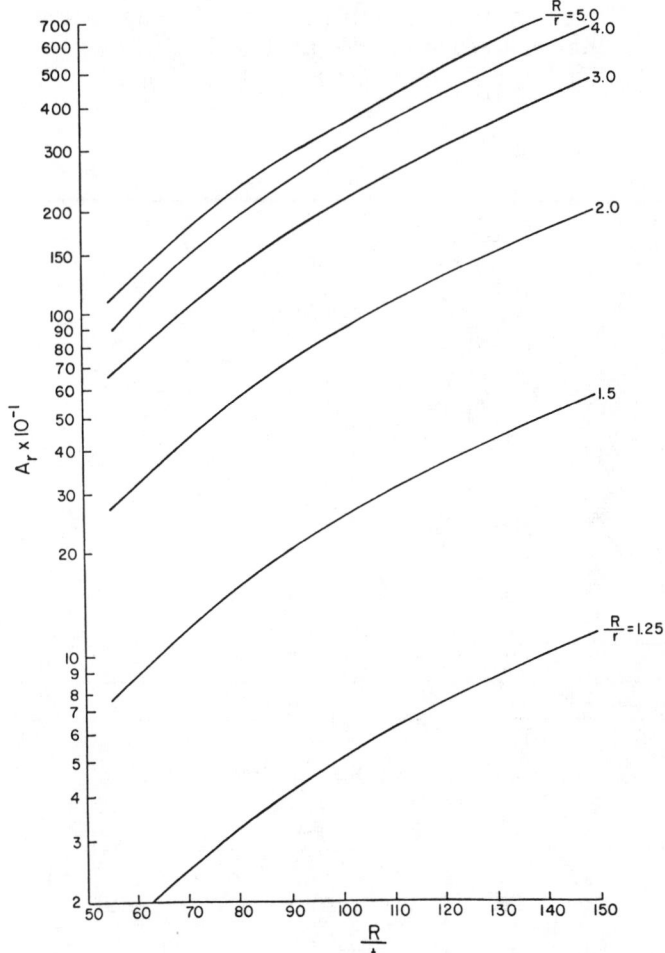

Case C10

Circular Plate with hole simply supported at the inner edge and loaded with $\overline{W}$ linearly distributed along the outer edge

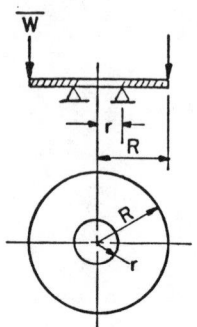

Max. Stress $f_b = B_r \dfrac{\overline{W}}{t^2}$

$W_{max} = A_r \dfrac{\overline{W}}{Et}$

R/r	1.25	1.50	2.0	3.0	4.0	5.0
B_r	1.10	1.26	1.48	1.88	2.17	2.34

			A_r			
R/r R/t	1.25	1.50	2.0	3.0	4.0	5.0
55	1032	1570	2033	2220	2190	2130
60	1228	1868	2419	2642	2606	2534
65	1441	2193	2839	3101	3059	2974
70	1671	2543	3293	2597	3548	3450
75	1918	2919	3780	4129	4073	3960
80	2182	3322	4301	4698	4634·	4506
85	2464	3750	4855	5303	5231	5068
90	2762	4204	5443	5945	5864	5702
95	3078	4684	6065	6624	6534	6354
100	3410	5190	6720	7340	7240	7040
110	4126	6280	8131	8881	8760	8518
120	4910	7474	9677	10570	10426	10138
130	5763	8771	11357	12405	12236	11898
140	6684	10172	13171	14386	14190	13798
150	7673	11678	15120	16515	16290	15840

Case C10 continued

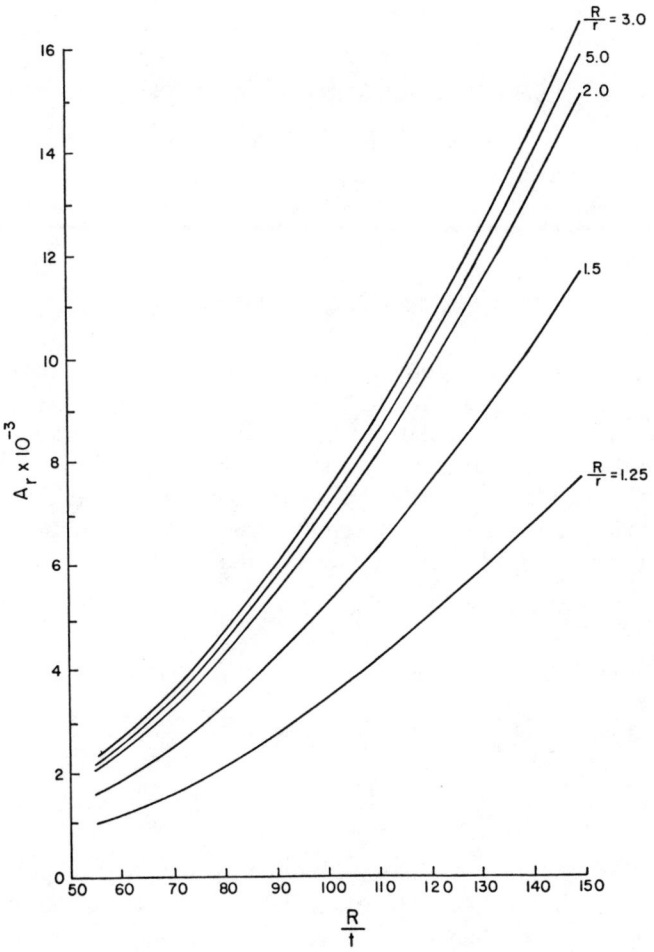

Chapter 8

Rings, Curved Bars, and Arches

BENDING OF RINGS

Consider the ring in Figure 8-1 subjected to concentrated loads as shown. Let R be the initial radius of curvature; the deformation of the ring is expressed by the radial displacement w. The radial displacement w is a function of the angle θ and is assumed to be a small quantity compared to R. It is taken as positive when the displacement is directed toward the center. The relation between the magnitude of bending moment and the change in curvature is expressed by the following equation [1,2]:

$$\frac{EI}{R^2}\left(w + \frac{d^2 w}{d\theta^2}\right) = -M$$

$$\text{or, } w + \frac{d^2 w}{d\theta^2} = -\frac{MR^2}{EI} \tag{8-1}$$

This is the differential equation of the deformed ring. It also applies for a deflected curve of a thin bar with a circular center line. For an infinitive R, Equation 8-1 coincides with that for a straight bar.

The bending moment at any cross section of the ring can be found from Figure 8-2, which is cut from Figure 8-1. Let M_0 be the bending moment at A and B; then the bending moment at any cross section C is

$$M = M_0 + \frac{WR}{2}(1 - \cos\theta) \tag{8-2}$$

The moment M_0 can be found by using *Castigliano's theorem*, which states that the partial derivative of the strain energy with respect to the bending moment M_0 is equal to the angle of rotation at the cross section A. In this case the angle of rotation is zero at A, or

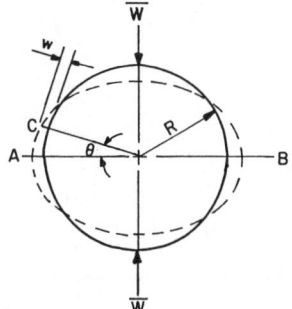

Figure 8-1. A deformed ring.

Figure 8-2. Free body of a ring.

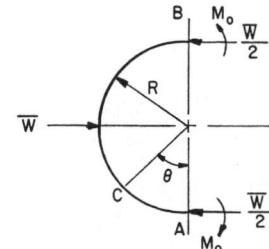

$$\frac{\partial U}{\partial M_0} = 0 \tag{8-3}$$

The strain energy U for the ring is expressed by

$$U = \int_0^{2\pi} \frac{M^2 R d\theta}{2EI} \tag{8-4}$$

Substituting Equation 8-2 into 8-4 and from Equation 8-3, we obtain

$$M_0 = WR \left(\frac{1}{\pi} - \frac{1}{2} \right) = -0.18169 \, WR \tag{8-5}$$

Substituting Equations 8-2 and 8-5 into Equation 8-1 leads to

$$\frac{d^2w}{d\theta^2} + w = -\frac{WR^3}{EI} \left(\frac{1}{\pi} - \frac{1}{2} \right) - \frac{WR^3}{2EI} (1 - \cos\theta) \tag{8-6}$$

The general solution of Equation 8-6 is

$$w = A \sin\theta + B \cos\theta - \frac{WR^3}{EI} \left(\frac{1}{\pi} - \frac{1}{2} \right)$$

$$- \frac{WR^3}{2EI} + \frac{WR^3}{4EI} \theta \sin\theta \qquad (8\text{-}7)$$

The constants A and B can be determined from the following boundary conditions:

$$\frac{dw}{d\theta} = 0 \text{ at } \theta = 0$$

$$\frac{dw}{d\theta} = 0 \text{ at } \theta = \frac{\pi}{2}$$

We obtain

$$A = 0$$

$$B = \frac{WR^3}{4EI}$$

Substituting A and B into Equation 8-7, we obtain the expression for radial deflection:

$$w = \frac{WR^3}{4EI} \left(\cos\theta + \theta \sin\theta - \frac{4}{\pi} \right) \qquad (8\text{-}8)$$

From this expression, the radial deflection at Point A or D can be calculated.

$$w_{\theta=0} = - \frac{WR^3}{EI} \left(\frac{1}{\pi} - \frac{1}{4} \right) = - 0.06831 \frac{WR^3}{EI}$$

The lengthening of diameter AB is $0.13662 \, WR^3/EI$

$$w_{\theta = \frac{\pi}{2}} = \frac{WR^3}{EI} \left(\frac{\pi}{8} - \frac{1}{\pi} \right)$$

and the shortening of diameter DE is $0.14878 \, WR^3/EI$.

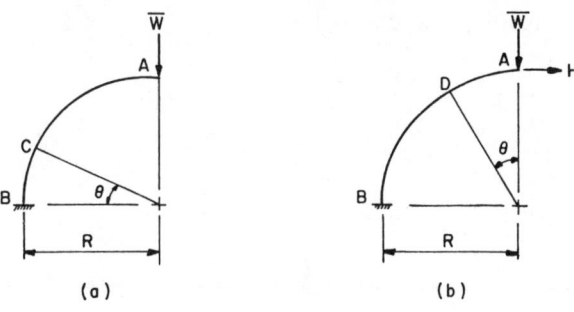

Figure 8-3. Curved beam for Example 8-1.

Example 8-1

A curved beam is in the form of a quadrant of a thin circular ring, as shown in Figure 8-3a; find the vertical and horizontal displacements at the loading point A.

Moment at any section C is M:

$$M = WR \cos\theta$$

$$\frac{\partial M}{\partial W} = R \cos\theta$$

From Castigliano's theorem, the vertical displacement at A is

$$\delta_v = \frac{\partial U}{\partial W} = \int_0^{\frac{\pi}{2}} \frac{M \frac{\partial M}{\partial W} R}{EI} d\theta = \frac{1}{EI} \int_0^{\frac{\pi}{2}} WR \cos\theta (R \cos\theta) R d\theta$$

$$= \frac{WR^3}{EI} \int_0^{\frac{\pi}{2}} \cos^2\theta \, d\theta = \frac{WR^3}{EI} \left[\frac{\theta}{2} + \frac{1}{4} \sin2\theta \right]_0^{\frac{\pi}{2}} = \frac{WR^3}{4EI}$$

In order to find the horizontal displacement at A, let us apply a dummy load at A as shown in Figure 8-3b. The bending moment at any section D is

$$M = WR \sin\phi - H(R - R \cos\phi)$$

$$\frac{\partial M}{\partial H} = R(1 - \cos\phi)$$

The horizontal displacement at A is

$$\delta_H = \frac{\partial U}{\partial H} = \frac{1}{EI} \int_0^{\frac{\pi}{2}} M \frac{\partial M}{\partial H} R \, d\phi$$

$$\delta_H = \frac{1}{EI} \int_0^{\frac{\pi}{2}} (WR \sin\phi) R(1 - \cos\phi) R \, d\phi = \frac{WR^3}{2EI}$$

Example 8-2

Calculate the reactions and the vertical deflection at A for a thin semi-circular ring that is loaded as shown in Figure 8-4.

The moment at any section B is

$$M = \frac{W}{2} R(1 - \cos\theta) - HR \sin\phi \text{ for } 0 < \theta < \frac{\pi}{2}$$

$$\frac{\partial M}{\partial H} = - R \sin\theta$$

$$\delta_H = \frac{\partial U}{\partial H} = 0$$

$$\frac{\partial U}{\partial H} = \frac{2}{EI} \int_0^{\frac{\pi}{2}} \left[\frac{W}{2} R(1 - \cos\theta) - HR \sin\theta \right] (-R \sin\theta) R d\theta = 0$$

Solving this equation we obtain

$$H = \frac{W}{\pi}$$

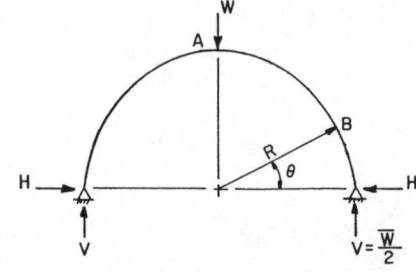

Figure 8-4. Semicircular ring for Example 8-2.

Deflection at $A = \delta_v$:

$$\delta_v = \frac{\partial U}{\partial W} = \frac{2}{EI} \int_0^{\frac{\pi}{2}} M \frac{\partial M}{\partial W} R \, d\theta$$

$$\delta_v = \frac{2}{EI} \int_0^{\frac{\pi}{2}} \left[\frac{WR}{2} (1 - \cos\theta) - \frac{WR}{\pi} \sin\theta \right]$$

$$\left[\frac{R}{2} (1 - \cos\theta) - \frac{R}{\pi} \sin\theta \right] R \, d\theta$$

$$= \frac{2WR^3}{EI} \int_0^{\frac{\pi}{2}} \left[\frac{1}{4} (1 - \cos\theta)^2 - \frac{1}{\pi} \sin\theta(1 - \cos\theta) + \frac{1}{\pi^2} \sin^2\theta \right] d\theta$$

$$= \frac{WR^3}{EI} \left[\frac{3\pi}{8} - 1 - \frac{1}{2\pi} \right]$$

BENDING STRESS OF CURVED BEAMS [3,5]

For a straight beam the bending stress is My/I, where y is the distance from the neutral axis (which passes through the centroid) to the fiber on which the stress acts. In a curved beam, the neutral axis does not pass through the centroid, and the bending stresses are not proportional to the distances from the neutral axis. Figure 8-5 shows an element cut from a curved beam. We assume that planes remain planar after bending and that Hooke's law holds for the curved beam. The elongation of a fiber at a distance y from the neutral axis is $y(\Delta d\theta)$, and the strain is expressed as

$$\epsilon = \frac{y(\Delta d\theta)}{(R + y) \, d\theta} \tag{8-9}$$

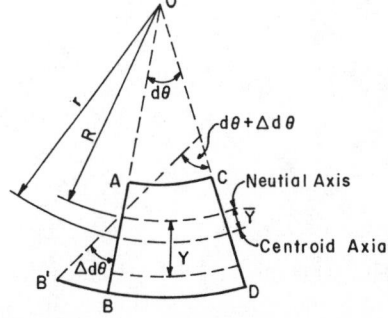

Figure 8-5.   Element of a curved beam.

where R is the radius of curvature of the neutral axis. The normal stress is expressed as

$$f_b = \frac{Ey(\Delta d\theta)}{(R + y)\, d\theta} \tag{8-10}$$

and the neutral axis is located by the fact that the resultant normal force over the cross section is zero. Thus

$$\int_A f_b \, dA = \int_A \frac{Ey(\Delta d\theta)}{(R + y)d\theta} = 0$$

That is,

$$\int_A \frac{y}{R + y} \, dA = 0$$

Let $U = R + y$

$$\int_A \frac{y}{R + y} \, dA = \int \frac{U - R}{U} \, dA = \int_A dA - \int_A \frac{R}{U} \, dA = 0$$

$$R = \frac{A}{\displaystyle\int_A \frac{dA}{U}} \tag{8-11}$$

Equation 8-11 locates the neutral axis. Note that the neutral axis always lies between the centroid axis and the center of curvature.

The bending moment must equal the sum of the moments of the normal forces on the fibers. That is,

$$M = \int_A f_b \, y \, dA = \frac{E(\Delta d\theta)}{d\theta} \int_A \frac{y^2 dA}{R + y}$$

$$\int_A \frac{y^2}{R + y} \, dA = \int_A y \, dA - R \int_A \frac{y}{R + y} \, dA$$

The first integral is the static moment of the cross-sectional area about the neutral axis, and the second integral vanishes.

$$M = \frac{E(\Delta d\theta)}{d\theta} (A\overline{y})$$

From Equation 8-10 we obtain

$$f_b = \frac{My}{A\overline{y}(R + y)} \qquad (8\text{-}12)$$

Equation 8-12 indicates that the stress distribution across the depth of the curved beam is hyperbolic. The maximum stress always occurs at the outer fibers on the concave side of the beam.

$$\text{max. } f_b = \frac{M(R-R_i)}{A\overline{y}\, R_i}$$

where R_i is the inside radius of the curved beam. The relation between the radius of curvature of the neutral axis and the centroid axis is

$$R = r - \overline{y}$$

For a given cross section of a curved beam, the relation between r/h and R/h can be established. Figures 8-6 and 8-7 show this relationship for rectangular and circular cross sections, respectively.

The radius of curvature of the neutral axis R can also be calculated from Equation 8-11. For a rectangular cross section this is expressed as

$$R = \frac{A}{\displaystyle\int_A \frac{dA}{U}} = \frac{bh}{\displaystyle\int \frac{b\,dU}{U}} = \frac{h}{[\ln U]_{R_i}^{R_0}} = \frac{h}{\ln\left(\dfrac{R_0}{R_i}\right)}$$

Note that for a thinner beam, the ratio R_0/R_i is close to unity; it is better to proceed as follows:

$$\ln\left(\frac{R_0}{R_i}\right) = \ln\left(\frac{r + \dfrac{h}{2}}{r - \dfrac{h}{2}}\right) = \ln\left(\frac{1 + \dfrac{h}{2r}}{1 - \dfrac{h}{2r}}\right)$$

$$= \frac{h}{r}\left[1 + \frac{1}{3}\left(\frac{h}{2r}\right)^2 + \frac{1}{5}\left(\frac{h}{2r}\right)^4 + \ldots\right] \approx \frac{h}{r}\left(1 + \frac{h^2}{12r^2}\right)$$

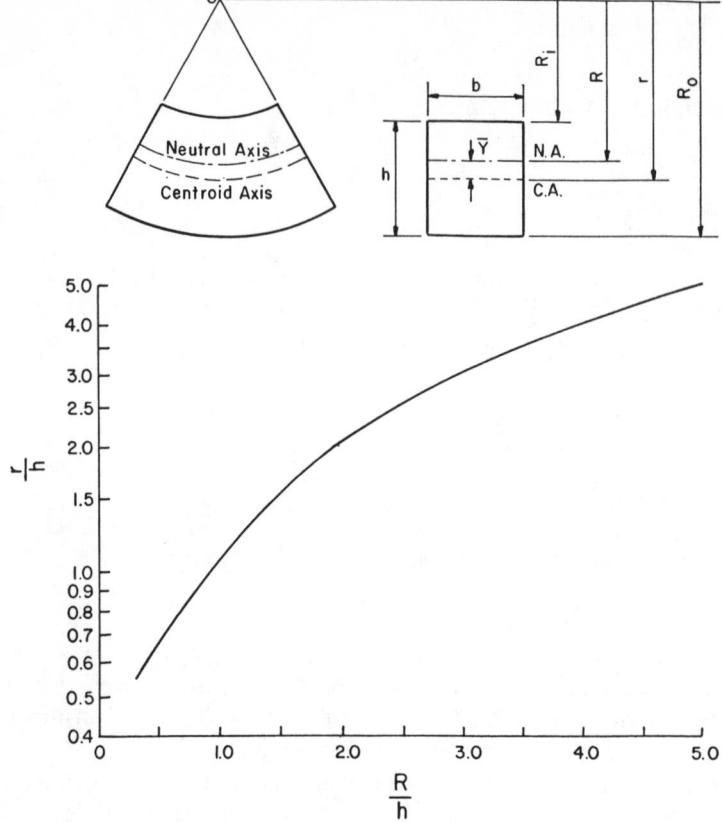

Figure 8-6. Relation of r/h and R/h for a rectangular curved beam.

$$R = \frac{h}{\ln\left(\dfrac{R_0}{R_i}\right)} = \frac{h}{\dfrac{h}{r}\left(1 + \dfrac{h^2}{12r^2}\right)} = \frac{r}{1 + \dfrac{h^2}{12r^2}}$$

$$\bar{y} = r - R \approx \frac{h^2}{12r}$$

Example 8-3

A rectangular curved beam with a cross section of 3 by 6 in. is loaded with a moment of 120 kip-in. Determine the maximum tensile and compressive stresses if the inside and outside radii of the beam are 4.5 in. and 10.5 in., respectively, as shown in Figure 8-8.

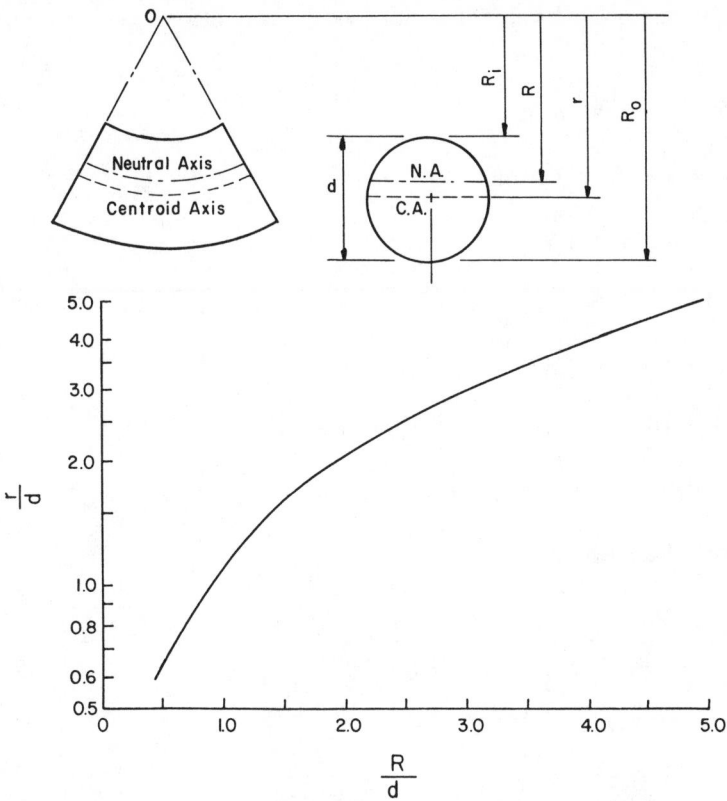

Figure 8-7. Relation of r/h and R/h for a circular curved beam.

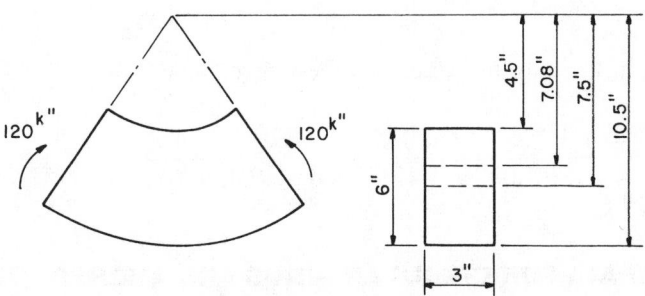

Figure 8-8. Rectangular curved beam for Example 8-3.

It is necessary to locate the neutral axis. From the given data we calculate

$$A = 6 \times 3 = 18 \text{ in.}^2$$

$$r = 4.5 + 3 = 7.5 \text{ in.}$$

$$r/h = 7.5/6 = 1.25$$

From Figure 8-6

$$R/h = 1.18$$

$$R = 1.18 \text{ h} = 1.18 \times 6 = 7.08 \text{ in.}$$

$$\overline{y} = R - r = 7.5 - 7.08 = 0.42 \text{ in.}$$

Or, from Equation 8-11

$$R = \frac{A}{\displaystyle\int_A \frac{dA}{du}} = \frac{bh}{b\displaystyle\int_A \frac{du}{u}} = \frac{h}{[\ln U]_{R_i}^{R_0}} = \frac{h}{\ln\left(\dfrac{R_0}{R_i}\right)}$$

$$R = \frac{6}{\ln(10.5/4.5)} = \frac{6}{0.8473} = 7.0813 \text{ in.}$$

The stress in the extreme fibers on the concave side is calculated from Equation 8-12:

$$f_i = \frac{120 \times (7.08 - 4.5)}{18 \times 0.42(7.08 - 2.58)} = \frac{120 \times 2.58}{18 \times 0.42 \times 4.5} = 9.1 \text{ kips/in.}^2$$

The stress in the extreme fibers on the convex side is

$$f_0 = \frac{120 \times (10.5 - 7.08)}{18 \times 0.42(7.08 + 3.42)} = \frac{120 \times 3.42}{18 \times 0.42 \times 10.5}$$
$$= 5.17 \text{ kips/in.}^2$$

DATA FOR CIRCULAR RINGS AND ARCHES

Moments and deflections of a ring can be derived from Equation 8-1. The differential equation can be expressed as

$$\frac{d^2w}{ds^2} + \frac{w}{R^2} = \frac{-M}{EI}$$

This equation coincides with that for a straight bar when R is infinitely large. So a closed ring may be regarded as a statically indeterminate beam. In solving the ring problem it is assumed that:

1. The radius of the ring is large in comparison with its radial thickness.
2. The ring cross section is uniform and the deformation is not so severe that its circular shape is lost.
3. The deflection is due solely to bending; the effects of axial and shear stresses are neglected.
4. The stress in a ring is always within the elastic limit.

Based on these assumptions, the formulas for bending moments and vertical and horizontal deflection of rings loaded and supported in various ways have been derived and given in several publications [4,6,7]. Most of the formulas consist of the algebraic sum of a number of terms, and calculations should be made with great care.

Formulas and data for selected cases of rings and arches have been developed or modified and presented in this section. Dimensionless data are provided in tabular or graphic forms which are designed for convenient application. Note that ordinates of some graphs have been factored for easier reading. In Case 7, for example, $K_d \times 50 = 1.35$; that is, $K_d = 1.35/50 = 0.027$. The following notations are used in the presented cases:

R = radius of a circular ring or arch
θ = angle to locate the cross section under consideration
W = concentrated load
M = bending moment; it is positive when it produces a decrease in the initial curvature
EI = flexural rigidity of the ring or arch
K_m = dimensionless moment coefficient
K_d = dimensionless deflection coefficient
w = radial displacement, taken as positive when it is directed outward from the center
K_A, K_B = moment coefficients at points A and B
p = uniform load per unit length
M = externally applied bending moment in ring, or fixed moment in built-in arch
H_0 = horizontal thrust in pin-joint arch
H_H = horizontal thrust in built-in arch
K_H = dimensionless thrust coefficient
y_A = maximum deflection of arch

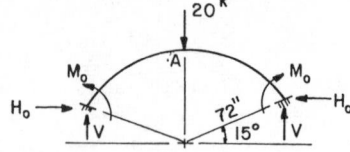

Figure 8-9. Supporting circular arch for Example 8-4

K'_m = dimensionless moment coefficient for arch
K_x, K_y = dimensionless deflection coefficients in x and y directions
α = angle defining length of arc for circular arch

Example 8-4

A supporting circular arch is fixed at its ends and loaded as shown in Figure 8-9. Calculate the reactions and the maximum moment and deflection if the moment of inertia is 20 in.[4]

$W = 20$ kips

$\alpha = 15$ degrees

From Case 16, $K_H = 0.6026$ $K'_m = -0.1251$
$K_m = 0.0888$ $K_d = 0.00644$

$H_0 = WK_H = 20 \times 0.6026 = 12.05$ kips

$$y_A = \frac{WR^3}{EI} K_d = \frac{20 \times 72^3}{29 \times 10^3 \times 20} \times 0.00644 = 0.0829 \text{ in.}$$

$$M_{max} = WRK'_m = -20 \times 72 \times 0.1251 = -180.144 \text{ kip-in.}$$

REFERENCES

1. Timoshenko, S., and Gere, J., *Theory of Elastic Stability,* McGraw-Hill Book Company, New York, 1964.
2. Den Hartog, J. P., *Advanced Strength of Materials,* McGraw-Hill Book Company, 1952.
3. Hopkings, R., *Design Analysis of Shaft and Beams,* McGraw-Hill Book Company, 1970.
4. Blake, A., "How to Find Deflection and Moment of Rings and Accurate Beams," *Production Engineering,* Vol. 34 No. 1, 1963.
5. Nash, W., *Theory and Problems of Strength of Materials,* Second Edition, McGraw-Hill Book Company, 1972.
6. Roark, R. J., *Formulas for Stress and Strain,* Fourth Edition, McGraw-Hill Book Company, 1965.
7. Roark, R. J. and Young, W., *Formulas for Stress and Strain,* Fifth Edition, McGraw-Hill Company, 1975.

Cases for Chapter 8

Case 1

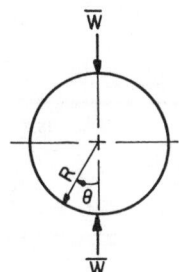

$M = K_m \overline{W} R$

$w = K_d \dfrac{\overline{W}R^3}{EI}$

$K_m = 0.3183 - 0.5 \sin\theta$

$K_d = 0.3183 - 0.25 \sin\theta - (0.3927 - 0.25\theta)\cos\theta$

θ	0°	15°	30°	45°	60°	75°	90°
K_m	0.3183	0.1889	0.0683	−0.0353	−0.1147	−0.1647	−0.1817
K_d	−0.0744	−0.0625	−0.0334	0.0027	0.0363	0.0599	0.0683

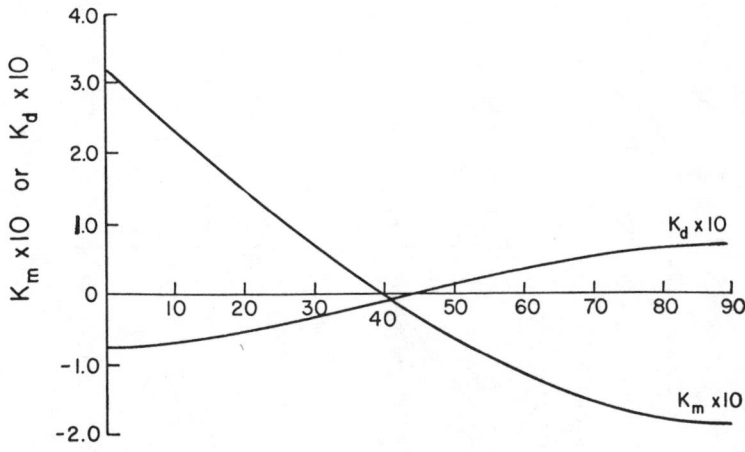

θ (Degree)

Case 2

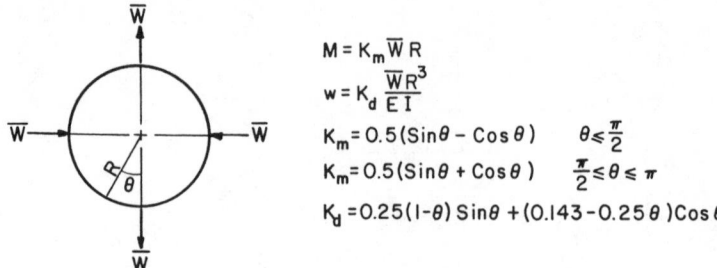

$$M = K_m \overline{W} R$$
$$w = K_d \frac{\overline{W} R^3}{E I}$$
$$K_m = 0.5(Sin\theta - Cos\theta) \qquad \theta \leqslant \frac{\pi}{2}$$
$$K_m = 0.5(Sin\theta + Cos\theta) \qquad \frac{\pi}{2} \leqslant \theta \leqslant \pi$$
$$K_d = 0.25(1-\theta) Sin\theta + (0.143 - 0.25\theta) Cos\theta$$

θ	0°	15°	30°	45°	60°	75°	90°
K_m	−0.50	−0.3536	−0.183	0.0	0.0183	0.3536	0.50
K_d	0.1427	0.1224	0.0698	0.0	−0.0698	−0.1224	−0.1427

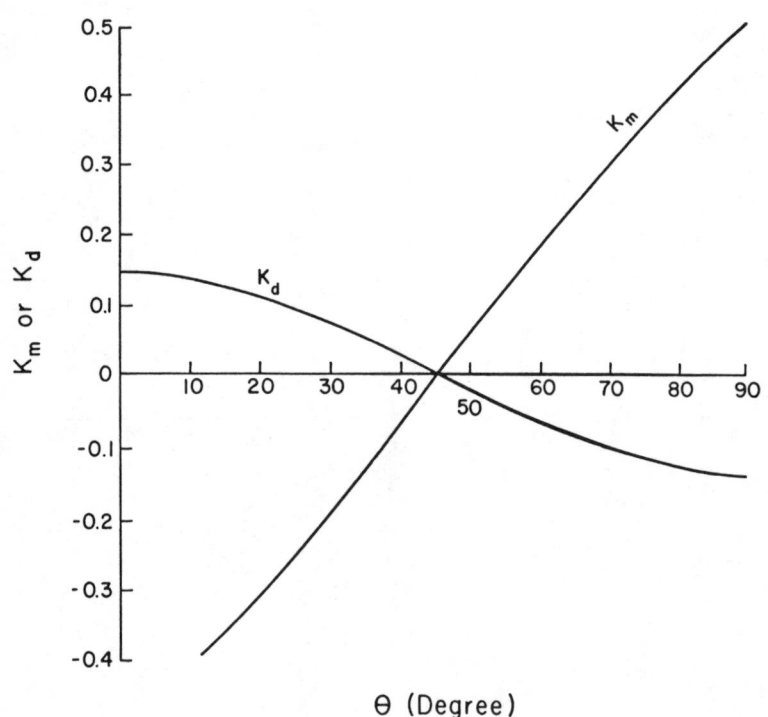

Θ (Degree)

Case 3

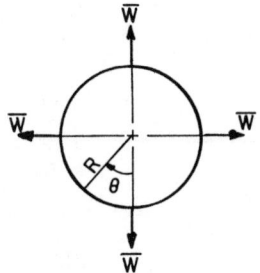

$$M = K_m \overline{W} R$$
$$w = K_d \frac{\overline{W} R^3}{E I}$$

$$K_m = 0.5(\sin\theta + \cos\theta) - 0.6366$$

$$K_d = 0.25(1+\theta)\sin\theta + (0.643 - 0.25\theta)\cos\theta - 0.637$$

θ	0°	15°	30°	45°	60°	75°	90°
K_m	−0.1366	−0.0243	0.0464	0.0705	0.0464	−0.0243	−0.1366
K_d	0.0061	0.00261	−0.0029	−0.0054	−0.0029	0.0026	0.0061

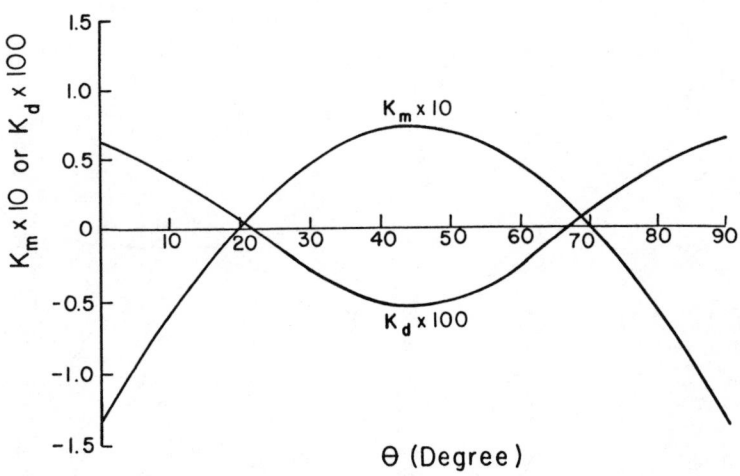

Case 4

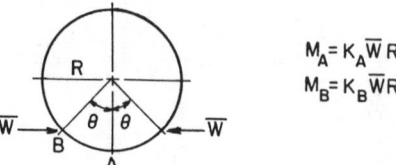

$$M_A = K_A \overline{W} R$$
$$M_B = K_B \overline{W} R$$

θ	15°	30°	45°	60°	75°	90°
K_A	−0.0284	−0.0903	−0.1540	−0.1955	−0.2045	−0.1817
K_B	0.0055	0.0398	0.1123	0.2068	0.2869	0.3183

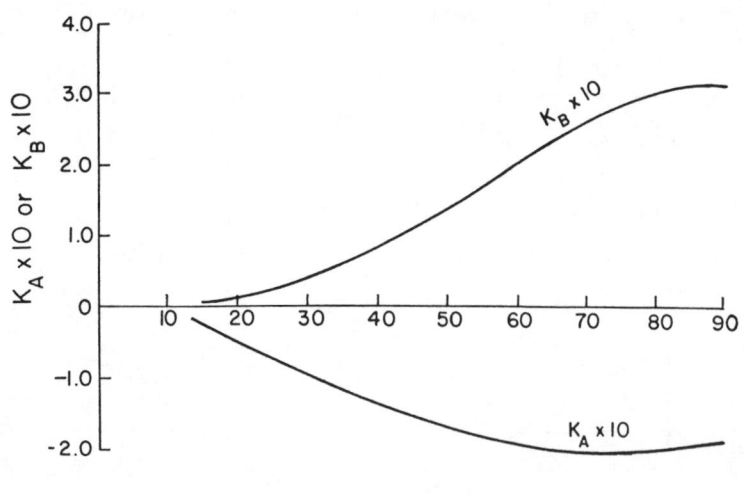

θ (Degree)

Case 5

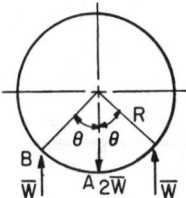

$$M_A = -\overline{W}RK_A$$
$$M_B = \overline{W}RK_B$$

θ	15°	30°	45°	60°	75°	90°
K_A	0.2268	0.3797	0.4644	0.4978	0.5024	0.50
K_B	0.0313	0.1096	0.1961	0.2489	0.2434	0.1817

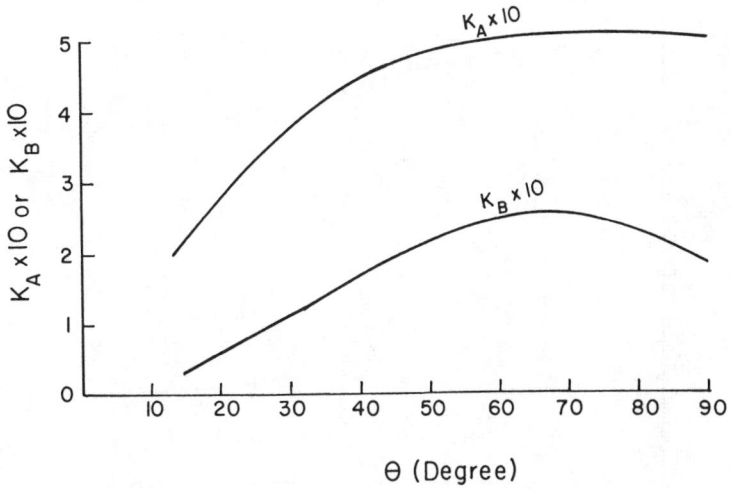

Case 6

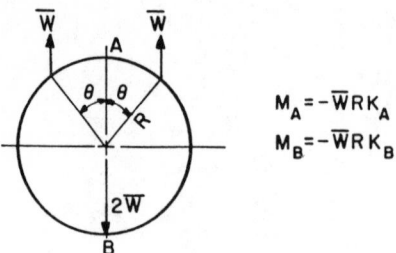

$$M_A = -\overline{W}RK_A$$
$$M_B = -\overline{W}RK_B$$

θ	15°	30°	45°	60°	75°	90°
K_B	0.6260	0.5977	0.5610	0.5274	0.5062	0.500
K_A	0.4098	0.2569	0.1722	0.1389	0.1343	0.1366

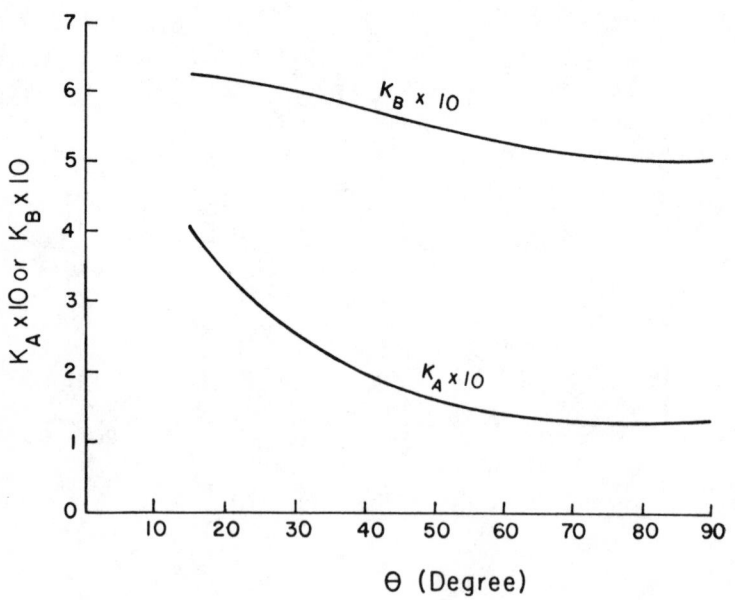

Case 7

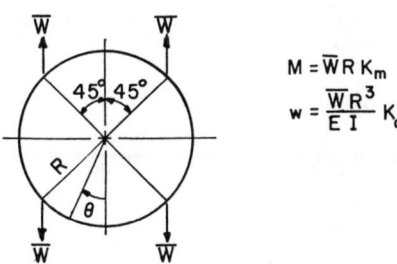

$$M = \overline{W}R\,K_m$$

$$w = \frac{\overline{W}R^3}{EI}\,K_d$$

θ	0°	15°	30°	45°	60°	75°	90°
K_m	−0.0966	−0.0966	−0.0966	−0.0966	0.0623	0.1622	0.1963
K_d	0.0461	0.0412	0.0270	0.0043	−0.0233	−0.0454	−0.0537

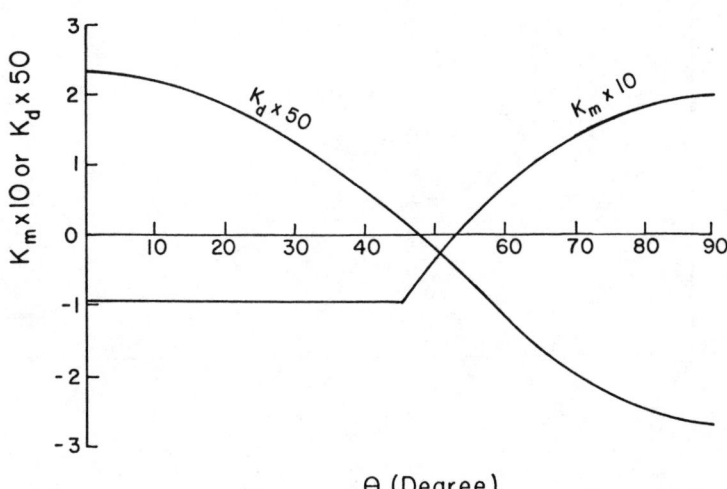

Θ (Degree)

Case 8

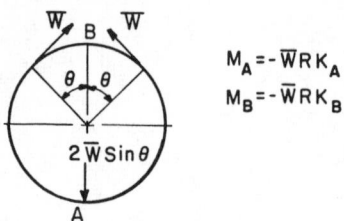

$$M_A = -\overline{W}RK_A$$
$$M_B = -\overline{W}RK_B$$

θ	30°	45°	50°	60°	75°	90°	105°	120°
K_A	0.3110	0.4268	0.4563	0.5000	0.5245	0.5000	0.4324	0.3333
K_B	0.2067	0.2305	0.2209	0.2180	0.1826	0.1366	0.0904	0.0513

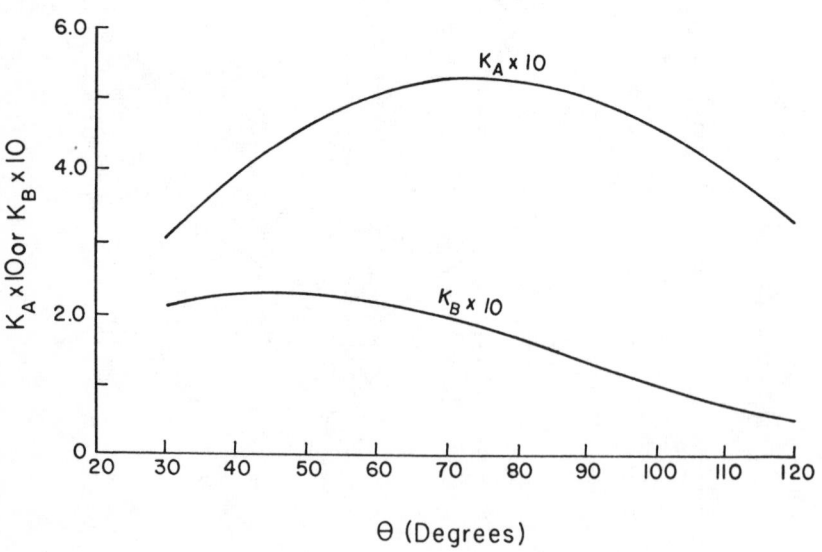

Case 9

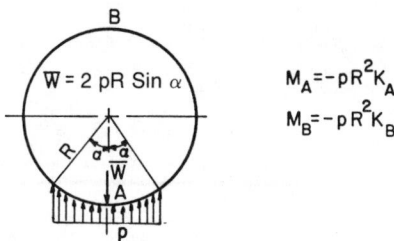

$$W = 2\,pR\,\sin\alpha$$

$$M_A = -pR^2 K_A$$
$$M_B = -pR^2 K_B$$

α	15°	30°	40°	50°	60°	75°	90°
K_A	0.0307	0.1050	0.1640	0.2211	0.2700	0.3201	0.3372
K_B	0.0009	0.0065	0.0138	0.0230	0.0329	0.0449	0.0494

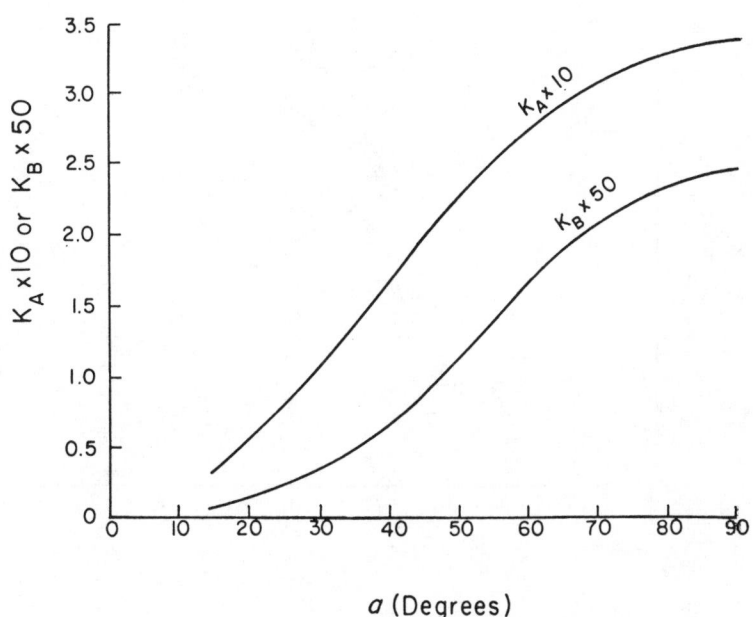

a (Degrees)

Case 10   l unit length of pipe filled with liquid of specific gravity p

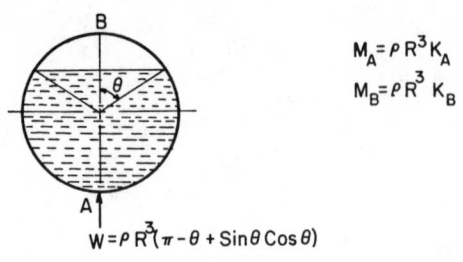

$$M_A = p R^3 K_A$$
$$M_B = p R^3 K_B$$

$$W = p R^3 (\pi - \theta + \sin\theta \cos\theta)$$

θ	0°	15°	30°	40°	50°	60°	75°	90°
K_A	0.7500	0.7463	0.7216	0.6861	0.6335	0.5649	0.4403	0.3067
K_B	0.2500	0.2468	0.2289	0.2006	0.1787	0.1466	0.0981	0.0567

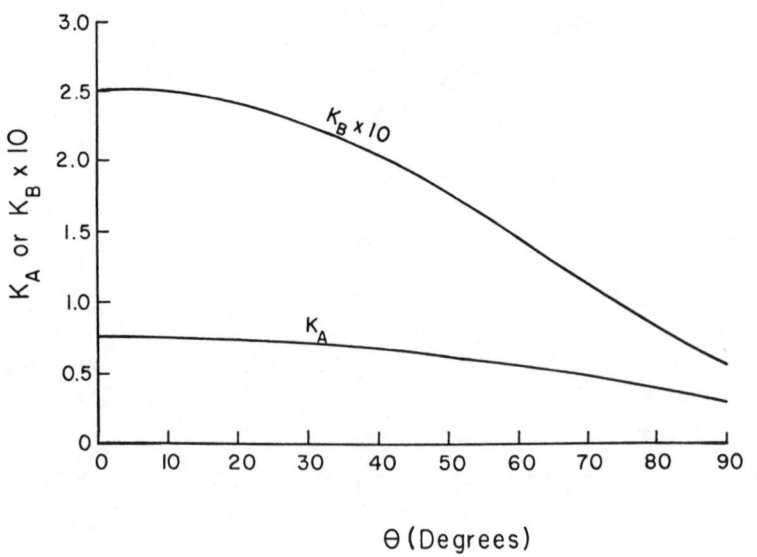

θ (Degrees)

Case 11

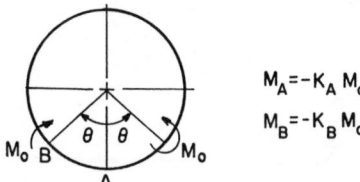

$$M_A = -K_A M_o$$
$$M_B = -K_B M_o$$

θ	15°	30°	45°	60°	75°	90°
K_A	0.7519	0.5150	0.2999	0.1154	−0.0316	−0.1366
K_B	0.7575	0.5577	0.4318	0.3910	0.4242	0.500

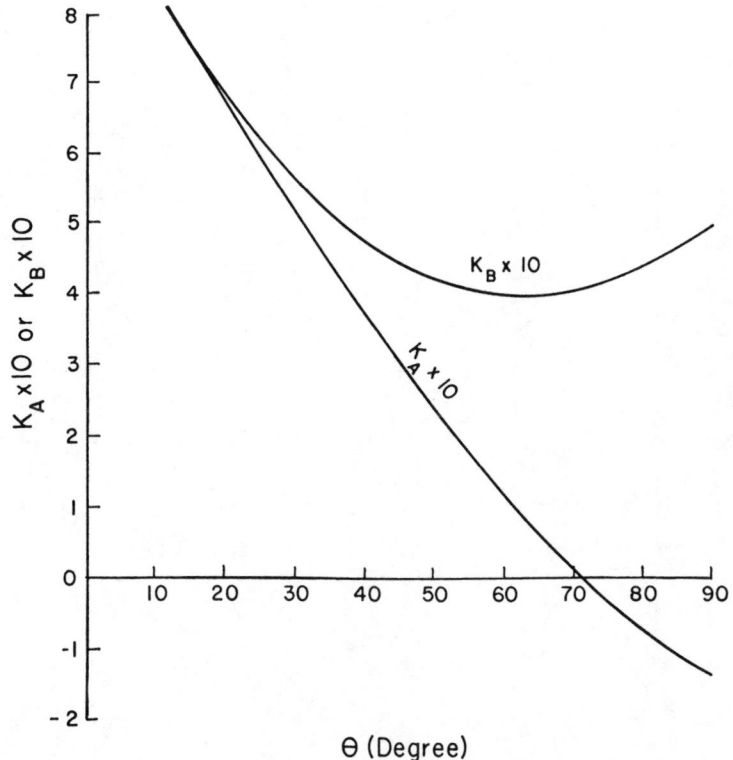

Case 12

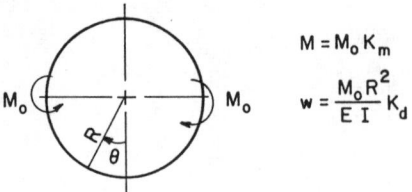

$$M = M_o K_m$$

$$w = \frac{M_o R^2}{E I} K_d$$

θ	0°	15°	30°	45°	60°	75°	90°
K_m	−0.1366	−0.1149	−0.0513	0.0499	0.1817	0.3352	0.500
K_d	0.0225	0.0172	0.0032	−0.0144	−0.0274	−0.0260	0.0

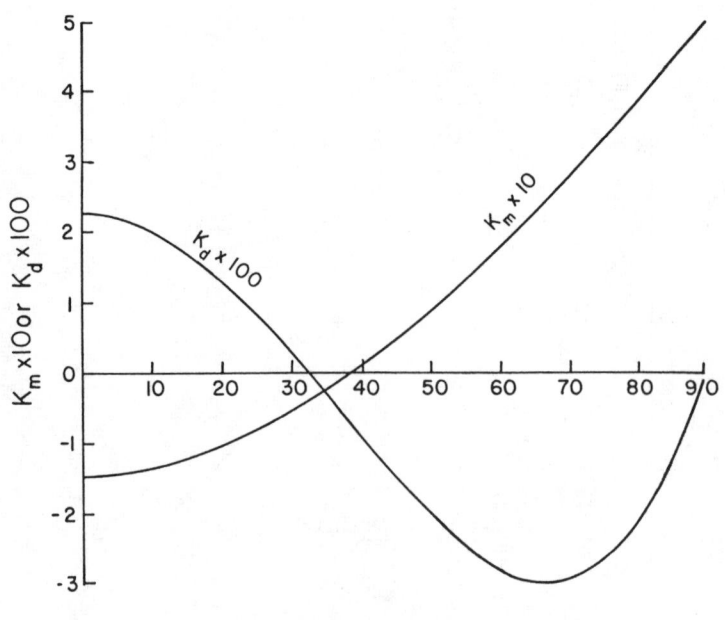

Θ (Degree)

Case 13

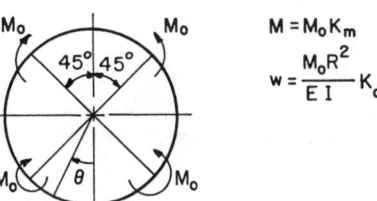

$$M = M_o K_m$$

$$w = \frac{M_o R^2}{E I} K_d$$

θ	15°	30°	45°	60°	75°	90°	105°	120°	135°
K_m	0.1304	0.2203	0.3634	−0.4501	−0.2330	0.0	−0.2330	−0.4501	−0.6365
K_d	−0.0242	−0.0088	0.0225	0.0432	0.0306	0.0	−0.0306	−0.0433	−0.0250

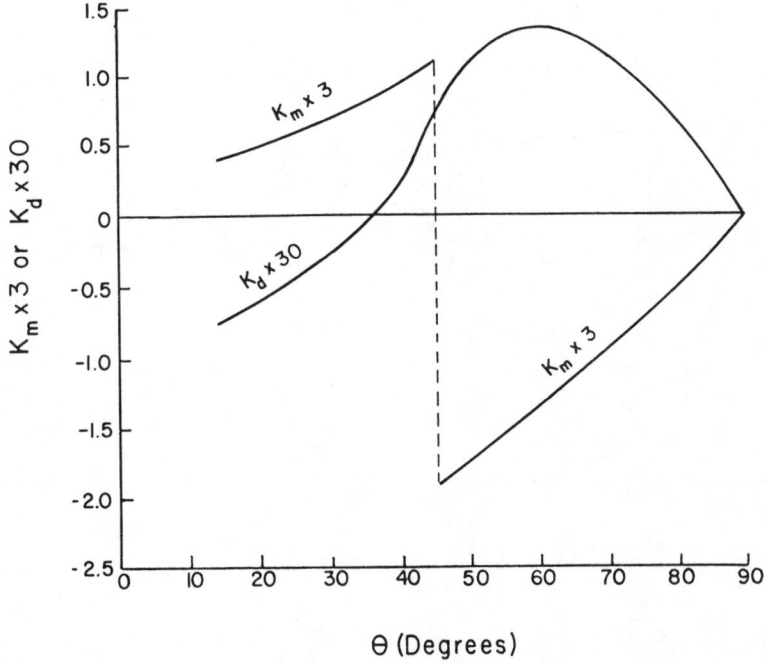

Case 14   Simply supported Circular arch

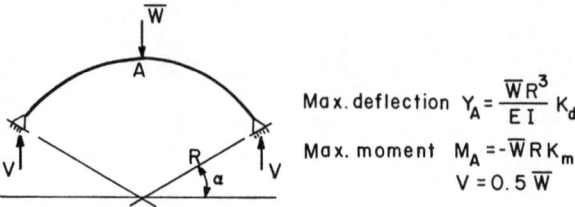

Max. deflection $Y_A = \dfrac{\overline{W} R^3}{E I} K_d$

Max. moment $M_A = -\overline{W} R K_m$

$V = 0.5 \overline{W}$

α	0°	15°	25°	30°	35°	40°	45°	50°	55°	60°
K_d	0.178	0.1595	0.1305	0.1132	0.0953	0.0775	0.0606	0.0453	0.032	0.0211
K_m	0.500	0.483	0.4532	0.433	0.4096	0.383	0.3536	0.3214	0.2868	0.250

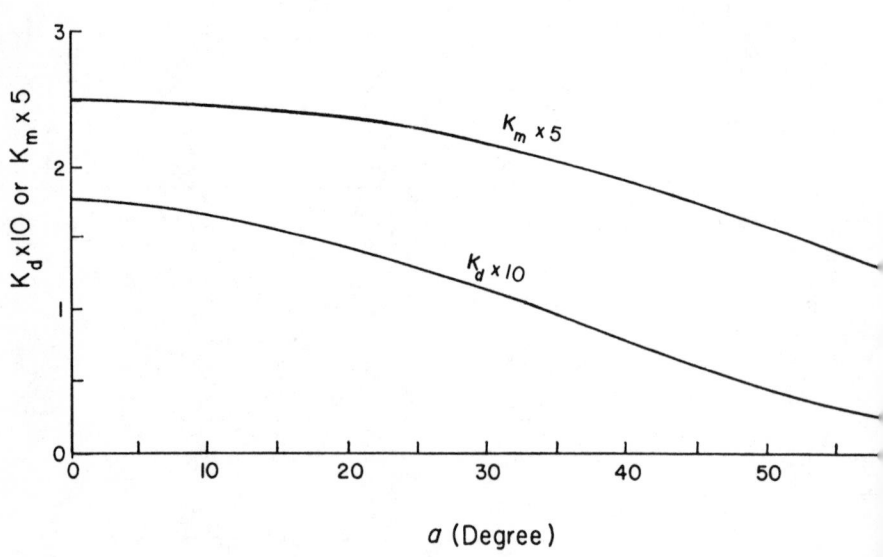

Case 15   Hinged Circular Arch

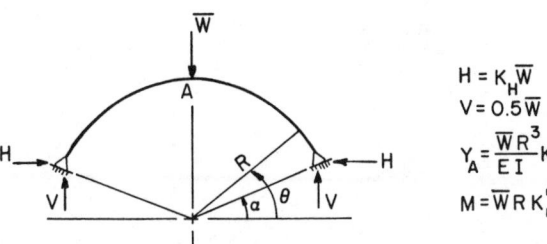

$$H = K_H \overline{W}$$
$$V = 0.5\overline{W}$$
$$Y_A = \frac{\overline{W}R^3}{EI}K_d$$
$$M = \overline{W}R K'_m$$

α	0°	7°	15°	22°	30°	37°	45°	52°	60°
K_H	0.3183	0.3760	0.4508	0.5271	0.6313	0.7442	0.9099	1.1068	1.4355
K_d	0.0190	0.0134	0.0102	0.0072	0.0052	0.0033	0.0020	0.0015	0.0006

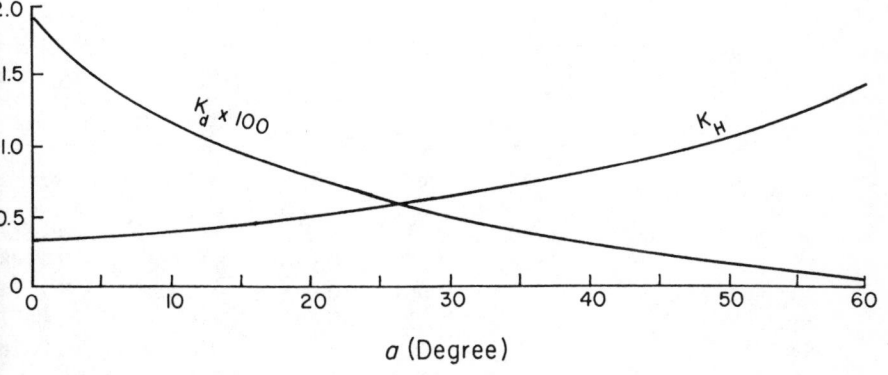

α (Degree)

Case 15 continued

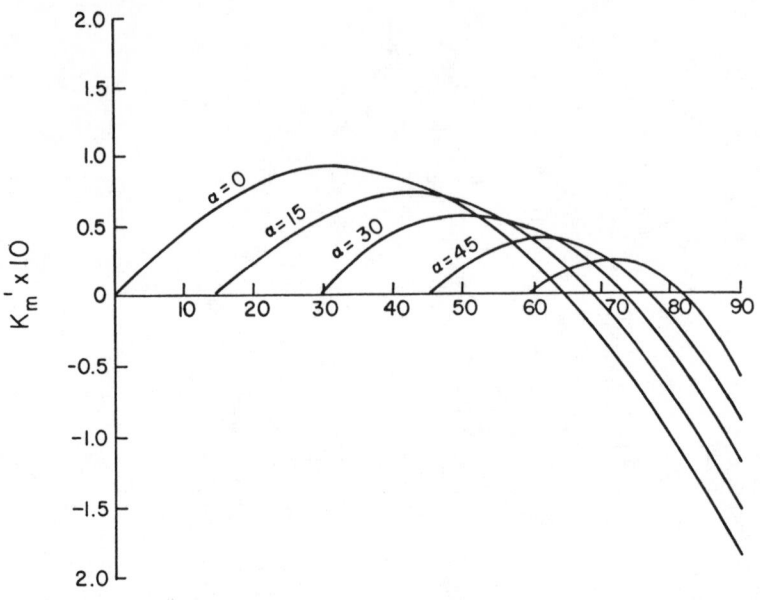

θ (Degree)

θ α	0°	15°	30°	45°	60°	75°	90°
0°	0	0.0654	0.0922	0.0786	0.0257	−0.0631	−0.1817
15°		0	0.0588	0.0727	0.0408	−0.0348	−0.1488
30°			0	0.0513	0.0481	−0.0095	−0.1174
45°				0	0.0411	0.0114	−0.0871
60°					0	0.0229	−0.0576

Case 16   Fixed Circular Arch

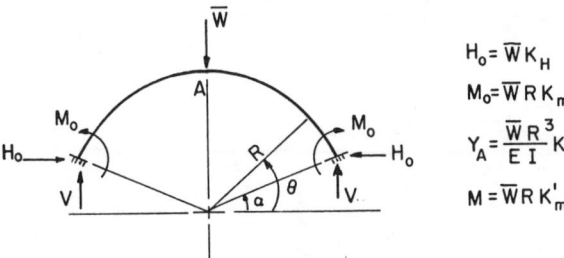

$$H_o = \overline{W} K_H$$

$$M_o = \overline{W} R K_m$$

$$Y_A = \frac{\overline{W} R^3}{E I} K_d$$

$$M = \overline{W} R K'_m$$

α	0°	15°	30°	45°	60°
K_H	0.4591	0.6026	0.8035	1.1240	1.770
K_m	0.1106	0.0888	0.0690	0.0510	0.0480
K_d	0.01168	0.00644	0.00345	0.0016	0.0019

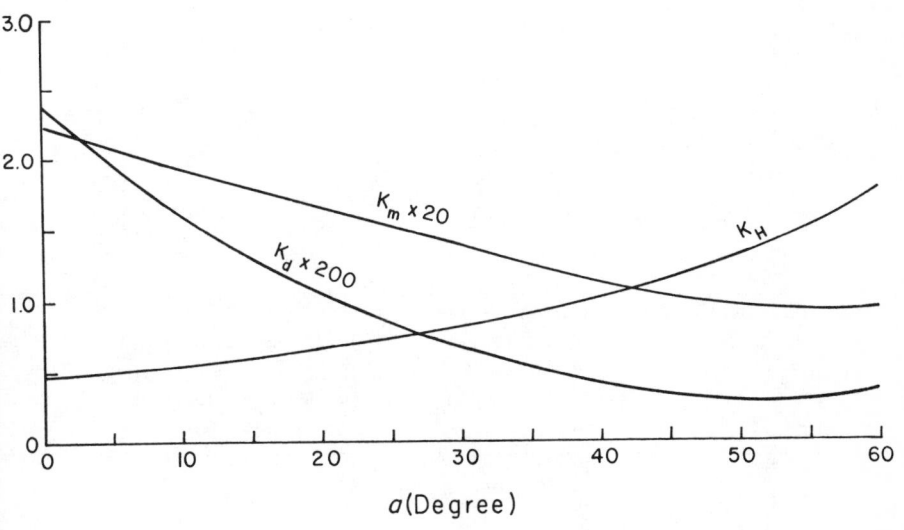

Case 16 continued

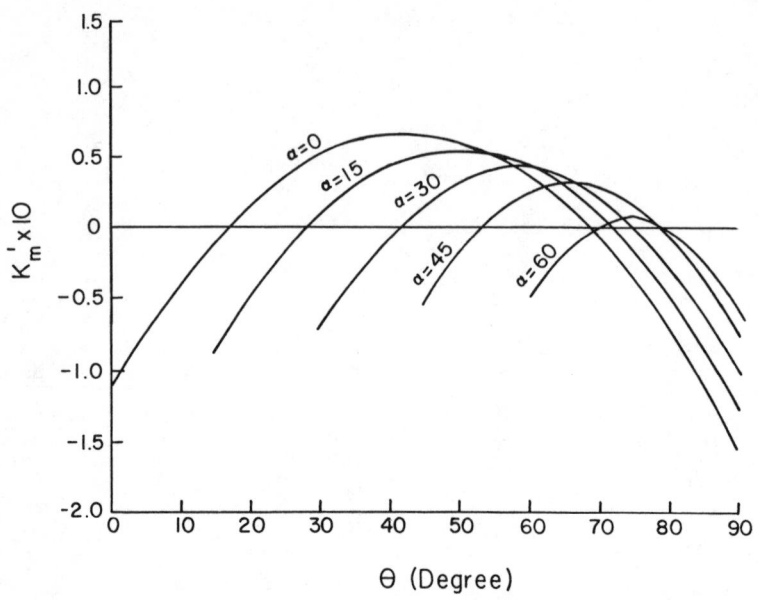

$$\mathbf{K_m'}$$

θ α	0°	15°	30°	45°	60°	75°	90°
0°	−0.1106	−0.0088	0.0520	0.0676	0.0370	−0.0377	−0.1515
15°		−0.0888	0.0066	0.0519	0.04415	−0.0162	−0.1251
30°			−0.0690	0.0180	0.0421	0.0018	−0.1002
45°				−0.051	0.0241	0.0158	−0.075
60°					−0.048	0.0083	−0.0608

Case 17

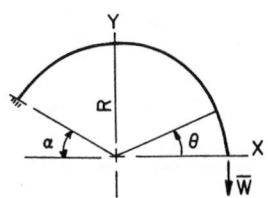

Deflection at free end

$$Y = -\frac{\overline{W} R^3}{EI} K_Y$$

$$X = -\frac{\overline{W} R^3}{EI} K_X$$

$$M = -\overline{W} R (\cos \theta - 1)$$

α	0°	15°	30°	45°	60°	75°	90°
K_y	4.7124	3.6770	2.7105	1.8701	1.1930	0.6920	0.3562
K_x	2.000	1.9324	1.7410	1.4571	1.1250	0.7923	0.500

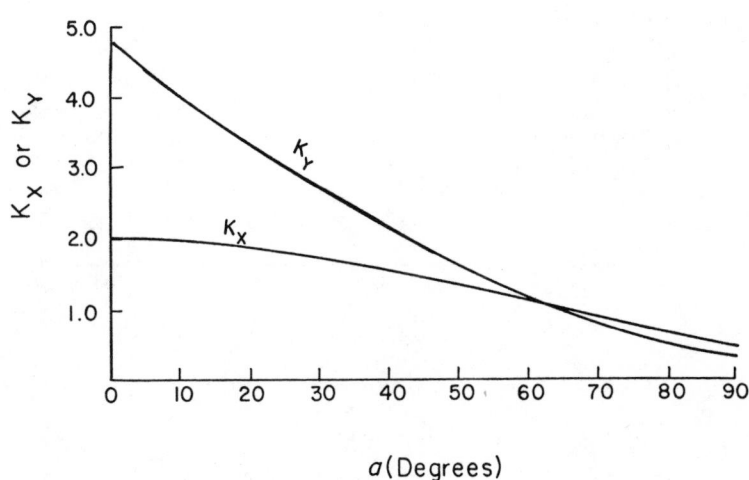

a(Degrees)

Case 18

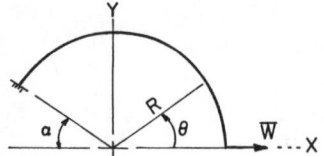

Deflection at free end   ($\theta = 0$)

$$Y = \frac{\overline{W} R^3}{EI} K_Y$$

$$X = \frac{\overline{W} R^3}{EI} K_X$$

$$M = \overline{W} R \sin \theta$$

α	0°	15°	30°	45°	60°	75°	90°
K_y	2.000	1.9324	1.7410	1.4571	1.1250	0.7923	0.50
K_x	1.5708	1.5649	1.5255	1.4281	1.2637	1.0413	0.7854

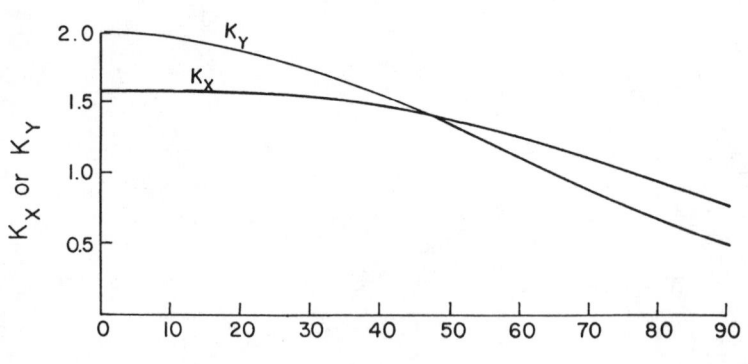

a (Degrees)

Case 19

Deflection at free end $(\theta = 0)$

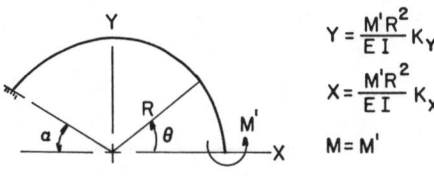

$$Y = \frac{M'R^2}{EI} K_Y$$

$$X = \frac{M'R^2}{EI} K_X$$

$$M = M'$$

α	0°	15°	30°	45°	60°	75°	90°
K_y	3.1416	2.6210	2.1180	1.6491	1.2284	0.8667	0.5708
K_x	2.000	1.9659	1.8660	1.7071	1.5000	1.2588	1.0000

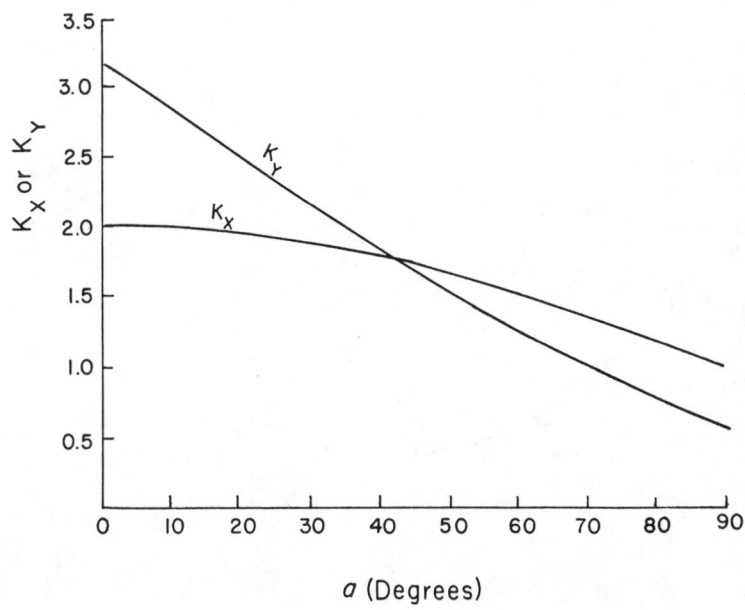

a (Degrees)

Case 20

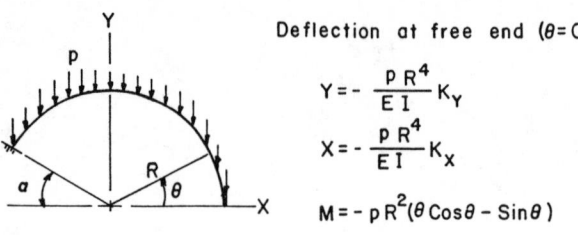

Deflection at free end $(\theta=0)$

$$Y = -\frac{pR^4}{EI}K_Y$$

$$X = -\frac{pR^4}{EI}K_X$$

$$M = -pR^2(\theta\cos\theta - \sin\theta)$$

α	0°	15°	30°	45°	60°	75°	90°
K_y	6.4674	4.8496	3.3822	2.1720	1.2668	0.6582	0.2961
K_x	2.3562	2.2509	1.9610	1.5531	1.1102	0.7070	0.3927

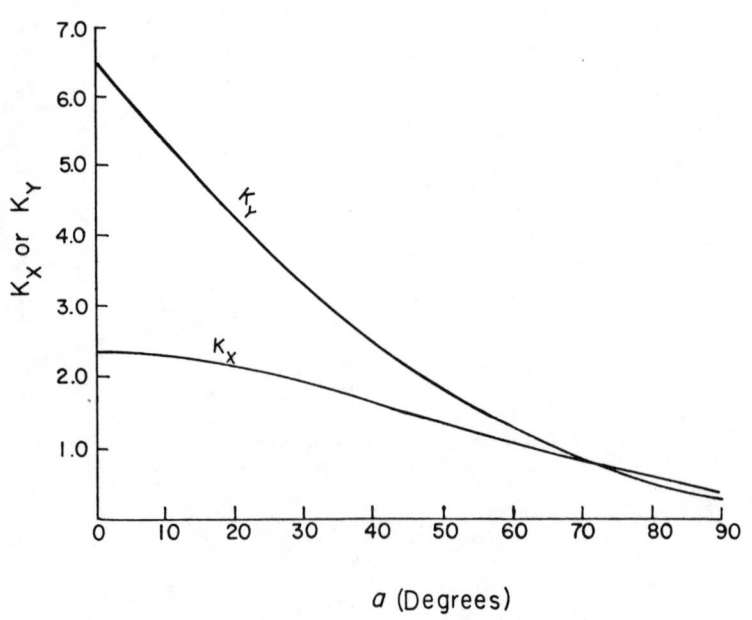

a (Degrees)

Chapter 9

Torsion

TORSION OF STRAIGHT MEMBERS WITH CIRCULAR CROSS SECTIONS [1,4,5]

A shaft of circular cross section is the most common member subjected to torsion. When the member twists, each section rotates about the longitudinal axis. At any point of a section there is a shear stress s_s. The magnitude of s_s is proportional to the distance from the center of the section, and its direction is perpendicular to the radius drawn through the point. If a round bar or tube is under constant torsional moment M_T, the cross sections rotate with respect to each other through an angle ϕ:

$$\phi = \frac{M_T \ell}{GJ} = \frac{2M_T \ell}{\pi R^4 G} \tag{9-1}$$

and the maximum shear stress is given by

$$s_s = \frac{M_T R}{J} = \frac{2M_T}{\pi R^3} \tag{9-2}$$

where M_T = twisting moment
$\quad$ G = modulus of rigidity
$\quad$ J = polar moment of inertia
$\quad$ ℓ = length of the member
$\quad$ R = radius of the circular section
$\quad$ ϕ = angle of twist in radians
$\quad$ s_s = shear stress

If the cross section is hollow, then

$$\phi = \frac{2M_T \ell}{\pi(R_1^4 - R_2^4)G} \tag{9-3}$$

$$S_s = \frac{2M_T R_1}{\pi(R_1^4 - R_2^4)} \tag{9-4}$$

where R_1 and R_2 are the outer and inner radii, respectively. In the derivation of these equations, it was assumed that:

1. the bar is straight and the section is uniform.
2. the bar is loaded within the elastic limit.
3. the planar cross sections in the untwisted state remain planar when the twisting moment is applied.
4. the cross sections remain undistorted in their own plane.

TORSION OF SHAFTS WITH NON-CIRCULAR CROSS SECTIONS

When a twisting shaft is not circular in cross section, it is no longer possible to prove that planar cross sections remain planar or that they remain undistorted in their own plane. The study of twist of non-circular cross sections was first published by Saint-Venant and is described in detail in Chapter 1 of *Advanced Strength of Materials* [1]. Let x and y be the perpendicular coordinates in the plane of a normal cross section with the origin in the center of twist; let z be the longitudinal coordinate and the section of reference be placed at z = 0. Saint-Venant assumed that the deformation of a section at z from the reference section could be written as

$$u = \theta\, zy \tag{9-5a}$$

$$v = -\theta z x \tag{9-5b}$$

$$w = f(x,y) \tag{9-5c}$$

where u, v, and w are the displacements of a point x, y, z from an untwisted state to a twisted state in the x, y, and z directions, respectively, and θ is the angle of twist per unit length of the shaft. Equations 9-5 express that there are only two strains γ_{xz} and γ_{yz}, since there is no distortion in the normal cross section. Thus $e_x = e_y = 0$ and also $\gamma_{xy} = 0$. The shear strains γ_{xz} and γ_{yz} are expressed as

$$\gamma_{xz} = \frac{\partial u}{\partial z} + \frac{\partial w}{\partial x}$$

$$\gamma_{yz} = \frac{\partial v}{\partial z} + \frac{\partial w}{\partial y}$$

Substituting Equations 9-5 into the above and applying Hooke's law the shear stresses can be written as

$$(s_s)_{xz} = G\left(\theta y + \frac{\partial w}{\partial x}\right) \tag{9-6a}$$

$$(s_s)_{yz} = G\left(-\theta x + \frac{\partial w}{\partial y}\right) \tag{9-6b}$$

The shear stresses are not constant across a normal cross section xy but differ from point to point. When all shear stresses on the faces of an element dx dy dz are set in equilibrium, the following equation of equilibrium can be derived:

$$\frac{\partial}{\partial x}(s_s)_{xz} + \frac{\partial}{\partial y}(s_s)_{yz} = 0 \tag{9-7}$$

Equation 9-7 is a partial differential equation in terms of $(s_s)_{xz}$ and $(s_s)_{yz}$, both depending on variables x and y. Saint-Venant assumed a function $\Phi(x,y)$ such that the stresses can be found from it by differentiation.

$$(s_s)_{yz} = \frac{\partial \Phi}{\partial x} \tag{9-8a}$$

$$(s_s)_{xz} = -\frac{\partial \Phi}{\partial y} \tag{9-8b}$$

The function $\Phi(x,y)$ is Saint-Venant's torsion function. The value Φ can be plotted vertically on an xy base and forms a curved surface. Equations 9-8 state that the shear stress component in the y direction is the slope of the Φ surface in the x direction. We can generalize the statement that the shear stress component in any direction equals the slope of the Φ surface in the perpendicular direction. Equations 9-6 now become

$$-\frac{\partial \Phi}{\partial y} = G\left(\theta y + \frac{\partial w}{\partial x}\right) \tag{9-9a}$$

$$\frac{\partial \Phi}{\partial x} = G\left(\theta x + \frac{\partial w}{\partial y}\right) \tag{9-9b}$$

These two equations can be combined to form the following:

$$\frac{\partial^2 \Phi}{\partial x^2} + \frac{\partial^2 \Phi}{\partial y^2} = -2G\theta \qquad (9\text{-}10)$$

Equation 9-10 is an equation of continuity. If one can find a Φ function which satisfies Equation 9-10, and the boundary condition Φ is constant along the periphery, then the stresses derived from that function are the solution to the torsion problem of the given cross section. It has been proved that the torque transmitted by a shaft equals twice the volume under the Φ surface, provided that $\Phi = 0$ at the boundary; that is,

$$M_T = 2 \int\int \Phi \, dA \qquad (9\text{-}11)$$

Torsion problems solved using Saint-Venant's method often involve complicated mathematics. Some practical sections such as I beams and channels may not be reducible to mathematical formulas, so approximate solutions are needed. The best approximate method is the membrane analog method developed by Prandtl in 1903. He observed that Equation 9-10 for the stress function is the same as the differential equation for the shape of a stretched membrane blown up from flat by air pressure. A membrane is originally flat in the xy plane and blown up to ordinate z by air pressure p; if $z = 0$ at the periphery of the membrane when the interior is blown up, the following equilibrium equation can be derived:

$$\frac{\partial^2 z}{\partial x^2} + \frac{\partial^2 z}{\partial y^2} = -\frac{p}{T} \qquad (9\text{-}12)$$

Equation 9-12 indicates that the sum of the curvature in two perpendicular directions is a constant for all points of the membrane. If we adjust the membrane tension T or the air pressure p so that $p/T = 2G\theta$, then Equation 9-12 of the membrane is identical to Equation 9-10 of the torsional stress function. Experiments have been made using rubber sheets or soap films for membranes. Blowing up a large membrane creates two cross sections, the one to be investigated and a purely circular one. The p/T value is the same for both sections since the same membrane is used, and the corresponding $2G\theta$ should also be the same. If the volumes and slopes are measured then

$$\frac{\text{volume}}{\text{slope}} = K \frac{M_T}{s_s}$$

K is a constant which is the same for both sections. K can be easily calculated since we know all about the circular section, and the torque stress ratio for the other section can be found. We also have the following relationship:

$$\frac{V_c}{V_o} = \frac{C_c}{C_o}$$

where V_c = volume of membrane hill of circular section
V_o = volume of membrane hill of other section
C_c = torsional stiffness of circular section
C_o = torsional stiffness of other section

By membrane analog we can translate from the membrane to torsion by

$$p/T \rightarrow 2G\theta$$

$$\text{slope} \rightarrow s_s$$

$$\text{volume} \rightarrow M_T/2$$

Using the membrane analog method, the following solutions can be obtained for a rectangular section of b by t:

$$s_s = K_1 \frac{3M_T}{bt^3} \tag{9-13}$$

$$\frac{M_T}{\theta} = K_2 \frac{Gbt^3}{3} \tag{9-14}$$

where K_1 and K_2 are factors depending on the b/t ratio. For a very narrow section (b >> t), both K_1 and K_2 approach unity, as shown in Figure 9-1. Note that the maximum stress occurs at the point closest to the center of the section, and at corners the shear stress is zero. Equation 9-14 can be written as

$$\theta = \frac{M_T}{GJ} \tag{9-15}$$

where J is a torsional constant. For a thin rectangular section, $J = bt^3/3$. Equation 9-15 is the unit angle of twist for all non-circular cross sections.

Equations 9-13 and 9-14 are good for open sections that can be built of rectangles, angles, T shapes, and slit tubes, if we interpret b as the total length of the cross section. For open sections with corners, the stress at

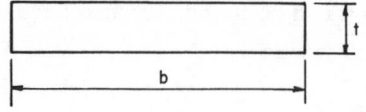

b/t	1	1.5	2	2.5	3	5	10	∞
K_1	1.600	1.433	1.353	1.287	1.247	1.147	1.067	1.0
K_2	0.423	0.588	0.687	0.747	0.789	0.873	0.936	1.0

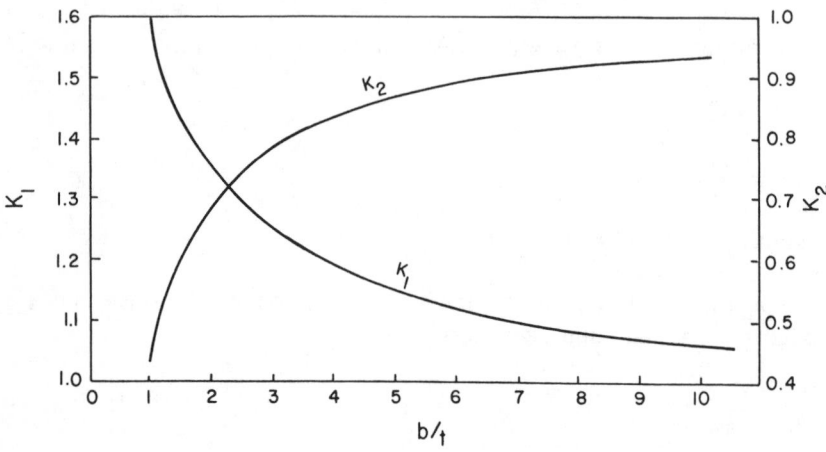

Figure 9-1. K_1 and K_2 factors for a rectangular section.

the reentrant corner is always greater than that given in Equation 9-13. The stress concentration depends greatly on the local fillet radius of the reentrant corner. The maximum shear stress is obtained by

$$(s_s)_{max} = K_3 s_s$$

where K_3 is the stress concentration factor. An approximate solution for the stress concentration factor was found by Trefftz in 1922. It is a function of t/r. As shown in Figure 9-2, t is the wall thickness of the web, and r is the fillet radius of the 90 degree corner.

TORSION OF MEMBERS WITH HOLLOW SECTIONS [1,3]

Thin-walled, closed tubes (circular, elliptic, or square) made of aluminum in aircraft construction, and the steel box girders used in building construction are some of the most important practical sections. For a

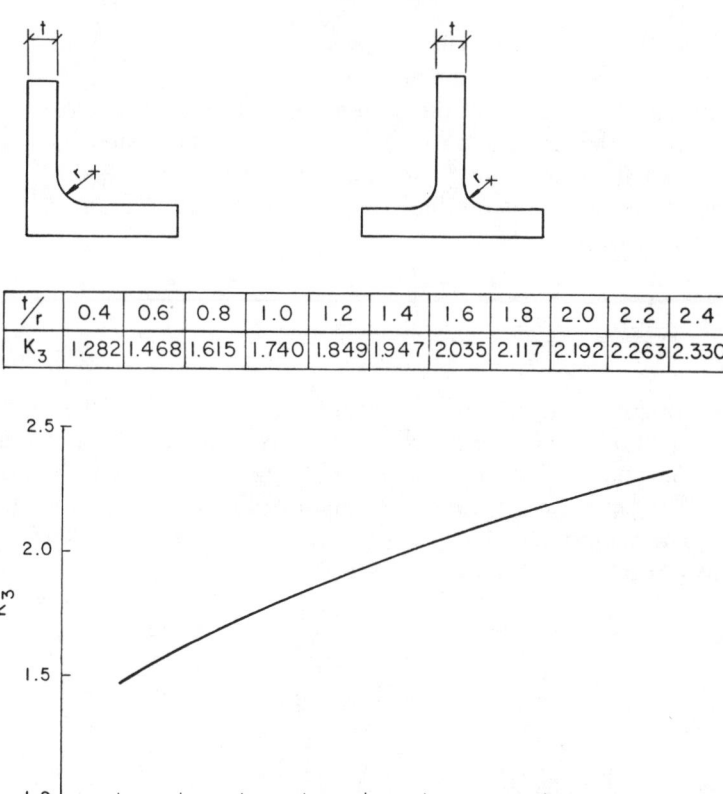

Figure 9-2. Stress concentration factor.

thin-walled, closed tube it is reasonable to assume that the shearing stress is constant across the thickness, and also that it is tangent to the centerline between the inner and outer boundaries. For a thin-walled tube of arbitrary cross section, the shearing force per unit length of the periphery of the tube is termed *shear flow*, denoted by q_f. If t is the thickness of the wall, then

$$q_f = s_s t$$

Let us consider the cross section shown in Figure 9-3a. The moment about an arbitrary point o of the shear force $q_f ds$ on the infinitesimal element ds is $q_f n ds$. The resultant moment of such a shearing force over the cross section is

$$M_T = \int q_f \, n \, ds = 2Aq_f = 2A \, s_s \, t \tag{9-16}$$

The product n ds is twice the area of the triangle; thus the integral is merely twice the area enclosed by the centerline of the tube. This expression is called *Bredt's law*. Note that A is the area enclosed by the centerline of the tube, it is not the cross-sectional area of the material. The shear stress is

$$s_s = \frac{M_T}{2At} \tag{9-17}$$

The torsional stiffness of the section can be found by the membrane analog method. The general section with its membrane is shown in Figures 9-3a and b. The blown up membrane must be sensibly straight across the gap of width t, and the height of the membrane must be constant. The membrane slope is h/t. Letting the upward push equal the downward pull, we obtain

$$pA = \int T \, ds \, h/t$$

Translating to torsion by setting slope $h/t = s_s$ and $p/T = 2G\theta$, we obtain

$$2G\theta = \frac{1}{A} \int s_s \, ds \tag{9-18}$$

From Equations 9-17 and 9-18

$$\theta = \frac{M_T}{4GA^2} \int \frac{ds}{t} \tag{9-19}$$

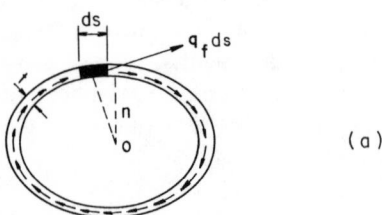

(a)

Figure 9-3.   Torsion of a hollow section.

(b)

$$C = \frac{M_T}{\theta} = \frac{4GA^2}{\int (ds/t)} = GJ \qquad (9\text{-}20)$$

When the wall thickness is constant, $J = 4A^2t/L$, where L is the peripheral length of the wall. We obtain the following values for J:

Square tube with constant t, $J = tb^3$

Rectangular tube with constant t, $J = 2tb^2 h^2/(b + h)$

Rectangular tube t is not constant, $J = 4h^2b^2/(\Sigma b/t_b + \Sigma h/t_h)$

Note that Equation 9-17 applies only where the shear stress is smooth; if there are sharp corners, the stress concentration factor K_3, shown in Figure 9-2, should be applied.

Example 9-1

Determine the torque transmitted if the maximum shear stress is 6 kips/in.[2], disregarding stress concentration in the corners; and the angle of twist per unit length of the thin-walled tube of rectangular cross section shown in Figure 9-4, if $G = 12 \times 10^3$ kips/in.[2]

$$A = (6 - 0.5)(6 - 0.25) = 31.625 \text{ in.}^2$$

From Equation 9-17,

$$M_T = s_s 2At = 6 \times 2 \times 31.625 \times 0.25 = 94.875 \text{ kip-in.}$$

$$\theta = M_T/GJ$$

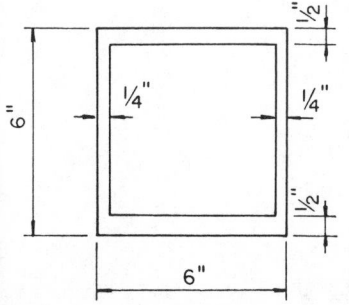

Figure 9-4. Rectangular thin-walled tube for Example 9-2.

$$J = 4h^2b^2/(\Sigma h/t_h + \Sigma b/t_b) = \cfrac{4 \times 5.5^2 \times 5.75^2}{\cfrac{5.5 \times 2}{0.25} + \cfrac{5.75 \times 2}{0.5}} = 59.7$$

$\theta = 94.875/(12,000 \times 59.7) = 1.324 \times 10^{-4}$ radian/in.
$= 0.091°/\text{ft}$

Example 9-2

The two-cell section shown in Figure 9-5 has a maximum shear stress of 5 kips/in.^{2}. Disregarding the stress concentration, calculate the torque transmitted and the angle of twist per unit length under that torque (use $G = 12 \times 10^3$ kips/in.^{2}).

Let h_1 and h_2 be the heights of the blown up membrane for cells one and two. The membrane slope for the inside wall will be $\Delta h/t$, where Δh is the difference between h_1 and h_2. We can write the equation of vertical equilibrium for each cell:

$$p \times 12^2 = T \times 12 \left(\frac{h_1}{0.5t} + \frac{h_1}{0.5t} + \frac{h_1 - h_2}{0.25t} + \frac{h_1}{0.5t} \right)$$

$$p \times 12^2 = T \times 12 \left(\frac{h_2 - h_1}{0.25t} + \frac{h_2}{0.5t} + \frac{h_2}{0.5t} + \frac{h_2}{0.5t} \right)$$

Solving these two equations we obtain

$$h_1 = h_2 = 2p/T$$

The central leg has zero slope in its membrane and hence has zero stress. We translate from the membrane to torsion by

$$p/T = 2G\theta$$

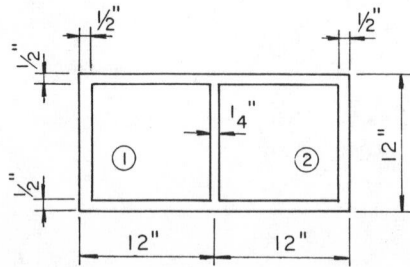

Figure 9-5. Two-cell section for Example 9-2.

slope $= s_s$

volume $= M_T/2$

$(s_s)_{max} = 5$ kips/in.$^2 = h_1/0.5 = 4p/T = 8G\theta$

$\theta = 5/(8 \times 12,000) = 5.208 \times 10^{-5}$ radian/in.

The volume under the membrane is

$$V = \left(12 \times 12 \frac{2p}{T}\right) \times 2 = 576p/T = 1,152 \ G\theta$$

$M_T/2 = 1,152 \ G\theta$

$M_T = 2,304 \times 5/8 = 1,440$ kip-in.

THE EFFECT OF WARPING RESTRAINT [2]

In the preceding section we discussed the case of torsion in which the cross sections of the shaft are free to warp. Under such conditions, warping is the same for all cross sections and takes place without any axial strain in the longitudinal direction. If one or both ends of the shaft are fixed so that warping is prevented, the stresses and angle of twist produced by a given torque are affected. For compact sections (rectangular, elliptical, etc.) the effect is slight, but for open tubes or flanged sections the effect may be considerable.

Consider a cantilever I beam subjected to an end torsion. The flange at the free end will displace a relative distance of u in the x direction, as shown in Figure 9-6. The moment and shear force introduced at a flange are M_f and V_f, respectively:

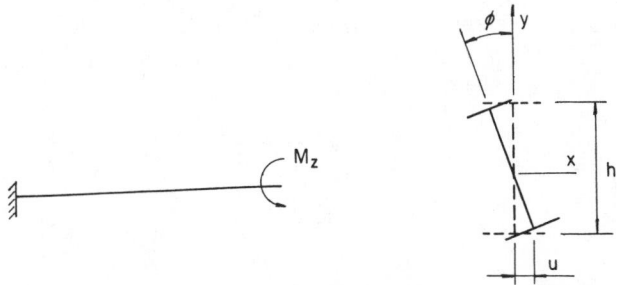

Figure 9-6. Cantilever I beam with at-end torsion.

$$M_f = -EI_f \frac{d^2u}{dz^2} \tag{9-21}$$

$$V_f = -EI_f \frac{d^3u}{dz^3} \tag{9-22}$$

where I_f is the moment of inertia of a flange ($I_f = I_y/2$), and the z axis is along the beam. Note that $u = (h/2)\phi$; hence, Equations 9-21 and 9-22 become

$$M_f = -EI_f \frac{h}{2} \phi''$$

$$V_f = -EI_f \frac{h}{2} \phi'''$$

The bending stresses caused by M_f can be calculated by the flexure formula, $f_b = M_f c/I_f$, and the shear stresses caused by M_f can be calculated from the beam shear formula, $f_s = V_f Q_f/I_f t$. Note that I_f and Q_f refer to the y axis of the beam and are for one flange only.

The shear forces V_f in the two flanges form a resisting couple which is expressed as follows:

$$M_w = V_f h = -EI_f \frac{h^2}{2} \phi''' = -EC_w\phi'''$$

C_w is defined as the *warping constant*. The total twisting moment is

$$M_Z = M_T + M_w = GJ \frac{d\phi}{dz} - EC_w \frac{d^3\phi}{dz^3} \tag{9-23}$$

Equation 9-23 takes into consideration both St. Venant torsion and warping torsion. This is *Timoshenko's differential equation*. It is a linear differential equation, and the solution is

$$\phi = \frac{M_Z}{GJ} \left[z + \frac{1}{\alpha} (e^{-\alpha z} - 1) \right] \tag{9-24}$$

where

$$\alpha = \sqrt{\frac{GJ}{EC_w}}$$

From Equation 9-24 we find ϕ'' and ϕ''', and the bending moment in the flange is

$$M_f = EI_f \frac{h}{2} \phi'' = \frac{EI_f h M_w}{2GJ} \alpha e^{-\alpha z}$$

The maximum value of M_f occurs at $z = 0$, hence

$$\text{max. } M_f = \frac{M_w}{\alpha h}$$

The shear force in the flange is

$$V_f = EI_f \frac{h}{2} \phi''' = -\frac{EI_f h M_w}{2GJ} \alpha^2 e^{-\alpha z}$$

The maximum value of ϕ''' also occurs at $z = 0$:

$$\text{max. } V_f = -\frac{M_w}{h}$$

Hence, at $z = 0$ the entire twisting moment is transmitted by transverse shear forces in the flanges, and nothing is transmitted there in the manner of St. Venant's torsional shear.

Values of ϕ', ϕ'', and ϕ''' for various loading and boundary conditions are available in Reference 6.

REFERENCES

1. Hartog, J. P., *Advanced Strength of Materials*, McGraw-Hill Book Company, 1952.
2. Timoshenko, S. and Gere, J., *Theory of Elastic Stability*, McGraw-Hill Book Company, 1961.
3. Brockenbrough, R. and Johnston, B., *Steel Design Manual*, U.S. Steel Corporation, 1968.
4. Roark, R., *Formulas for Stress and Strain*, Fourth Edition, McGraw-Hill Book Company, 1965.
5. Roark, R. and Young, W., *Formulas for Stress and Strain*, Fifth Edition, McGraw-Hill Book Company, 1975.
6. Bethlehem Steel, *Torsion Analysis of Rolled Steel Sections*, 1963.

Chapter 10

Thin Shells

DEFINITIONS AND ASSUMPTIONS [1,2]

A thin shell is a curved slab whose thickness t is small compared with its other dimensions and with its radii of curvature r_x and r_y. The surface that bisects the shell thickness is called the middle surface. By specifying the form of the middle surface and the thickness, we can completely define the geometry of the shell. Analysis of a shell consists of establishing equilibrium of an infinitely small element cut from the shell, as shown in Figure 10-1. The principal radii of curvature lie in the xz and yz planes. If we neglect the small quantities z/r_x and z/r_y, the resultant forces per unit length of a normal section are

$$N_x = \int_{-\frac{t}{2}}^{\frac{t}{2}} \sigma_x \, dx = \frac{Et}{1 - \nu^2} (e_1 + \nu \, e_2) \tag{10-1}$$

$$N_y = \int_{-\frac{t}{2}}^{\frac{t}{2}} \sigma_y \, dz = \frac{Et}{1 - \nu^2} (e_2 + \nu \, e_1) \tag{10-2}$$

$$N_{xy} = \int_{-\frac{t}{2}}^{\frac{t}{2}} \tau_{xy} \, dz = \frac{\gamma tE}{2(1 + \nu)} = N_{yx} \tag{10-3}$$

$$M_x = \int_{-\frac{t}{2}}^{\frac{t}{2}} \sigma_x z \, dz = -D(\chi_x + \nu\chi_y) \tag{10-4}$$

$$M_y = \int_{-\frac{t}{2}}^{\frac{t}{2}} \sigma_y z \, dz = -D(\chi_y + \nu\chi_x) \tag{10-5}$$

$$M_{xy} = -M_{yx} = - \int_{-\frac{t}{2}}^{\frac{t}{2}} \tau_{xy} z \, dz = D(1 - \nu) \, \chi_{xy} \tag{10-6}$$

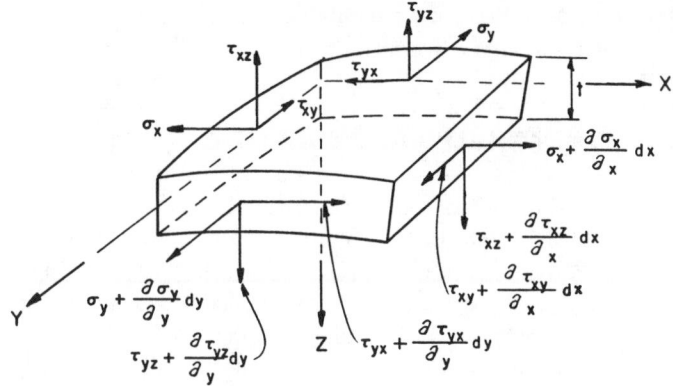

Figure 10-1. Forces on a shell element.

$$Q_x = \int_{-\frac{t}{2}}^{\frac{t}{2}} \tau_{xz} \, dz \tag{10-7}$$

$$Q_y = \int_{-\frac{t}{2}}^{\frac{t}{2}} \tau_{yz} \, dz \tag{10-8}$$

where e_1, e_2 = unit elongation of the middle surface in the x and y directions, respectively
  γ = shearing strain in the middle surface
  χ_x, χ_y = changes of curvatures
  χ_{xy} = twist of the middle surface
  E = modulus of elasticity
  D = flexural rigidity of shell
  ν = Poisson's ratio

Derivations of the equations shown above are available in Chapter 14 of *Theory of Plates and Shells* [1].

In many cases of analysis and design of shell structures, the bending stresses can be neglected and only the stresses due to strain in the middle surface are considered. If the bending can be neglected, the analysis is greatly simplified, since the moments M_x, M_y, M_{xy} and shearing forces Q_x, Q_y, vanish. The only unknowns are N_x, N_y, and N_{xy}, which can be determined from the conditions of equilibrium. The theory of shells neglecting bending stresses is called *membrane theory,* and N_x, N_y, and N_{xy} obtained in this manner are called *membrane forces.* The membrane theory can sometimes provide a reasonable basis for design if the loading is smoothly distributed over the surface and the displacements do not rise to appreciable bending.

THIN SHELLS OF REVOLUTION [1,5]

A shell that has the form of a surface of revolution is obtained by rotation of a plane curve about an axis lying in the plane of the curve. This curve is called the *meridian* and its plane is a *meridian plane*. In this case an element is defined by the angles θ and ϕ; the subscripts θ and ϕ are used in notation for stresses and resultant forces.

The shell element shown in Figure 10-2 is cut out by two meridians and two parallel circles. The meridional force N_ϕ, the hoop force N_θ, and the shearing force $N_{\phi\theta}$ can be determined from the conditions of equilibrium. Writing the equation of equilibrium of the element with the forces in the direction of the tangent to the meridian, we obtain

$$\frac{\partial}{\partial \phi} (r_0 \, N_\phi) + r_1 \frac{N_{\theta\phi}}{\partial \theta} - r_1 \, N_\theta \cos \phi + p_y \, r_0 \, r_1 = 0 \qquad (10\text{-}9)$$

The second equation is obtained by writing the equation of equilibrium of the forces in the direction of the tangent to the parallel circle. We obtain

$$\frac{\partial}{\partial \phi} (r_0 \, N_{\phi\theta}) + r_1 \frac{\partial N_\theta}{\partial \theta} + r_1 \, N_{\phi\theta} \cos\phi + p_x \, r_0 \, r_1 = 0 \qquad (10\text{-}10)$$

The third equation of equilibrium is obtained by summing up the projections of the forces in the z direction. We obtain

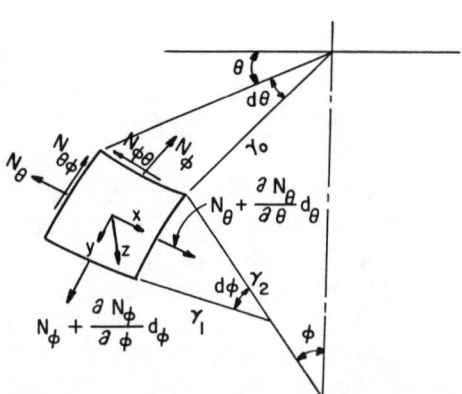

Figure 10-2. Membrane forces on a shell element.

$$N_\theta \ r_1 \ \sin\phi \ + \ N_\phi \ r_0 \ + \ P_z \ r_0 \ r_1 \ = \ 0 \qquad (10\text{-}11)$$

In many practical designs of shells, the loads are symmetrical with respect to their axes. The stresses are independent of θ, and all derivatives with respect to θ in Equations 10-9 and 10-10 vanish. For shells in the form of a surface of revolution and loaded symmetrically with respect to their axes, the equations of equilibrium are

$$\frac{d}{d\phi} \ (r_0 \ N_\phi) \ - \ r_1 \ N_\theta \ \cos\phi \ + \ p_y \ r_0 \ r_1 \ = \ 0 \qquad (10\text{-}9a)$$

$$\frac{N_\phi}{r_1} \ + \ \frac{N_\theta}{r_2} \ = \ -p_z \qquad (10\text{-}11a)$$

From the symmetry of loading and deformation it can be stated that there will be no shearing forces acting on the sides of the element. Solving Equations 10-9a and 10-11a we obtain

$$N_\phi \ = \ \frac{1}{r_2 \ \sin^2\phi} \left[- \int r_1 \ r_2 \ (p_z \ \cos\phi \ + \ p_y) \ \sin\phi \ d\phi \ + \ C \right] \qquad (10\text{-}12)$$

Equation 10-12 is a condition of equilibrium for the part of the shell above a parallel circle ϕ = constant. If the resultant of the total load on that portion of the shell is R, then

$$N_\phi \ = \ \frac{-R}{2\pi \ r_0 \ \sin\phi} \ = \ - \ \frac{R}{2\pi \ r_2 \ \sin^2\phi} \qquad (10\text{-}13)$$

$$N_\theta \ = \ \frac{R}{2\pi \ r_1 \ \sin^2\phi} \ - \ p_z \ r_2 \qquad (10\text{-}14)$$

Take a spherical dome, for example. The spherical shell shown in Figure 10-3 is submitted to a dead load p per unit area. We find that

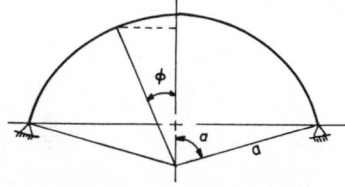

Figure 10-3. A spherical shell.

$$r_1 = r_2 = a$$

$$R = 2\pi \int_0^\phi a^2 \, p \, \sin\phi \, d\phi = 2\pi a^2 \, p(1 - \cos\phi)$$

$$N_\phi = \frac{-2\pi \, a^2 \, p(1 - \cos\phi)}{2\pi \, a \, \sin^2\phi} = \frac{-ap(1 - \cos\phi)}{\sin^2\phi} = \frac{-ap}{1 + \cos\phi}$$

$$p_y = p \, \sin\phi, \; p_z = p \, \cos\phi$$

$$N_\theta = \frac{2\pi \, a^2 \, p(1 - \cos\phi)}{2\pi \, a \, \sin^2\phi} - ap \, \cos\phi = ap\left(\frac{1}{1 + \cos\phi} - \cos\phi\right)$$

Note that N_ϕ is always negative. Thus there is a compression along the meridians. For small angles of ϕ, the hoop forces N_θ are also negative. For $\phi = 51.83°$, $N_\theta = 0$, and when $\phi > 51.83°$, N_θ is positive and there are tensile stresses in the direction perpendicular to the meridians.

CYLINDRICAL SHELLS WITH SYMMETRICAL LOADING [1,3]

The equations required to solve the problems of a circular cylindrical shell loaded symmetrically with respect to its axis have been derived in Chapter 15 of *Theory of Plates and Shells* [1] and Chapter 5 of *Stresses in Shell* [6]. The shell element shown in Figure 10-4 is defined by coordinate x and the angle ϕ. The forces acting on the sides of the element are shown. Note that the membrane shearing forces $N_{x\phi} = N_{\phi x}$ and the twisting moments $M_{x\phi} = M_{\phi x}$ vanish in this case, and that N_ϕ and M_ϕ are constant along the circumference. Assuming that the external forces are normal to the surface, the following three equations are obtained by projecting the forces on the x and z axes and by taking the moment of the forces about the y axis:

$$\frac{dN_x}{dx} R \, dx \, d\phi = 0 \tag{10-15a}$$

$$\frac{dQ_x}{dx} + \frac{N_\phi}{R} = -p_z \tag{10-15b}$$

$$\frac{dM_x}{dx} - Q_x = 0 \tag{10-15c}$$

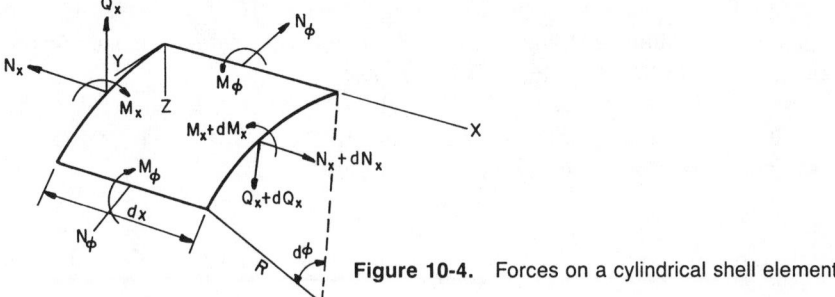

Figure 10-4. Forces on a cylindrical shell element.

Equation 10-15a indicates that the forces N_x are constant. The remaining two equations contain three unknowns, N_ϕ, Q_x and M_x. To solve the problem, the displacements of points in the middle surface of the shell are considered. Considering the displacements u and w in the x and z directions, Equations 10-15b and 10-15c can be reduced to the following single equation:

$$\frac{d^2}{dx^2}\left(D\,\frac{d^2w}{dx^2}\right) + \frac{Et}{R^2}\,w = p_z \tag{10-16a}$$

where $D = \dfrac{E\,t^3}{12(1 - \nu^2)}$

If the thickness of the shell t is constant, then

$$D\,\frac{d^4w}{dx^4} + \frac{Et}{R^2} = p_z \tag{10-16b}$$

let $\beta^4 = \dfrac{Et}{4R^2D} = \dfrac{3(1 - \nu^2)}{R^2t^2}$

Equation 10-16b is rewritten as

$$\frac{d^4w}{dx^4} + 4\,\beta^4\,w = \frac{p_z}{D} \tag{10-16c}$$

The general solution of Equation 10-16c is

$$w = e^{\beta x}(A\,\cos\beta x + B\,\sin\beta x) + e^{-\beta x}(C\,\cos\beta x + D\,\sin\beta x)$$
$$+ f(x) \tag{10-17}$$

where f(x) is a particular solution of Equation 10-16c, and A, B, C, and D are the constants of integration, which must be determined from the boundary conditions at the ends of the shell.

There is no change in curvature in the circumferential direction, and the curvature in the x direction is equal to $-d^2w/dx^2$. The three unknown quantities can be calculated as follows:

$$M_x = -D \frac{d^2w}{dx^2} \tag{10-18}$$

$$N_\phi = -\frac{Etw}{R} \tag{10-19}$$

$$Q_x = -D \frac{d^3w}{dx^3} \tag{10-20}$$

The bending moments M_ϕ, which are constant along the circumference, are calculated by

$$M_\phi = \nu M_x$$

Note that the deflection w is taken as positive toward the axis of the cylinder.

Example 10-1

A cylindrical tank with uniform wall thickness is filled with liquid as shown in Figure 10-5; derive the formulas for calculation of N_ϕ, M_0 and Q_0.

Let γ be the specific gravity of the liquid, then the liquid pressure at x is

$$p_z = -\gamma(d-x)$$

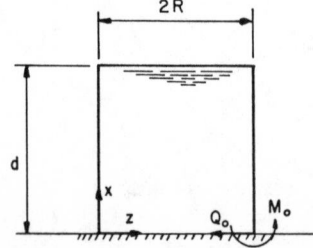

Figure 10-5. Cylindrical tank for Example 10-1.

Substituting p_z into Equation 10-16c we obtain

$$\frac{d^4w}{dx^4} + 4\beta^4 w = \frac{p_z}{D} \tag{10-21}$$

In most tanks where the thickness t is small in comparison with depth d and radius R, we may consider the cylindrical shell infinitely long. The constants A and B in Equation 10-17 are equal to zero. We obtain the following solution of Equation 10-21:

$$w = e^{-\beta x} (C \cos\beta x + D \sin\beta x) + f(x) \tag{10-22}$$

The particular solution $f(x) = -\gamma(d-x)R^2/Et$. The constants C and D can be obtained from the boundary conditions. The bottom of the tank is fixed to a rigid foundation. We have

$$w = 0 \text{ at } x = 0$$

$$dw/dx = 0 \text{ at } x = 0$$

$$C - \frac{\gamma R^2 d}{Et} = 0 \qquad C = \frac{\gamma R^2 d}{Et}$$

$$\beta(D-C) + \frac{\gamma R^2}{Et} = 0 \qquad D = \frac{\gamma R^2}{Et}\left(d - \frac{1}{\beta}\right)$$

Equation 10-22 becomes

$$w = -\frac{\gamma R^2}{Et}\left[d - x - de^{-\beta x}\cos\beta x - \left(d - \frac{1}{\beta}\right)e^{-\beta x}\sin\beta x\right]$$

From Equation 10-19 the force per unit length in the circumferential direction is

$$N_\phi = -\frac{Etw}{R}$$

$$= \gamma Rd\left[1 - \frac{x}{d} - e^{-\beta x}\cos\beta x - \left(1 - \frac{1}{\beta d}\right)e^{-\beta x}\sin\beta x\right] \tag{10-23}$$

From Equation 10-18, the bending moment per unit length is

$$M_x = -D \frac{d^2w}{dx^2}$$

$$= \frac{\gamma Rdt}{\sqrt{12(1 - \nu^2)}} \left[\left(1 - \frac{1}{\beta d} \right) e^{-\beta x} \cos\beta x - e^{-\beta x} \cos\beta x \right]$$

The bending moment at bottom $x = 0$ is

$$M_0 = (M_x)_{x=0} = \left(1 - \frac{1}{\beta d} \right) \frac{\gamma Rdt}{\sqrt{12(1 - \nu^2)}} \tag{10-24}$$

The particular solution $f(x) = -\gamma(d - x)R^2/Et$ is the radial expansion of a cylindrical shell with free edges under the action of hoop stresses. When $x = 0$,

$$f(x)_0 = -\frac{\gamma dR^2}{Et}, \qquad \left(\frac{df(x)}{dx} \right)_{x=0} = \frac{\gamma R^2}{Et}$$

The deflection and slope at the bottom will be eliminated by the combined action of Q_0 and M_0. The radial displacement and slopes by Q_0 and M_0 can be expressed as follows [1]:

$$(w)_{x=0} = \frac{-1}{2\beta^3 D} (Q_0 + \beta M_0)$$

$$\left(\frac{dw}{dx} \right)_{x=0} = \frac{1}{2\beta^2 D} (Q_0 + 2\beta M_0)$$

We have

$$f(x)_0 = -(w)_{x=0}$$

$$\left(\frac{df(x)}{dx} \right)_{x=0} = \left(\frac{dw}{dx} \right)_{x=0}$$

From these equations we obtain

$$Q_0 = \frac{\gamma Rdt}{\sqrt{12(1 - \nu^2)}} \left(2\beta - \frac{1}{d} \right) \tag{10-25}$$

Equations 10-23 through 10-25 have been derived in Chapter 15 of *Theory of Plates and Shells* [1].

DATA FOR THIN SHELLS [1,4,7]

When a shell is subjected to internal or external pressure, stresses are set up in the wall. At any point of the shell there is a meridional force acting parallel to the meridian, a circumferential or hoop force acting parallel to the circumference, and a radial force. In addition, there may be bending moments and shear forces. When the thickness of the shell is small (less than about one-tenth the radius), the stresses can be assumed to be uniformly distributed across the thickness, and the bending and radial stresses are negligibly small. The meridional and hoop stresses are the only important ones present. Tables and graphs for selected cases of thin shell are developed and provided for practical application. Here, Poisson's ratio ν has been taken as 0.3, the usual value for steel. In the tabulation, ν has been absorbed in the numerical coefficients. The coefficients K_ϕ, K_θ, K_w, etc., are dimensionless; they can be applied in metric or Imperial units. Note that ordinates of some graphs have been factored for easier reading. In Case 4, for example, $K_w \times 10^{-1} = 8.589$; that is, $K_w = 85.89$. The symbols in the equations are defined as follows:

N_ϕ, N_θ = meridional force and hoop force per unit length of the principal normal section of the shell

$N_{\phi\theta}$ = shear force per unit length of section

N_ϕ, N_θ = membrane factors per unit length of section in the hoop and axial directions, respectively, of a cylindrical shell

σ_t = membrane stress in the hoop direction (hoop stress)

σ_x = membrane stress in the axial direction of a cylindrical shell

q = load per unit area

p = pressure per unit area

M_0, Q_0 = edge moment and shear per unit length of section

R, t = radius and thickness of shell, respectively

w = radial displacement

γ = weight per unit volume of liquid

β = a notation defined by $\beta^4 = 3(1 - \nu^2)/(R\,t)^2$ where ν is Poisson's ratio

K_x, K_t = dimensionless stress factors

K_ϕ, K_θ = dimensionless force factors

K_M = dimensionless moment factor

$K_{\phi\theta}$ = dimensionless shear factor

K_w = dimensionless displacement factor

K_p, K_Q = dimensionless force and shear factors

Example 10-2

A spherical dome loaded with 50 lb/ft^2 dead load and 60 lb/ft^2 live load is supported on a ring as shown in Figure 10-6; calculate the reactions.

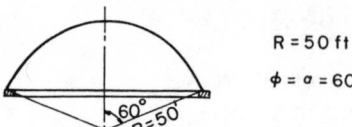

R = 50 ft

$\phi = \sigma = 60°$

60°
R=50

Figure 10-6. Spherical dome for Example 10-2.

$q = 50 + 60 = 110 \text{ lb/ft}^2$

From Case 1

$N_\phi = -RqK_\phi \qquad K_\phi = 0.667$
$N_\phi = -50 \times 110 \times 0.667 = -3,669 \text{ lb/ft}$
$N_\theta = RqK_\theta \qquad K_\theta = 0.167$
$N_\theta = 50 \times 110 \times 0.167 = 918 \text{ lb/ft}$
$F_H = N_\phi \sin 30 = -3,669 \times 0.5 = -1,835 \text{ lb/ft}$
$F_v = N_\phi \cos 30 = -3,669 \times 0.866 = -3,177 \text{ lb/ft}$

The horizontal components of the forces N_ϕ are taken by the supporting ring which undergoes a uniform circumferential extension. The vertical components of the forces N_ϕ are imposed on the dome by supports.

Example 10-3

A cylindrical tank filled with water is built into a rigid foundation as shown in Figure 10-7; calculate the bending moment and shear force at the bottom.

$$\beta^4 = \frac{3(1-\nu^2)}{R^2\,t^2} = \frac{3 \times 0.91}{240^2 \times 12^2} = 3.219 \times 10^{-7}$$

$$\beta = 0.02395 \text{ in.}^{-1}$$

$$\beta d = 0.02395 \times 25 \times 12 = 7.185$$

From Case 22

$$K_M' = 0.2597 \qquad K_Q = 4.046$$

$$M_0 = Rd\,\gamma\,tK_M' = 20 \times 25 \times 62.4 \times 1 \times 0.2597 = 8,103 \text{ lb-ft/ft}$$

$$Q_0 = R\,\gamma\,tK_Q = 20 \times 62.4 \times 1 \times 4.046 = 5,050 \text{ lb/ft}$$

If $\nu = 0.25$ is used in the calculation, M_0 and Q_0 will decrease about 1.5%.

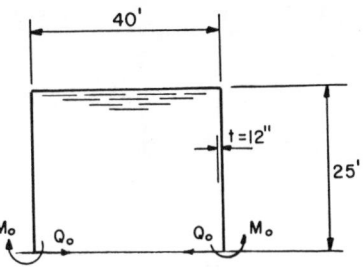

Figure 10-7. Cylindrical tank for Example 10-3.

REFERENCES

1. Timoshenko, S. and Woinowsky-Krieger, S., *Theory of Plates and Shells,* McGraw-Hill Book Company, 1959.
2. Billington, D., *Thin Shell Concrete Structures,* McGraw-Hill Book Company, 1965.
3. Den Hartog, J. P., *Advanced Strength of Materials,* McGraw-Hill Book Company, 1952.
4. Roark, R. J., *Formulas for Stress and Strain,* Fourth Edition, McGraw-Hill Book Company, 1965.
5. Timoshenko, S. and Gere, J. M., *Theory of Elastic Stability,* McGraw-Hill Book Company, 1964.
6. Flugge, W., *Stresses in Shell,* Springer-Verlag, New York, 1966.
7. Roark, R. J. and Young, W. C., *Formulas for Stress and Strain,* Fifth Edition, McGraw-Hill Book Company, 1975.

Cases for Chapter 10

Case 1   Spherical dome loaded by own weight q
per unit area

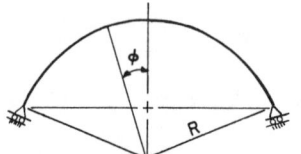

$$N_\phi = -Rq K_\phi$$
$$N_\theta = Rq K_\theta$$

ϕ	0°	10°	20°	30°	40°
K_ϕ	0.500	0.5038	0.5155	0.5359	0.5662
K_θ	−0.500	−0.4810	−0.4241	−0.3301	−0.1998

50°	60°	70°	80°	90°
0.6087	0.6667	0.7451	0.8520	1.000
−0.0341	0.1667	0.4031	0.6784	1.000

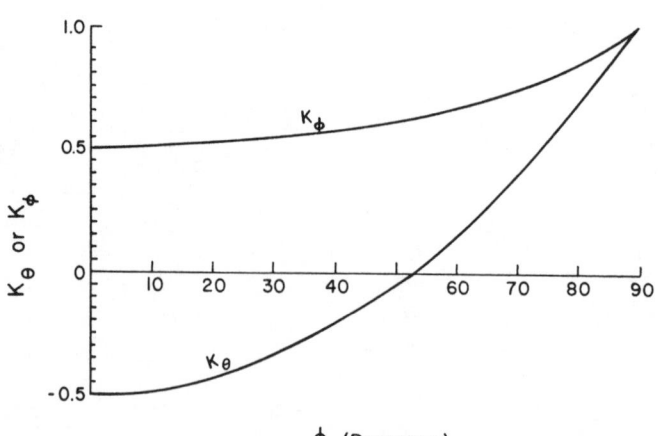

ϕ (Degrees)

Case 2   Spherical dome loaded with q per unit area of horizontal projection

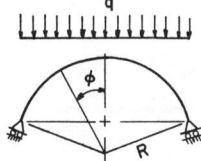

$$N_\phi = \frac{qR}{2}$$

$$N_\theta = \frac{qR}{2} K_\theta$$

ϕ	0°	10°	20°	30°	40°	45°	50°	60°	70°	80°	90°
K_θ	1.000	0.940	0.766	0.500	0.174	0	−0.174	−0.500	−0.766	0.940	−1.000

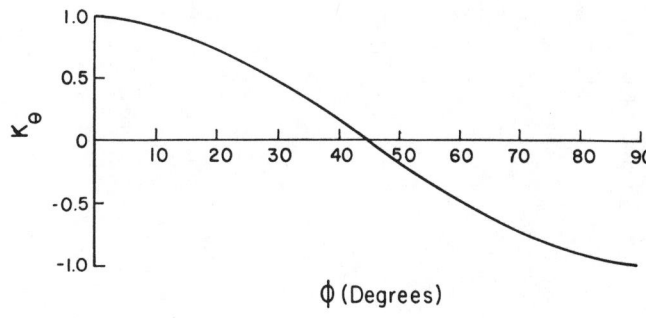

ϕ (Degrees)

Case 3   Spherical dome loaded with Q_0 per unit length

$$w = \frac{Q_0 R}{E t} K_w$$

$$(\sigma_x)_{max} = Q_0 \frac{\cos\phi}{t} \quad \text{at edge}$$

$$(\sigma_t)_{max} = -\frac{Q_0}{t} K_t \quad \text{at edge}$$

K_t

ϕ R/t	15°	30°	45°	60°	75°	90°
10	1.858	3.821	5.543	6.893	7.776	8.130
20	2.715	5.500	7.923	9.809	11.028	11.497
30	3.378	6.790	9.749	12.046	13.524	14.081
40	3.938	7.878	11.288	13.932	15.628	16.259
50	4.432	8.836	12.645	15.594	17.482	18.178
60	4.880	9.703	13.872	17.097	19.158	19.913
70	5.292	10.501	15.000	18.479	20.699	21.509
80	5.675	11.243	16.050	19.765	22.134	22.994
90	6.035	11.940	17.036	20.972	23.480	24.389
100	6.376	12.599	17.969	22.115	24.755	25.708

K_w

ϕ R/t	15°	30°	45°	60°	75°	90°
10	0.406	1.780	3.770	5.840	7.436	8.130
20	0.628	2.620	5.452	8.365	10.577	11.497
30	0.799	3.265	6.744	10.304	12.988	14.081
40	0.944	3.809	7.832	11.936	15.020	16.259
50	1.072	4.288	8.792	13.375	16.811	18.178
60	1.188	4.722	9.659	14.676	18.431	19.913
70	1.295	5.120	10.456	15.873	19.919	21.509
80	1.394	5.492	11.199	16.987	21.305	22.994
90	1.487	5.840	11.896	18.033	22.605	24.389
100	1.575	6.170	12.556	19.023	23.836	25.708

Case 3 continued

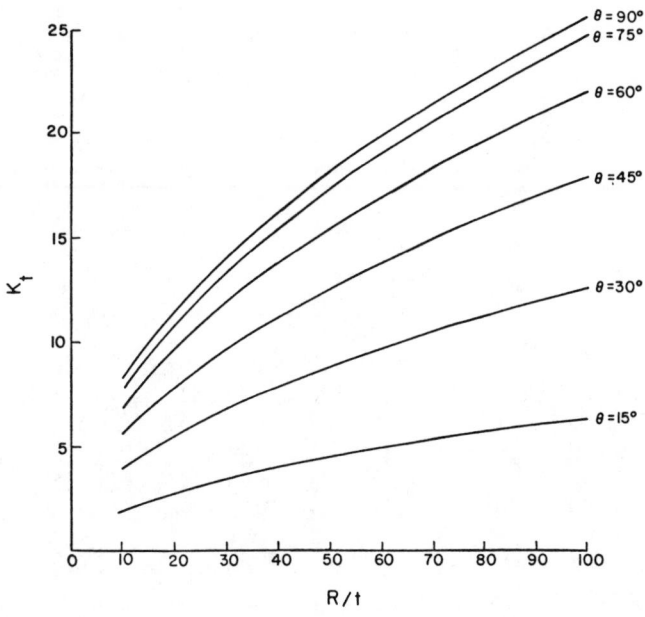

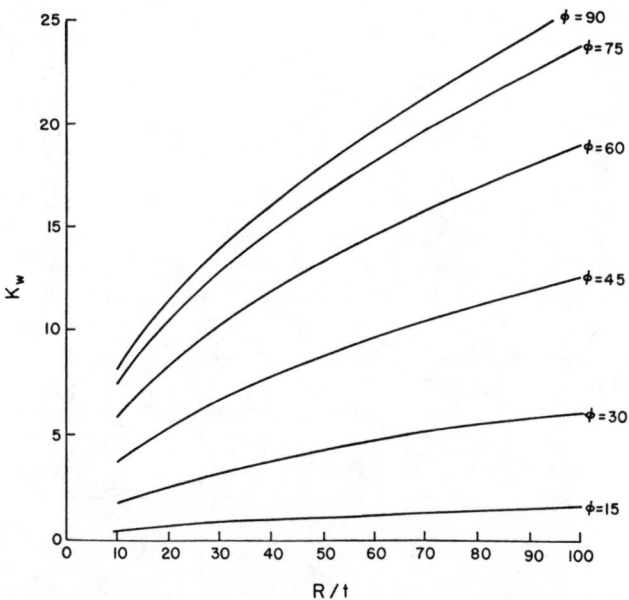

Case 4   Spherical shell loaded with uniform moment M_o per unit length at edge

$$w = \frac{M_o}{Et} K_w$$

$$(\sigma_t)_{max} = \frac{M_o}{tR} K_M$$

K_w

ϕ R/t	15°	30°	45°	60°	75°	90°
10	10.476	18.058	24.571	29.454	32.346	33.045
20	19.684	35.166	48.418	58.410	64.438	66.091
30	28.704	52.134	72.150	87.287	96.492	99.137
40	37.669	69.031	95.823	116.120	128.525	132.181
50	46.589	85.886	119.461	144.931	160.545	165.227
60	55.477	102.710	143.075	173.725	192.725	198.272
70	64.334	119.508	166.667	202.494	224.559	231.319
80	73.172	136.300	190.246	231.284	256.564	264.367
90	81.994	153.053	213.809	260.035	288.547	297.412
100	90.804	169.800	237.371	288.782	320.542	330.456

K_M

ϕ R/t	15°	30°	45°	60°	75°	90°
10	40.477	36.115	34.748	34.011	33.487	33.045
20	75.949	70.331	68.473	67.446	66.771	66.091
30	110.904	104.267	102.035	100.790	99.896	99.137
40	145.542	138.062	135.514	134.084	133.059	132.181
50	180.006	171.772	168.944	167.352	166.208	165.227
60	214.348	205.420	202.339	200.600	199.348	198.272
70	248.569	239.015	235.703	233.820	232.481	231.319
80	282.715	272.599	269.048	267.064	265.615	264.367
90	316.800	306.105	302.371	300.264	298.726	297.412
100	350.840	339.591	335.693	333.457	331.850	330.456

Case 4 continued

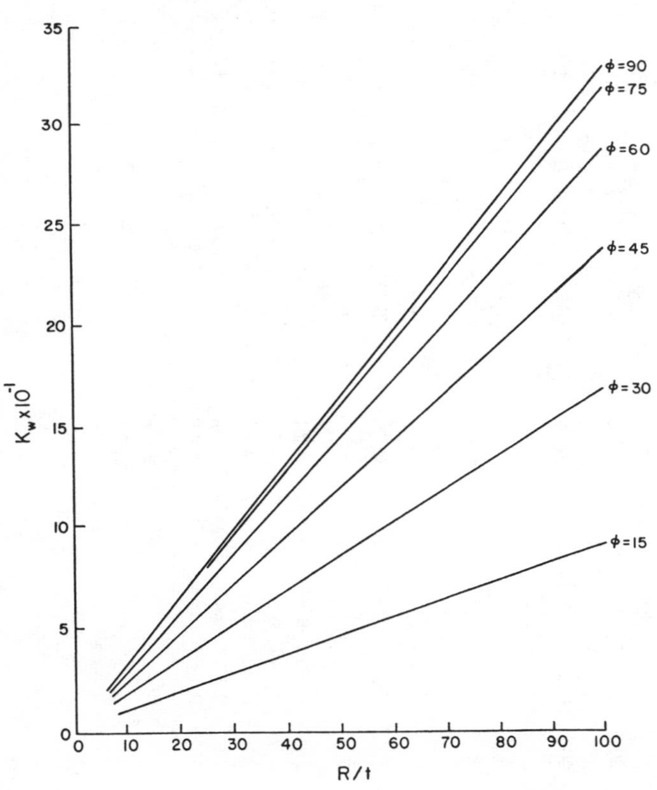

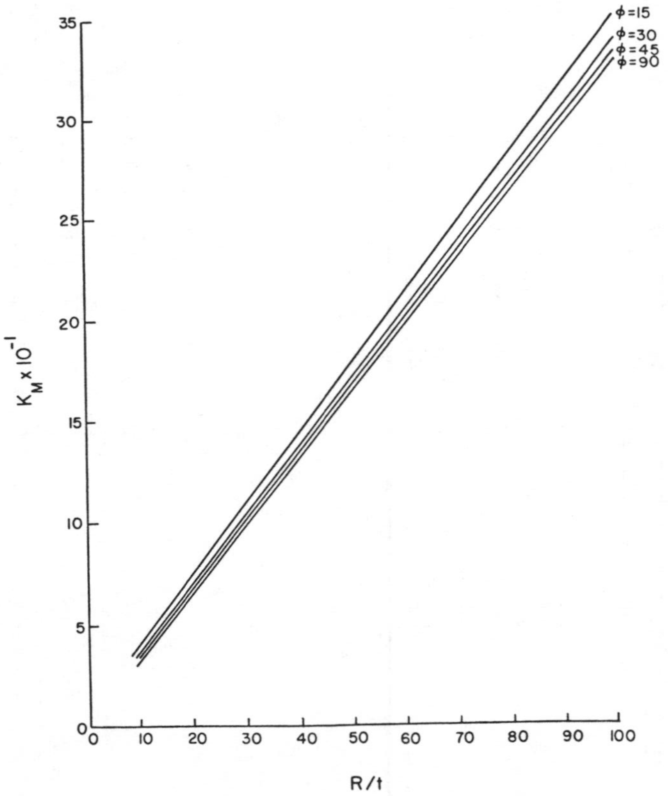

Case 5   Open spherical dome loaded with p per unit length and uniform load q per unit area (own weight)

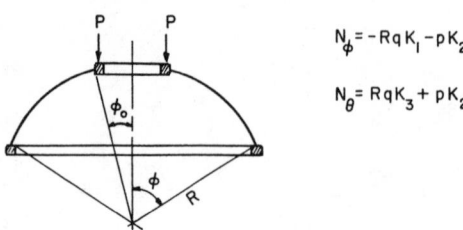

$$N_\phi = -RqK_1 - pK_2$$

$$N_\theta = RqK_3 + pK_2$$

K₁

φ φ₀	90°	85°	80°	75°	70°	65°	60°	55°	50°	45°
5°	0.9962	0.9160	0.8481	0.7903	0.7408	0.6983	0.6616	0.6298	0.6022	0.5782
10°	0.9848	0.9045	0.8364	0.7781	0.7279	0.6844	0.6464	0.6129	0.5828	0.5554
15°	0.9659	0.8855	0.8169	0.7579	0.7066	0.6614	0.6212	0.5847	0.5507	0.5176
20°	0.9397	0.8591	0.7899	0.7298	0.6768	0.6295	0.5863	0.5456	0.5060	0.4652
25°	0.9063	0.8254	0.7554	0.6940	0.6390	0.5889	0.5417	0.4959	0.4491	0.3984
30°	0.8660	0.7848	0.7139	0.6508	0.5934	0.5398	0.4880	0.4358	0.3804	0.3178

K₂

φ φ₀	90°	85°	80°	75°	70°	65°	60°	55°	50°	45°
5°	0.0872	0.0878	0.0899	0.0934	0.0987	0.1061	0.1162	0.1299	0.1485	0.1743
10°	0.1736	0.1750	0.1790	0.1861	0.1967	0.2114	0.2315	0.2588	0.2959	0.3473
15°	0.2588	0.2608	0.2669	0.2774	0.2931	0.3151	0.3451	0.3857	0.4411	0.5176
20°	0.3420	0.3446	0.3527	0.3666	0.3873	0.4164	0.4560	0.5097	0.5828	0.6840
25°	0.4226	0.4259	0.4358	0.4530	0.4786	0.5145	0.5635	0.6298	0.7202	0.8452
30°	0.5000	0.5038	0.5155	0.5359	0.5662	0.6087	0.6667	0.7451	0.8520	1.0000

Case 5 continued

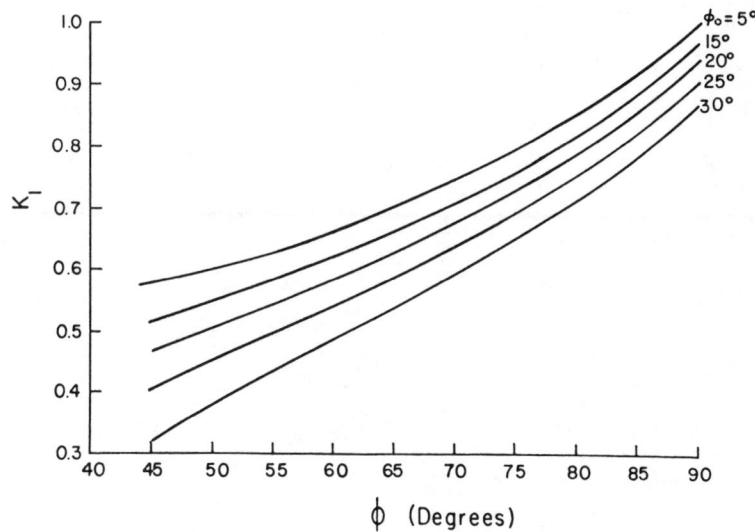

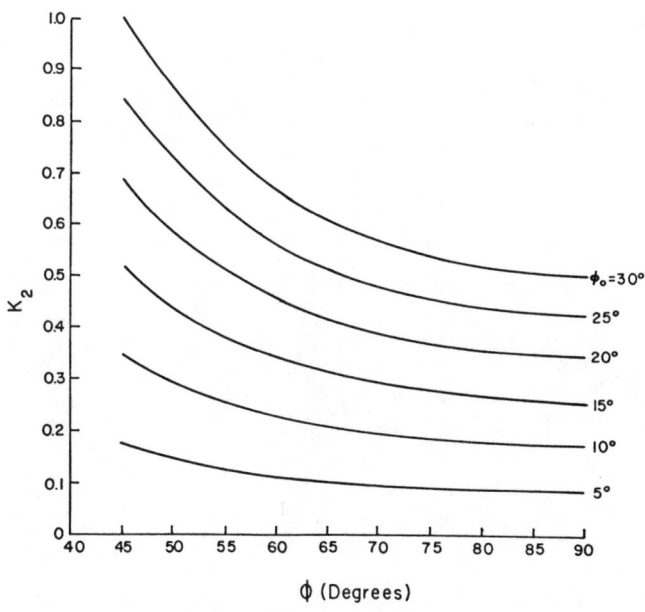

Case 5 continued

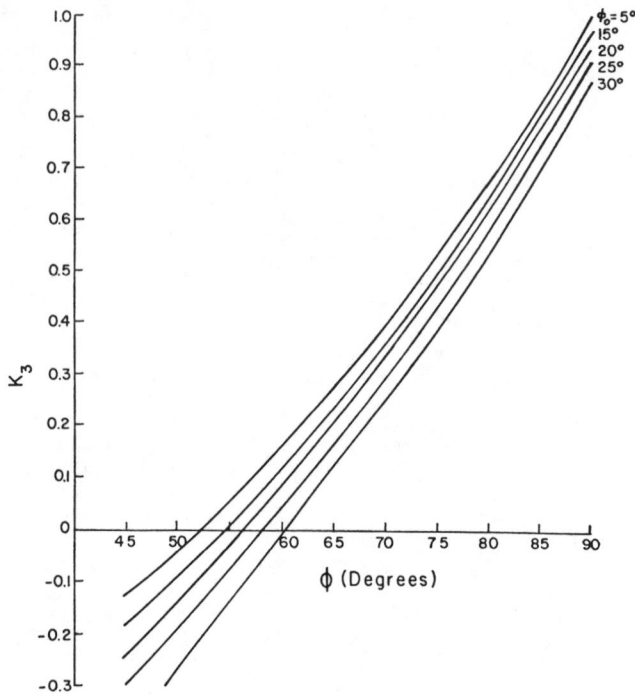

$$K_3$$

ϕ_0	ϕ 90°	85°	80°	75°	70°	65°	60°	55°	50°	45°
5°	0.9962	0.8288	0.6745	0.5315	0.3988	0.2757	0.1616	0.0562	−0.0406	−0.1289
10°	0.9848	0.8173	0.6627	0.5193	0.3859	0.2618	0.1464	0.0393	−0.0600	−0.1517
15°	0.9659	0.7983	0.6433	0.4991	0.3645	0.2388	0.1212	0.0114	−0.0921	−0.1895
20°	0.9379	0.7719	0.6162	0.4709	0.3348	0.2069	0.0863	−0.0280	−0.1368	−0.2419
25°	0.9063	0.7382	0.5818	0.4352	0.2970	0.1662	0.0417	−0.0777	−0.1937	−0.3087
30°	0.8660	0.6976	0.5403	0.3920	0.2514	0.1172	−0.0120	−0.1377	−0.2624	−0.3893

Case 6   Spherical dome with wind load

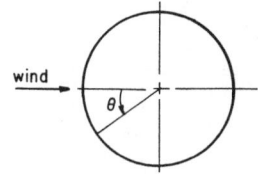

$$N_\phi = -\frac{pR}{3}K_\phi$$

$$N_{\phi\theta} = -\frac{pR}{3}K_{\phi\theta}$$

$$N_\theta = -\frac{pR}{3}K_\theta$$

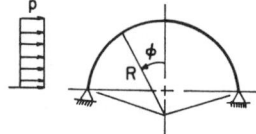

$\mathbf{K}_\phi$

θ \ ϕ	0°	15°	30°	45°	60°	70°	75°	80°	90°
0°	0	0.1919	0.3564	0.4645	0.4811	0.4179	0.3564	0.2699	0
30°	0	0.1662	0.3087	0.4022	0.4167	0.3619	0.3086	0.2337	0
45°	0	0.1357	0.2520	0.3284	0.3402	0.2955	0.2520	0.1908	0
60°	0	0.0960	0.1782	0.2322	0.2406	0.2090	0.1782	0.1349	0
90°	0	0	0	0	0	0	0	0	0

$\mathbf{K}_{\phi\theta}$

θ \ ϕ	0°	15°	30°	45°	60°	65°	70°	80°	90°
0°	0	0	0	0	0	0	0	0	0
30°	0	0.0993	0.2058	0.3285	0.4811	0.5424	0.6109	0.7772	1.000
45°	0	0.1404	0.2910	0.4645	0.6804	0.7671	0.8640	1.0990	1.4142
60°	0	0.1720	0.3564	0.5689	0.8333	0.9395	1.0582	1.3461	1.7321
75°	0	0.1918	0.3975	0.6345	0.9294	1.0479	1.1802	1.5013	1.9319
90°	0	0.1986	0.4115	0.6569	0.9624	1.0849	1.2219	1.5543	2.000

$\mathbf{K}_\theta$

θ \ ϕ	0°	15°	30°	45°	60°	65°	70°	80°	90°
0°	0	0.5845	1.1436	1.6569	2.1170	2.2604	2.4011	2.6846	3.000
30°	0	0.5063	0.9904	1.4349	1.8333	1.9576	2.0794	2.3249	2.5981
45°	0	0.4133	0.8086	1.1716	1.4969	1.5984	1.6979	1.8983	2.1213
60°	0	0.2923	0.5718	0.8284	1.0585	1.1302	1.2006	1.3423	1.500
75°	0	0.1513	0.2960	0.4288	0.5479	0.5850	0.6215	0.6948	0.7765
90°	0	0	0	0	0	0	0	0	0

Case 6 continued

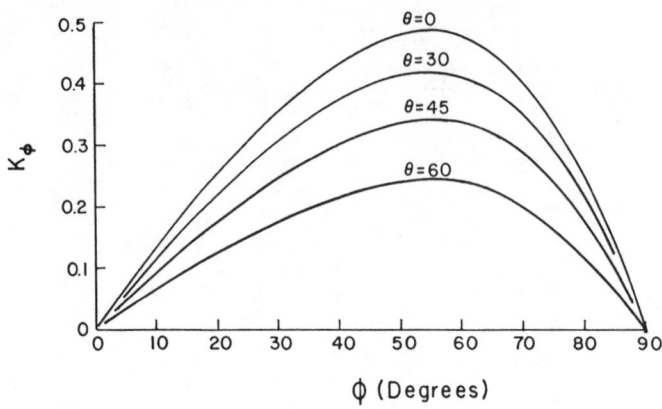

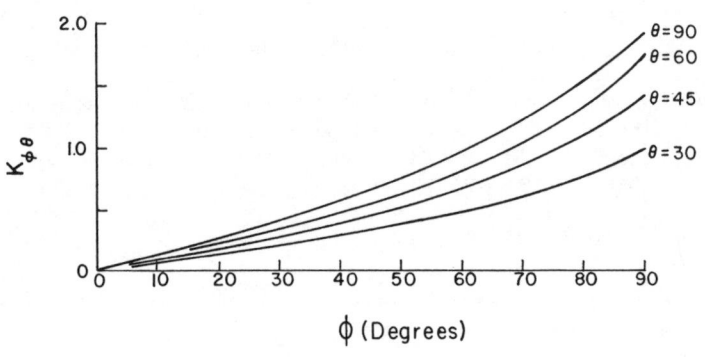

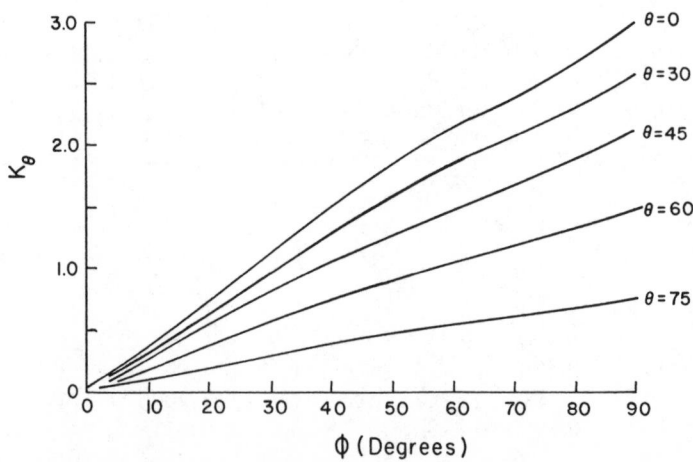

Case 7   Spherical tank filled with liquid

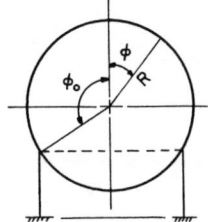

$$N_\phi = \gamma R^2 K_\phi$$
$$N_\theta = \gamma R^2 K_\theta$$

$$\phi \leq \phi_0$$

ϕ	0°	15°	30°	45°	60°	75°	90°	105°	120°	135°	150°
K_ϕ	0	0.0085	0.0327	0.0690	0.1111	0.1489	0.1667	0.1365	0	−0.4024	−1.6994
K_θ	0	0.0256	0.1013	0.2239	0.3889	0.5923	0.8333	1.1223	1.500	2.1095	3.5654

$$\phi > \phi_0$$

ϕ	90°	105°	120°	135°	150°	165°	180°
K_ϕ	0.8333	0.8511	0.8889	0.9310	0.9673	0.9915	1.000
K_θ	0.1667	0.4078	0.6111	0.7761	0.8987	0.9714	1.000

Case 7 continued

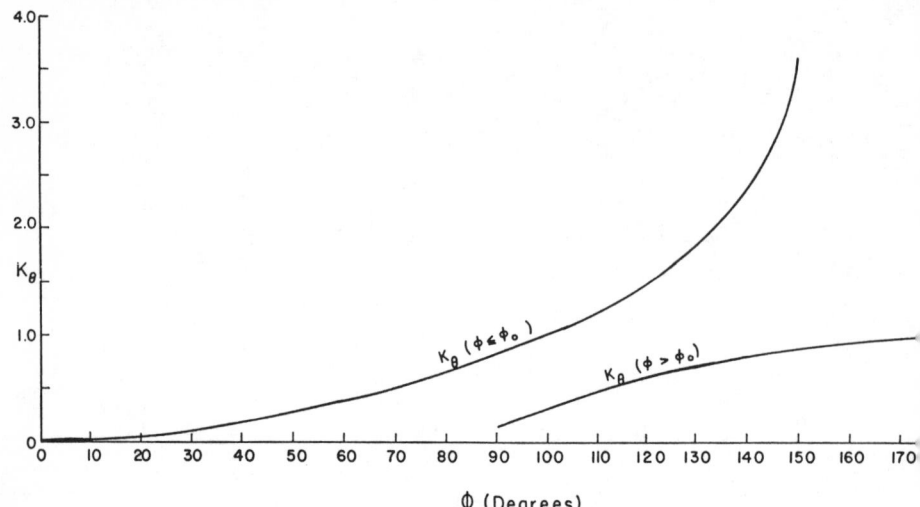

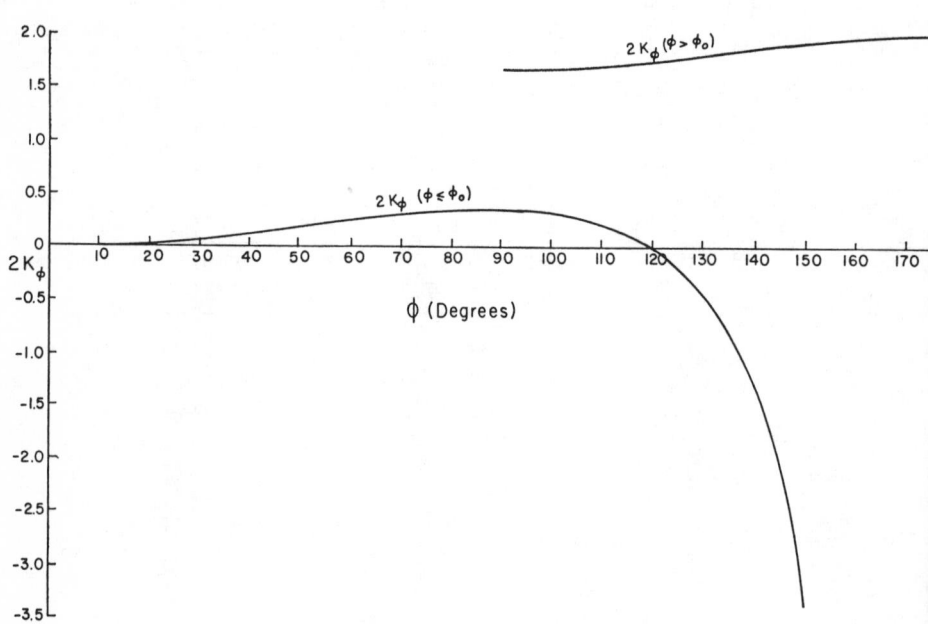

Case 8   Conical shell filled with liquid

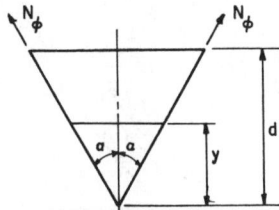

$$N_\phi = \frac{\gamma d^2}{10} K_\phi$$

$$N_\theta = \frac{\gamma d^2}{10} K_\theta$$

$$C = \frac{y}{d}$$

K_ϕ

C \ α	10°	20°	30°	40°	50°
0.2	0.155	0.336	0.578	0.949	1.607
0.4	0.263	0.568	0.978	1.607	2.719
0.6	0.322	0.697	1.200	1.972	3.337
0.8	0.334	0.723	1.244	2.045	3.461
1.0	0.298	0.646	1.111	1.827	3.090

K_θ

C \ α	10°	20°	30°	40°	50°
0.2	0.286	0.620	1.067	1.753	2.966
0.4	0.430	0.930	1.600	2.629	4.450
0.6	0.430	0.930	1.600	2.629	4.450
0.8	0.286	0.620	1.067	1.753	2.966
1.0	0	0	0	0	0

Case 8 continued

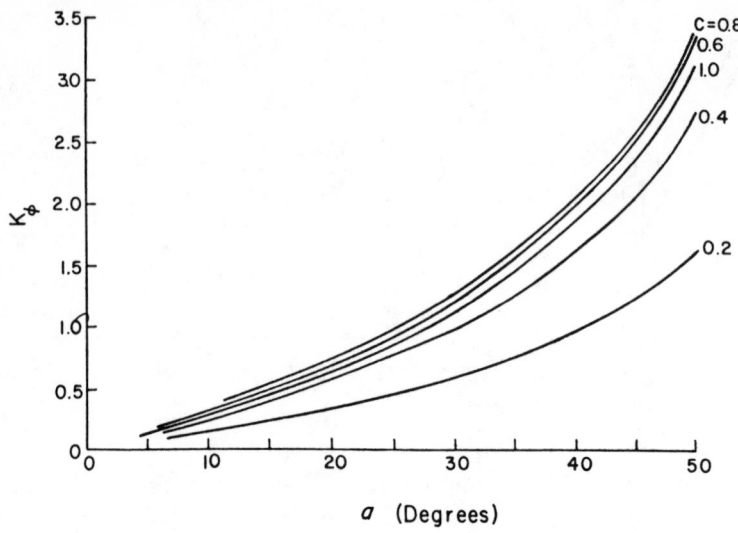

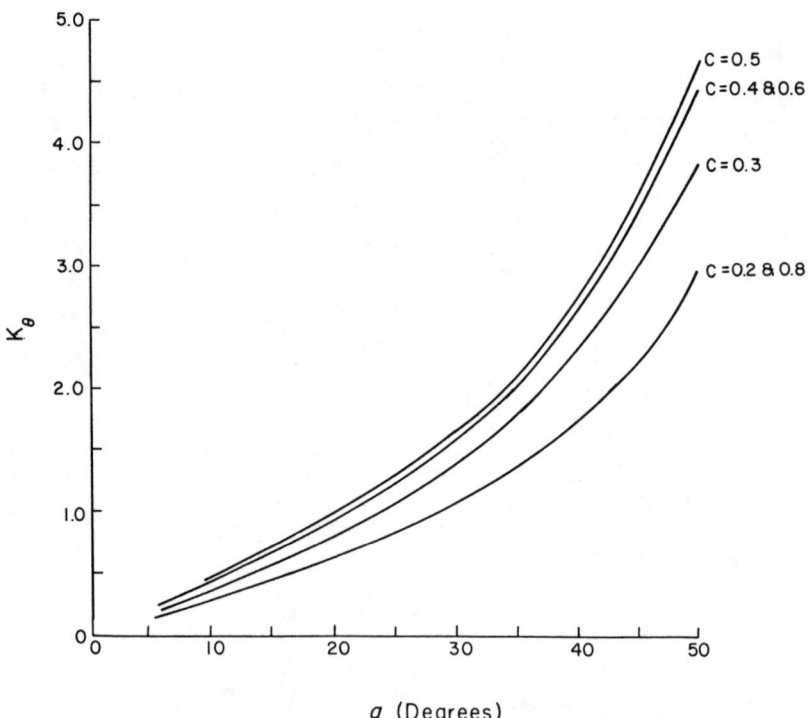

Case 9 E llipsoid boiler end

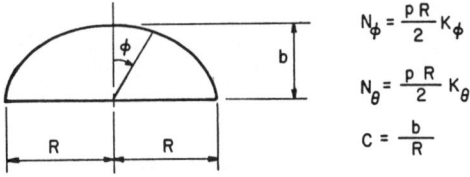

$$N_\phi = \frac{pR}{2} K_\phi$$

$$N_\theta = \frac{pR}{2} K_\theta$$

$$C = \frac{b}{R}$$

$\mathbf{K}_\phi$

C \ ϕ	0°	10°	20°	30°	40°	50°	60°	70°	80°	90°
0.2	5.000	3.808	2.562	1.890	1.513	1.287	1.147	1.061	1.015	1.000
0.4	2.500	2.323	1.968	1.644	1.404	1.238	1.125	1.053	1.013	1.000
0.5	2.000	1.915	1.721	1.512	1.336	1.204	1.109	1.047	1.012	1.000
0.6	1.667	1.624	1.516	1.387	1.266	1.166	1.091	1.040	1.010	1.000
0.7	1.429	1.407	1.349	1.273	1.195	1.126	1.071	1.031	1.008	1.000
0.8	1.250	1.240	1.211	1.170	1.126	1.084	1.048	1.022	1.006	1.000

$\mathbf{K}_\theta$

C \ ϕ	0°	10°	20°	30°	40°	50°	60°	70°	80°	90°
0.2	1.000	0.211	−0.926	−1.890	−2.699	−3.369	−3.900	−4.286	−4.521	−4.600
0.4	1.000	0.782	0.304	−0.206	−0.657	−1.030	−1.320	−1.532	−1.658	−1.700
0.5	1.000	0.871	0.558	0.189	−0.160	−0.458	−0.693	−0.863	−0.966	−1.000
0.6	1.000	0.922	0.721	0.462	0.202	−0.030	−0.218	−0.356	−0.439	−0.467
0.7	1.000	0.954	0.829	0.659	0.477	0.307	0.164	0.058	−0.007	−0.029
0.8	1.000	0.975	0.905	0.805	0.691	0.581	0.485	0.411	0.366	0.350

Case 9 continued

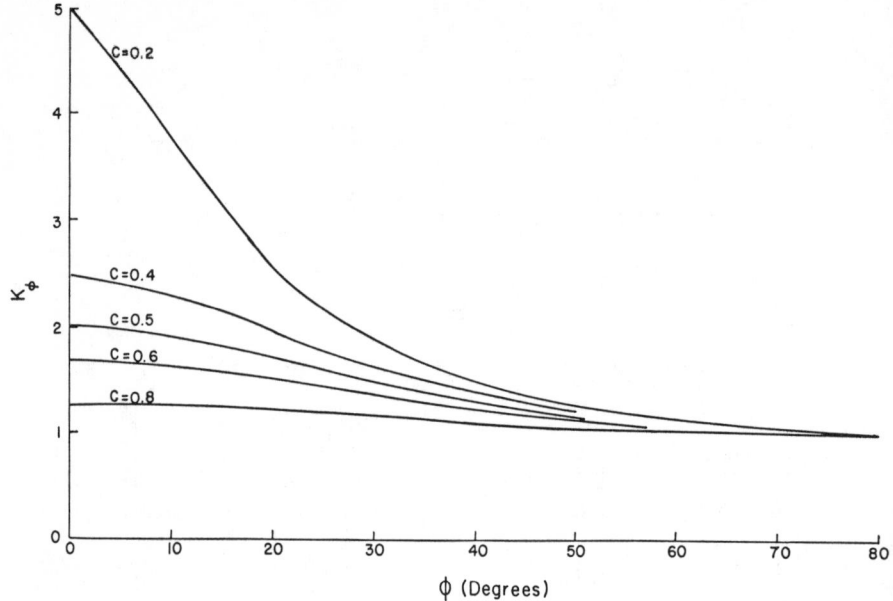

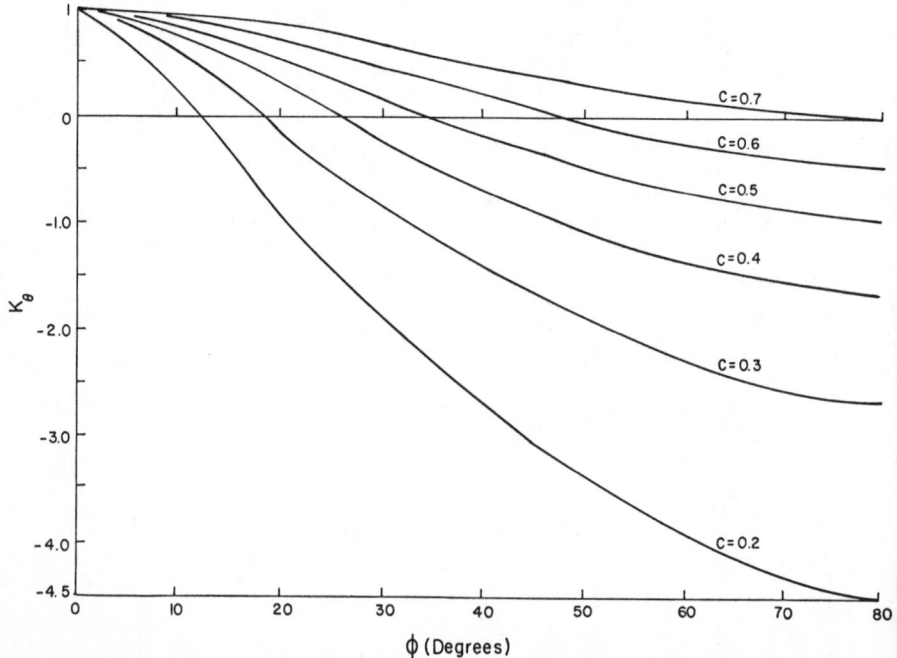

Case 10   Hemispherical tank filled with liquid

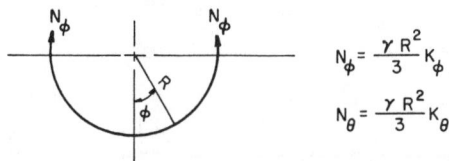

$$N_\phi = \frac{\gamma R^2}{3} K_\phi$$

$$N_\theta = \frac{\gamma R^2}{3} K_\theta$$

ϕ	1°	10°	20°	30°	40°
K_ϕ	1.500	1.4886	1.4552	1.4019	1.3323
K_θ	1.500	1.4658	1.3638	1.1962	0.9659

50°	60°	70°	80°	85°	90°
1.2515	1.1667	1.0872	1.0257	1.0070	1.000
0.6769	0.3333	−0.0611	−0.5047	−0.7455	−1.00

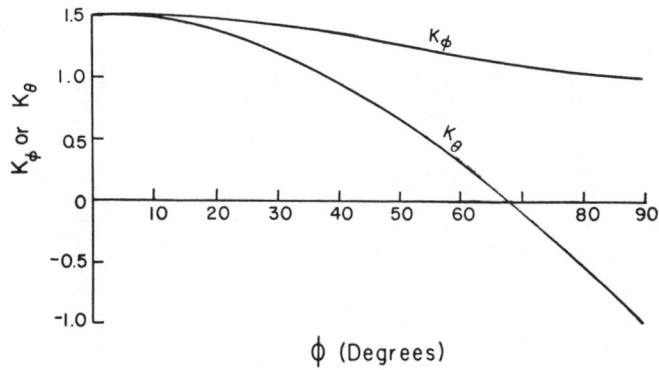

ϕ (Degrees)

Case 11 Long circular pipe uniformly loaded with p along the edge.

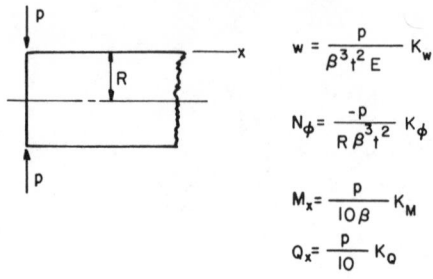

$$w = \frac{p}{\beta^3 t^2 E} K_w$$

$$N_\phi = \frac{-p}{R\beta^3 t^2} K_\phi$$

$$M_x = \frac{p}{10\beta} K_M$$

$$Q_x = \frac{p}{10} K_Q$$

βx	$K_w = K_\phi$	K_M	K_Q
0.0	5.460	0.000	10.000
0.2	4.381	1.627	6.398
0.4	3.371	2.610	3.564
0.6	2.473	3.099	1.431
0.8	1.710	3.223	−0.093
1.0	1.085	3.096	−1.108
1.2	0.596	2.807	−1.716
1.4	0.229	2.430	−2.011
1.6	−0.032	2.018	−2.077
1.8	−0.205	1.610	−1.985
2.0	−0.307	1.230	−1.794
2.2	−0.356	0.895	−1.548
2.4	−0.365	0.613	−1.282
2.6	−0.347	0.383	−1.019
2.8	−0.313	0.204	−0.777
3.0	−0.269	0.071	−0.563
3.2	−0.222	−0.024	−0.383
3.4	−0.176	−0.085	−0.237
3.6	−0.134	−0.121	−0.124
3.8	−0.097	−0.137	−0.040
4.0	−0.066	−0.139	−0.019

Case 11 continued

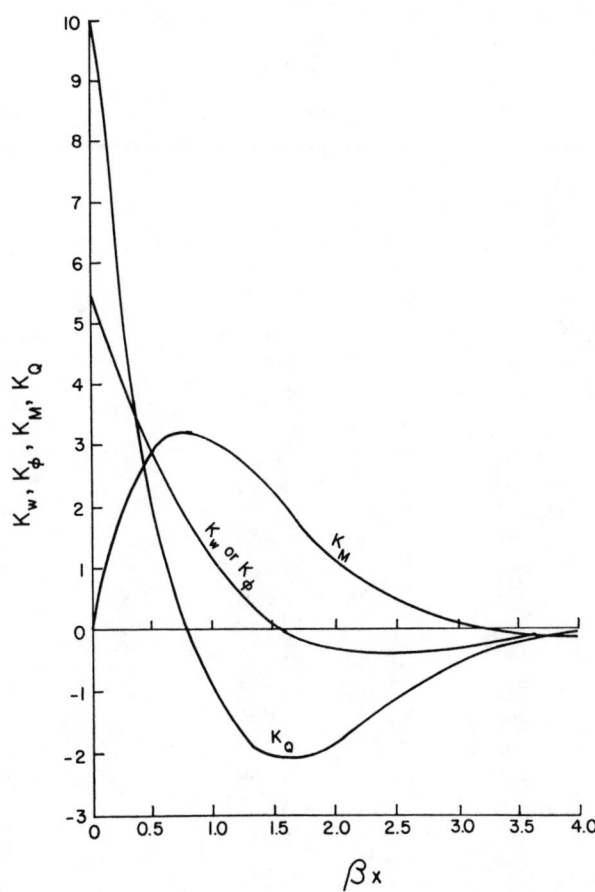

Case 12 Long circular pipe uniformly loaded with M_o at edge

$$w = \frac{M_o}{\beta^2 t^3 E} K_w$$

$$N_\phi = \frac{M_o}{R\beta^2 t^2} K_\phi$$

$$M_x = \frac{M_o}{10} K_M$$

$$Q_x = \frac{-\beta M_o}{5} K_Q$$

βx	$K_w = K_\phi$	K_M	K_Q
0	5.460	10.000	0.000
0.2	3.493	9.651	1.627
0.4	1.946	8.784	2.610
0.6	0.781	7.628	3.099
0.8	−0.051	6.354	3.223
1.0	−0.605	5.083	3.096
1.2	−0.937	3.899	2.807
1.4	−1.098	2.849	2.430
1.6	−1.134	1.959	2.018
1.8	−1.084	1.234	1.610
2.0	−0.980	0.667	1.230
2.2	−0.845	0.244	0.895
2.4	−0.700	−0.056	0.613
2.6	−0.556	−0.254	0.383
2.8	−0.420	−0.369	0.204
3.0	−0.307	−0.423	0.071
3.2	−0.209	−0.431	−0.024
3.4	−0.129	−0.408	−0.085
3.6	−0.068	−0.366	−0.121
3.8	−0.022	−0.314	−0.137
4.0	−0.010	−0.258	−0.139

Case 12 continued

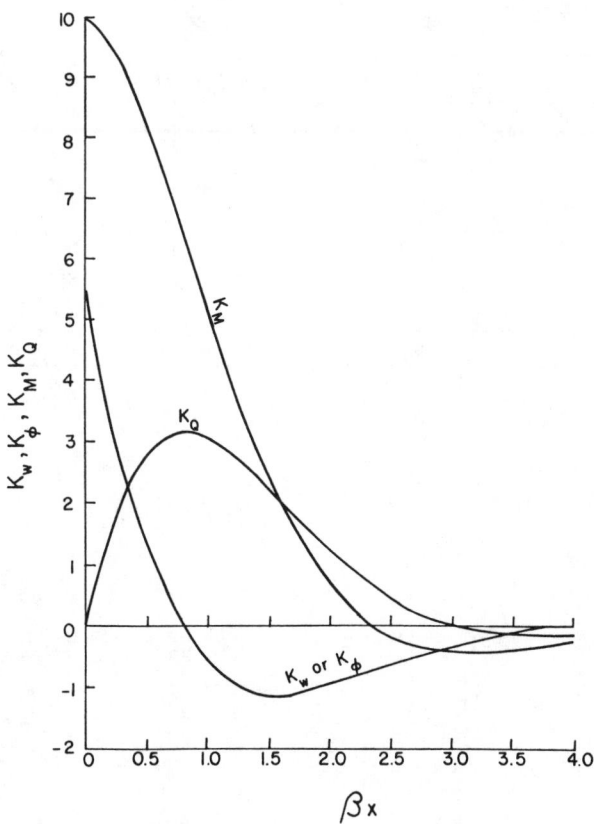

Case 13 Long circular pipe uniformly loaded along a circular section

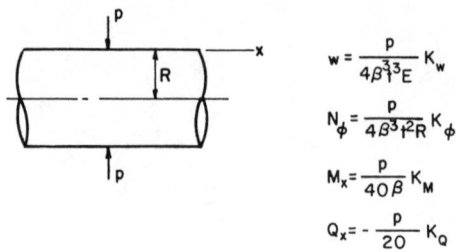

$$w = \frac{p}{4\beta^3 E} K_w$$

$$N_\phi = \frac{p}{4\beta^3 t^2 R} K_\phi$$

$$M_x = \frac{p}{40\beta} K_M$$

$$Q_x = -\frac{p}{20} K_Q$$

βx	$K_w = K_\phi$	K_M	K_Q
0	5.460	10.000	10.000
0.2	5.269	6.398	8.024
0.4	4.796	3.564	6.174
0.6	4.165	1.431	4.530
0.8	3.469	−0.093	3.131
1.0	2.775	−1.108	1.988
1.2	2.129	−1.716	1.091
1.4	1.556	−2.011	0.419
1.6	1.070	−2.077	−0.059
1.8	0.674	−1.985	−0.376
2.0	0.364	−1.794	−0.563
2.2	0.133	−1.548	−0.652
2.4	−0.031	−1.282	−0.669
2.6	−0.139	−1.019	−0.636
2.8	−0.202	−0.777	−0.573
3.0	−0.231	−0.563	−0.493
3.2	−0.235	−0.383	−0.407
3.4	−0.223	−0.237	−0.323
3.6	−0.200	−0.124	−0.245
3.8	−0.171	−0.040	−0.177
4.0	−0.141	−0.019	−0.120

Case 13 continued

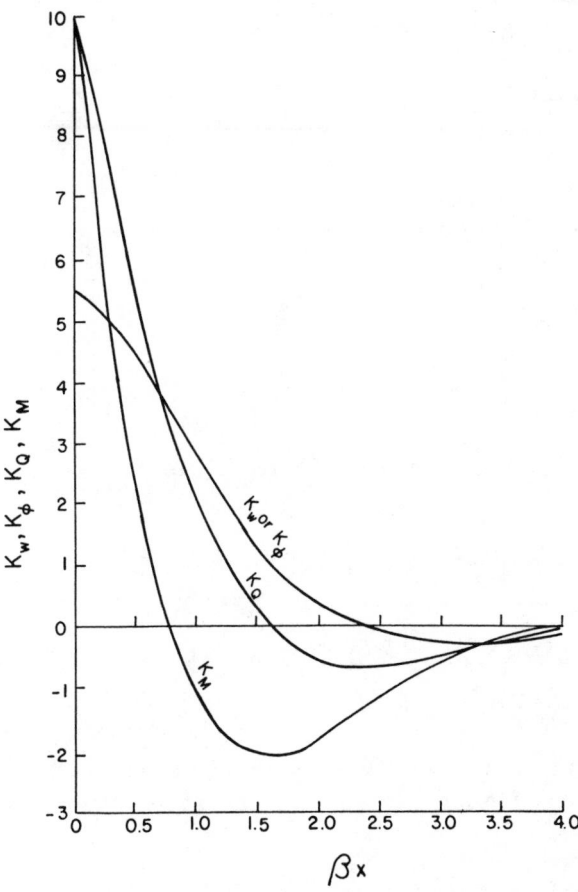

Case 14 Long circular pipe loaded with internal pressure

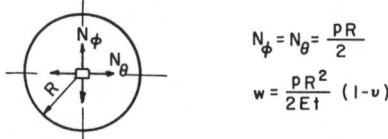

$$\sigma_t = \frac{N_\phi}{t} = p\,\frac{R}{t}$$

$$\sigma_x = \frac{N_x}{t} = \frac{pR}{2t}$$

$$w = \frac{pR^2}{Et}\left(1 - \frac{v}{2}\right)$$

Case 15 Spherical shell loaded with internal pressure

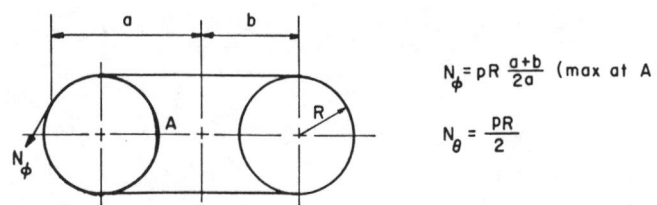

$$N_\phi = N_\theta = \frac{pR}{2}$$

$$w = \frac{pR^2}{2Et}(1 - v)$$

Case 16 Torus loaded with internal pressure

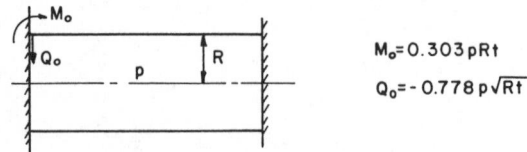

$$N_\phi = pR\,\frac{a+b}{2a} \quad (\text{max at A})$$

$$N_\theta = \frac{pR}{2}$$

Case 17 Cylindrical shell with sufficient length and loaded with internal pressure (edges fixed)

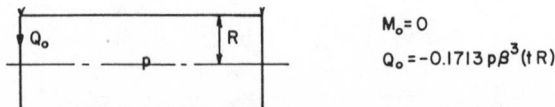

$$M_o = 0.303\,pRt$$

$$Q_o = -0.778\,p\sqrt{Rt}$$

Case 18 Cylindrical shell with sufficient length and simply supported edges loaded with internal pressure

$$M_o = 0$$

$$Q_o = -0.1713\,p\beta^3(tR)^2$$

Case 19   Long pipe reinforced with equidistant rings
loaded with internal pressure

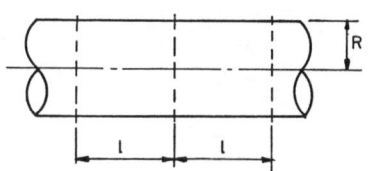

Reactive force (P) and Moment (M) per
unit length of the ring

$$P = \frac{P\sqrt{Rt}}{K_p}$$

$$M = \frac{pRt}{10} K_M$$

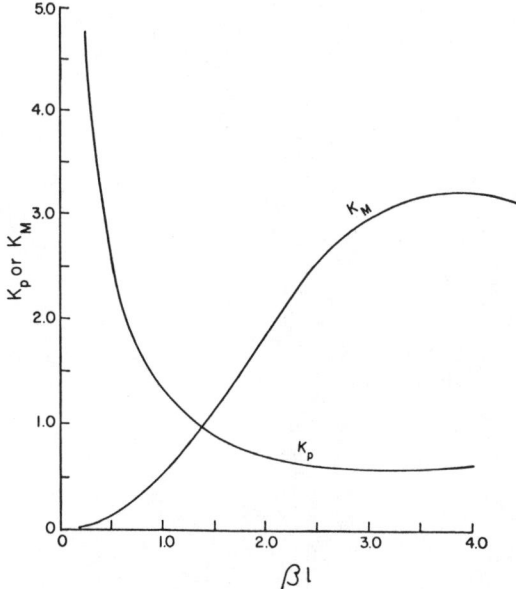

$\beta\ell$	K_p	K_M
0.2	6.425	0.021
0.4	3.213	0.082
0.6	2.143	0.182
0.8	1.610	0.322
1.0	1.292	0.505
1.2	1.083	0.717
1.4	0.938	0.959
1.6	0.832	1.235
1.8	0.753	1.528
2.0	0.698	1.816
2.5	0.615	2.487
3.0	0.585	2.957
3.5	0.588	3.177
4.0	0.606	3.202
4.5	0.630	3.147
5.0	0.631	3.117

Case 20   Pressure vessel with hemispherical ends

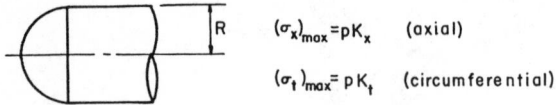

$(\sigma_x)_{max} = pK_x$ (axial)

$(\sigma_t)_{max} = pK_t$ (circumferential)

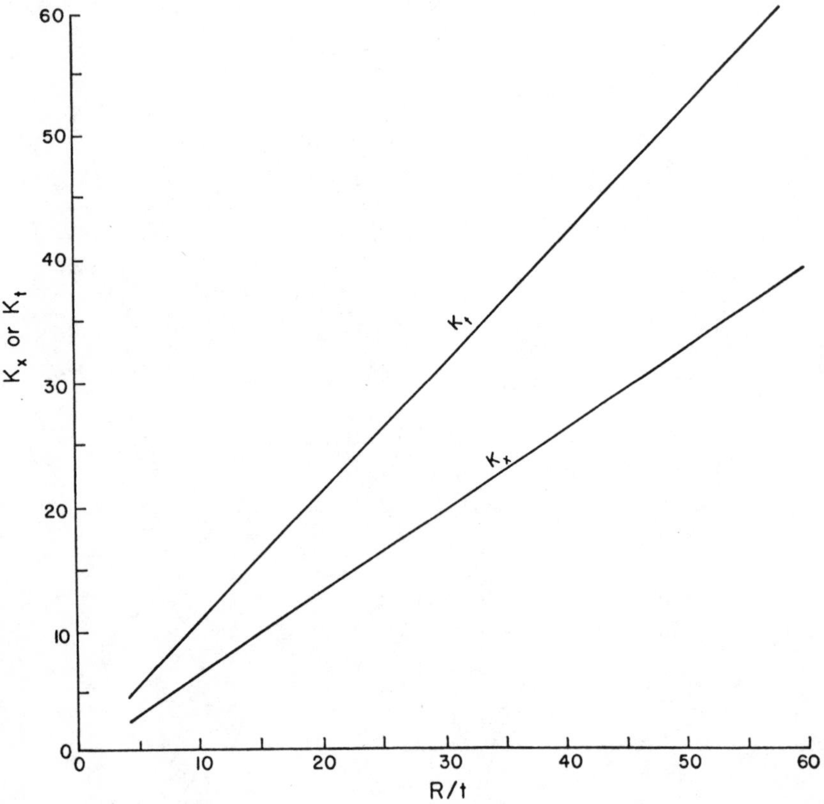

K_x or K_t

R/t

Case 21   Pressure vessel with elliptic ends

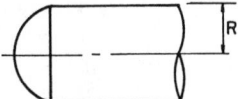

$(\sigma_x)_{max} = p\,K_x$ (axial)

$(\sigma_t)_{max} = p\,K_t$ (circumferential)

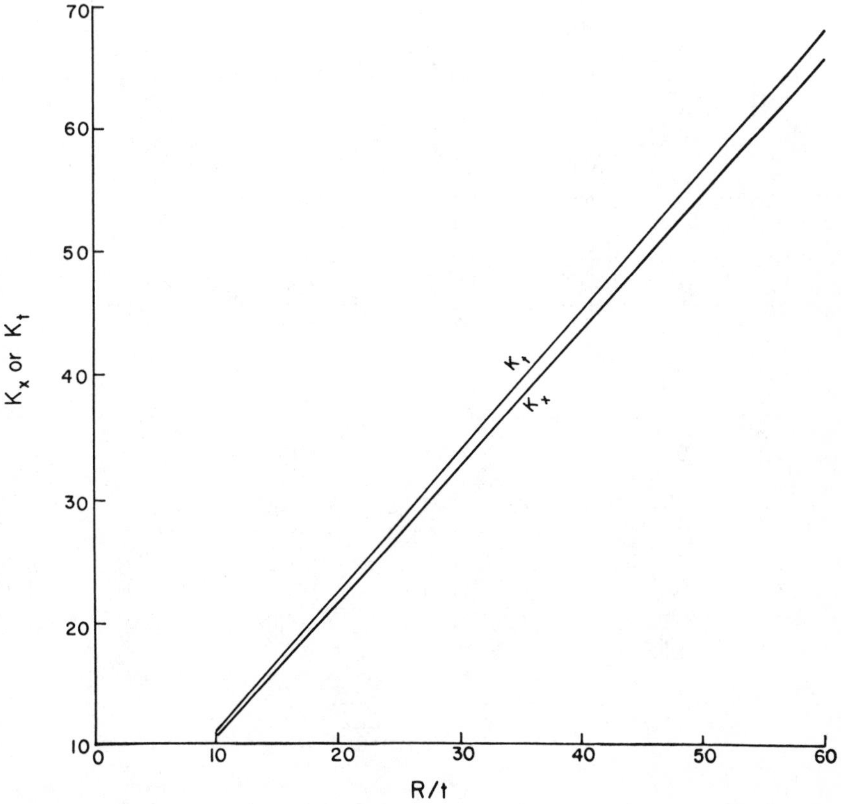

K_x or K_t

R/t

Case 22   Cylindrical tank filled with liquid

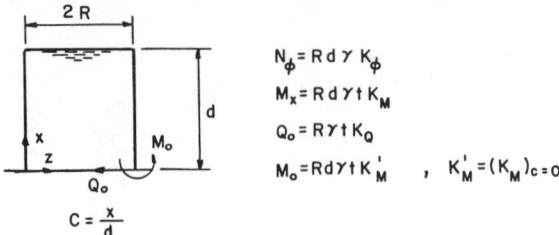

$N_\phi = R\, d\, \gamma\, K_\phi$

$M_x = R\, d\, \gamma\, t\, K_M$

$Q_o = R\, \gamma\, t\, K_Q$

$M_o = R\, d\, \gamma\, t\, K'_M$, $K'_M = (K_M)_{c=0}$

$C = \dfrac{x}{d}$

K_ϕ

C βd	0	0.2	0.4	0.6	0.8	1.0
2	0	0.0521	0.1258	0.1506	0.1050	−0.0052
4	0	0.2452	0.4546	0.4209	0.2425	0.0224
6	0	0.4570	0.6158	0.4346	0.2061	−0.0018
8	0	0.6293	0.6428	0.4065	0.1980	−0.0002
10	0	0.7456	0.6245	0.3982	0.2000	0.0001

K_M

C βd	0	0.2	0.4	0.6	0.8	1.0
2	0.1513	0.0144	−0.0502	−0.0684	−0.0620	−0.0457
4	0.2270	−0.0267	−0.0624	−0.0337	−0.0085	0.0015
6	0.2522	−0.0574	−0.0354	0.0025	0.0027	0.0008
8	0.2648	−0.0626	−0.0101	0.0027	0.0004	−0.0001
10	0.2724	−0.0526	0.0009	0.0009	0.0002	−0.0001
∞	0.3026	−	−	−	−	−

βd	2	4	6	8	10	20
K_Q	0.9078	2.1182	3.3286	4.5390	5.7494	11.801

Case 22 continued

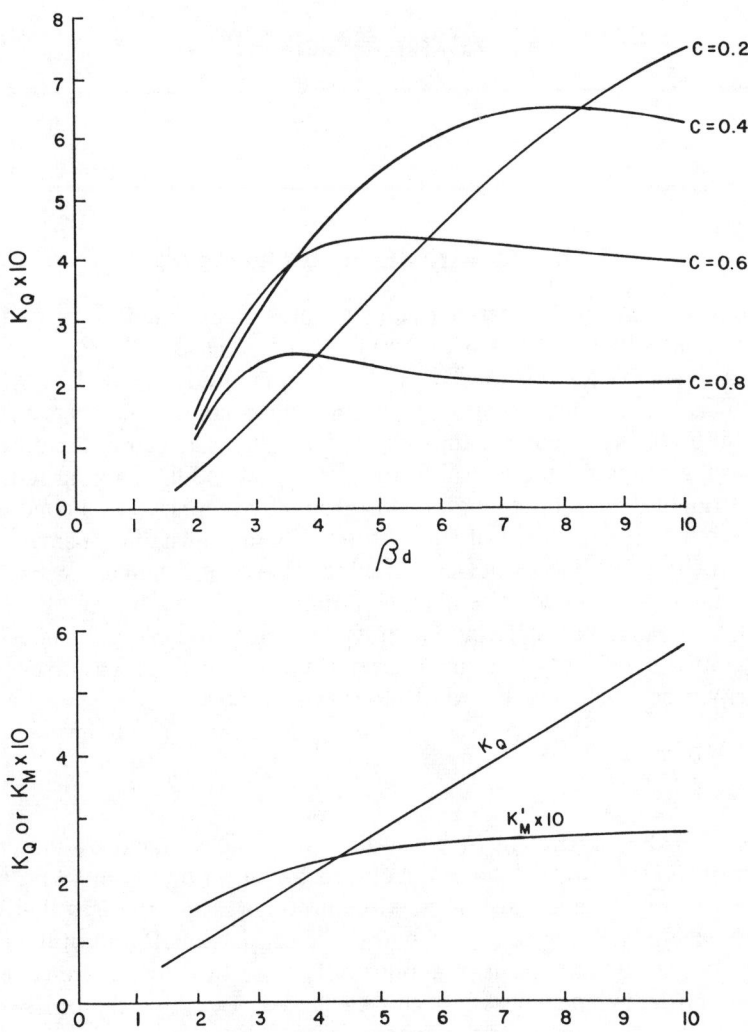

Chapter 11

Elastic Stability

ELASTIC BUCKLING OF BARS [4]

The maximum load a member can sustain is determined under certain circumstances by the stiffness of the member. As discussed in Chapter 5, when a slender column is loaded axially, it remains straight under a small load. This straight form of elastic equilibrium is stable. This means that if a lateral force is applied and the column is slightly deflected, the deflection disappears when the lateral force is removed and the column returns to its straight form. If the axial load is gradually increased to a condition in which the straight form of equilibrium becomes unstable, a small lateral force will produce a deflection which does not disappear when the lateral force is removed. The axial load that is sufficient to keep the column in such a slightly deflected form is defined as the *critical load*. The critical load of an ideal column, known as the *Euler load*, was discussed in Chapter 5. The critical load of a bar is given by

$$P_{cr} = \frac{C\pi^2 EI}{L^2} \tag{11-1}$$

where C is the coefficient of constraint and is determined by the ends conditions. The critical load is developed based on the assumption that the bar is very slender, so the maximum compressive stresses that occurred during buckling remained within the proportional limit of the material. To establish the limit of applicability of critical loads, we have to find the critical slenderness ratio for each case. Let A be the cross-sectional area of the bar; then the critical stress is

$$\sigma_{cr} = \frac{P_{cr}}{A} = \frac{\pi^2 E}{(KL/r)^2} \tag{11-2}$$

where K is the effective length factor. The critical stress depends on the modulus of elasticity E of the material and on the effective slenderness ratio KL/r. Equation 11-2 is valid as long as the stress σ_{cr} remains within the proportional limit. When the proportional limit and the modulus E of a material are known, the limiting value of the slenderness ratio KL/r can be calculated. Equation 11-2 can be represented graphically, as shown in Figure 11-1. The critical stress is plotted as a function of KL/r; the curve approaches the horizontal axis asymptotically, and σ_{cr} approaches zero as KL/r increases. The curve is also asymptotic to the vertical axis but is applicable only in the region where σ_{cr} remains below the proportional limit of the material. Values of C and K for bars with different boundary conditions are presented in Table 11-1.

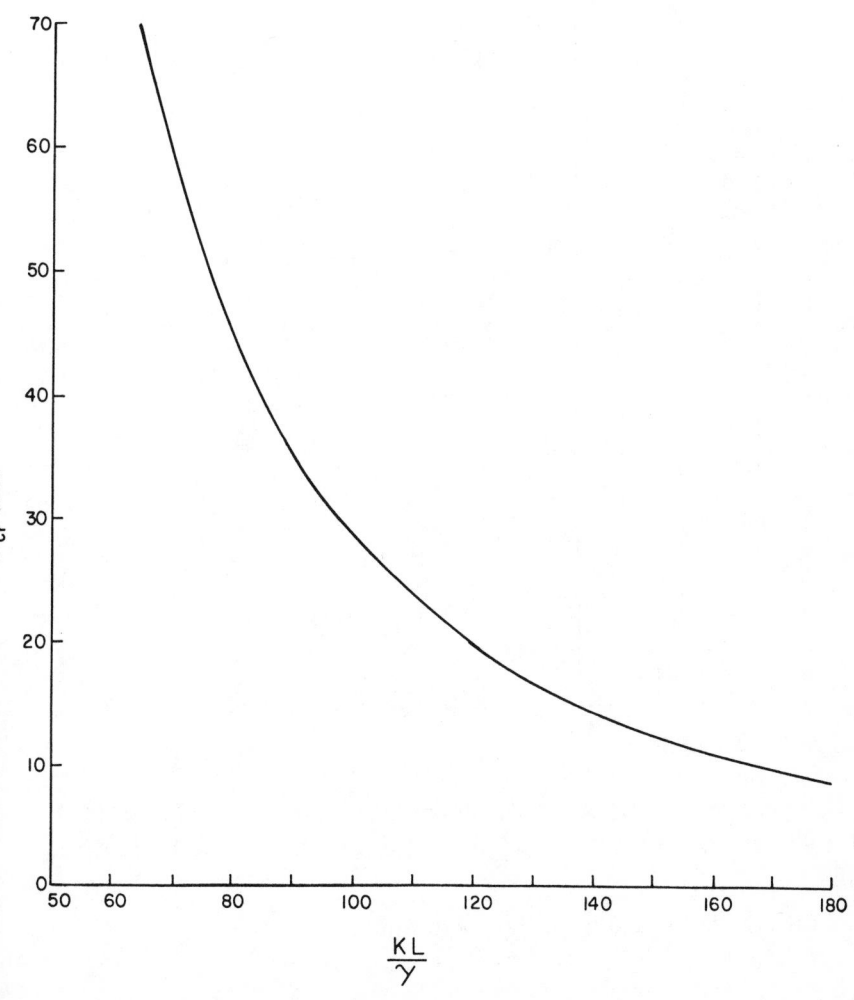

Figure 11-1.   Critical stress of a bar.

Table 11-1
Coefficient of Constraint and Effective Length Factor

	$P = \dfrac{C\pi^2 EI}{L^2}$		$\sigma_{cr} = \dfrac{\pi^2 E}{(KL/r)^2}$
	C	K	Remarks
	0.25	2	one end free one end fixed
	1	1	both ends pinned
	4	0.5	both ends fixed
	2.047	0.699	one end pinned one end fixed
	0.794	1.122	$P_{cr} = p_{cr}L$

Example 11-1 [2,3]

Calculate the coefficient of constraint C and effective length factor K of a column with one end fixed and the other end pinned, as shown in Figure 11-2.

The differential equation for the column is

$$\frac{d^4y}{dx^4} + k^2 \frac{d^2y}{dx^2} = 0 \tag{11-3}$$

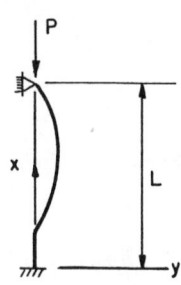

Figure 11-2. Column for Example 11-1.

where $k = \sqrt{P/EI}$

The general solution of Equation 11-3 is

$$y = A \sin kx + B \cos kx + Cx + D \qquad (11\text{-}4)$$

The boundary conditions of this column are

$$y = \frac{dy}{dx} = 0 \quad \text{at } x = 0$$

$$y = \frac{d^2y}{dx^2} = 0 \quad \text{at } x = L$$

Using these boundary conditions with Equation 11-4, we obtain the following equations:

$$B + D = 0$$

$$Ak + C = 0$$

$$CL + D = 0$$

$$A \sin KL + B \cos kL = 0$$

Solving these equations, we obtain

$$\tan kL = kL$$

$$kL = 4.493$$

$$k^2 = P/EI = 20.187/L^2$$

$$P_{cr} = \frac{20.187EI}{L^2} = \frac{2.045\pi^2EI}{L^2}$$

$$C = 2.045$$

$$K = 1/\sqrt{C} = 0.6992$$

This case is solved in detail in Chapter 2 of *Theory of Elastic Stability* [1].

ELASTIC BUCKLING OF FRAMES

Critical loads for rigid frames are discussed in Chapter 2 of *Theory of Elastic Stability* [1]. As shown in Figure 11-3, a frame will move laterally when loaded with P. The differential equation of the deflection curve of the column AB is

$$EI_c \frac{d^2y}{dx^2} = -Py \tag{11-5}$$

The solution of Equation 11-5 is

$$y = A \sin kx + B \cos kx \tag{11-6}$$

$$y = 0 \quad \text{at } x = 0, B = 0$$

$$y = \delta \quad \text{at } x = L, A = \delta/\sin kL$$

Equation 11-6 becomes

$$y = \frac{\delta}{\sin kL} \sin kx \tag{11-7}$$

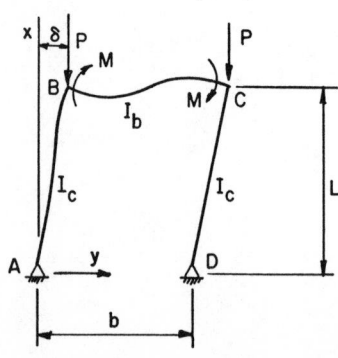

Figure 11-3. Rigid frame loaded with pinned supports.

The rotation at BC bar is

$$\theta_{BC} = \frac{Mb}{6\,EI_b} = \frac{Pb\delta}{6\,EI_b} \tag{11-8}$$

$$\theta_{BA} = \frac{dy}{dx} \text{ at } x = L$$

From Equation 11-7 we obtain

$$\left(\frac{dy}{dx}\right)_{x=L} = \frac{\delta k}{\sin kL} \cos kL \tag{11-9}$$

From Equations 11-8 and 11-9 we obtain

$$\frac{Pb}{6EI_b} = \frac{k \cos kL}{\sin kL} \tag{11-10}$$

Equation 11-10 can be represented in the following form:

$$kL \tan kL = \frac{6I_bL}{I_cb} \tag{11-11}$$

The critical value of P can be found for any ratio of $I_bL/(I_cb)$.

$$P_{cr} = \frac{C\pi^2 EI_c}{L^2}$$

C values for various ratios $I_bL/(I_cb)$ are presented in Figure 11-4. For example, if $I_bL/(I_cb) = 1$, then

$$kL \tan kL = 6$$

$$kL = 1.35$$

$$P_{cr} = \frac{0.184\,\pi^2 EI_c}{L^2}$$

The frame shown in Figure 11-5 is pinned at A; the coefficient of end restraint at A is zero, and the critical load P_{cr} is a function of the ratio $I_cb/(I_bL)$. For the case of $I_cb/(I_bL) = 1$, we find that

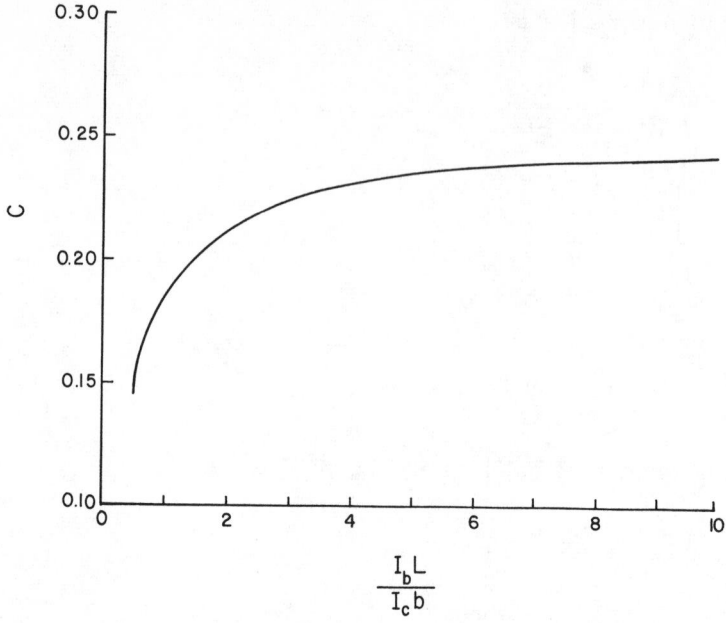

Figure 11-4. Coefficient of constraint for the rigid frame in Figure 11-3.

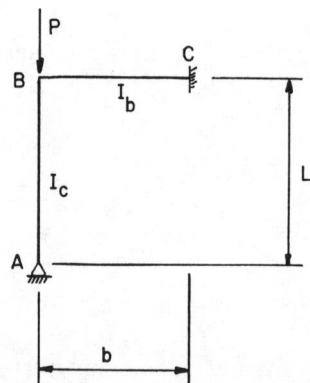

Figure 11-5. Rigid frame pinned at A and fixed at C.

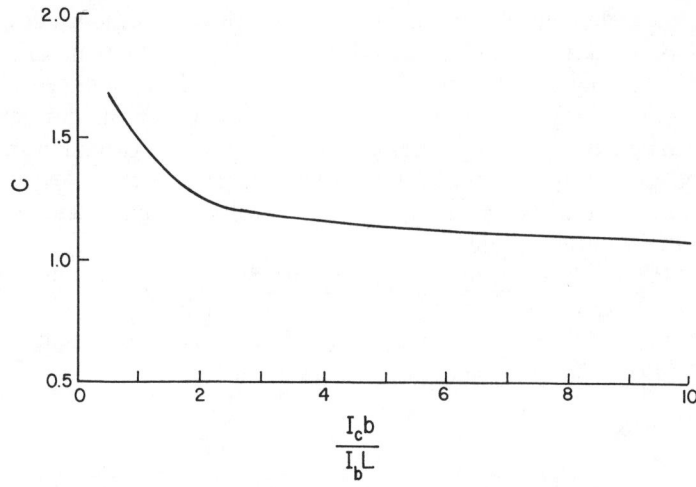

Figure 11-6.   Coefficient of constraint for rigid frame in Figure 11-5.

$$kL = 3.83$$

$$P_{cr} = \frac{C\pi^2 EI_c}{L^2} = \frac{1.486\pi^2 EI_c}{L^2}$$

In this case $C = 1.486$; values of C for various ratios $I_c b/(I_b L)$ are presented in Figure 11-6.

ELASTIC BUCKLING OF THIN PLATES [1,5]

The energy method can be used to study the buckling of thin plates. When a plate is stressed by force acting in the middle plane, it undergoes small lateral bending. If ΔU is the strain energy of the bending and ΔW is the work done by the forces acting in the middle plane of the plate, the critical value of forces can be expressed by the equation $\Delta W = \Delta U$. If $\Delta W < \Delta U$, the flat form of equilibrium of the plate is stable. If $\Delta W > \Delta U$, the plate is unstable and buckling occurs. The principal use of the energy method in problems of buckling is the approximation of critical loads for cases in which the exact solution of deflection curve is either unknown or too complicated. It begins by assuming a reasonable deflection curve or shape. The energy method gives the exact value of the critical load when the true shape of the deflection is known. The energy method always gives values for the critical load larger than the true value unless the assumed deflection shape is correct. This is because the true deflection shape is the only one that represents a deflection configuration for which each element of the plate is in equilibrium. An incorrect configuration of buckling requires additional constraints to maintain the shape and hence increase the rigidity of the plate, and the obtained critical load becomes larger than the true value. In assuming the deflection shape, it can be expressed in several parameters, and the deflection shape can then be altered by changing the parameter values. The most accurate critical load P_{cr} is the one corresponding to values of the parameters such that P_{cr} is the minimum value.

Figure 11-7 shows a simply supported rectangular plate uniformly compressed along the sides $x = 0$ and $x = a$. Let N_x be the force per unit length of the edge. The critical value of N_x, solved in Chapter 9 of *Theory of Elastic Stability* [1], can be expressed by

$$(N_x)_{cr} = \frac{\pi^2 D}{a^2} k' \tag{11-12}$$

where k' is a factor depending on the a/b ratio and a parameter m, which gives the number of half-waves into which the plate buckles. The k' factor is given as

$$k' = \left(m + \frac{a^2}{mb^2}\right)^2$$

If $m = 1$, Equation 11-12 can be written as

$$(N_x)_{cr} = \frac{\pi^2 D}{b^2} \left(\frac{a}{b} + \frac{b}{a}\right)^2 = \frac{\pi^2 D}{b^2} k \tag{11-13}$$

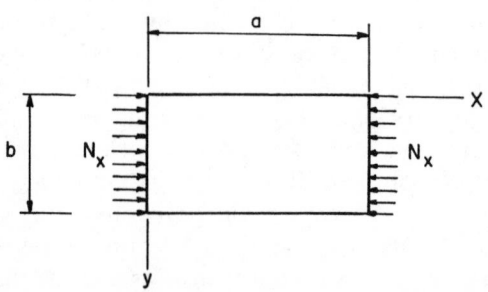

Figure 11-7. Uniformly compressed rectangular plate.

where k is a numerical factor depending on the ratio a/b. It can be seen that k acquires its minimum value when $a = b$; this means that for a plate with a given width, the critical load is the smallest when the plate is square. In this case $k = 4$ or $(N_x)_{cr} = 4\pi^2 D/b^2$. This is analogous to Euler's formula for the buckling of a column. Since $(N_x)_{cr}$ is the force per unit width and D is the flexural rigidity of the plate, we can conclude that because of the continuity of the plate, each longitudinal strip can carry a load four times larger than Euler's load for the same strip when it is isolated.

The number of half-waves into which the plate buckles is a function of a/b. The relation between m and a/b is expressed by

$$a/b = \sqrt{m(m + 1)} \qquad (11\text{-}14)$$

when m = 1, a/b = 1.414
     m = 2, a/b = 2.449
     m = 3, a/b = 3.464

At ratio $a/b = 1.414$, the plate buckles from one to two half-waves. The number of half-waves increases with the ratio a/b. When m is large, the plate is very long ($a >> b$) and will buckle in half-waves with length approximately equal to the width of the plate. When $a/b > 1.414$, the number of half-waves can be obtained from Equation 11-14, and the critical load can be calculated from Equation 11-13 by using a/m instead of a. The critical stress is obtained simply by dividing the critical load by plate thickness t.

$$\sigma_{cr} = \frac{\pi^2 D}{b^2 t} k = \frac{\pi^2 E k}{12(1 - \nu^2)} \left(\frac{t}{b}\right)^2$$

The critical loads or stresses obtained by theoretical formulas are usually greater than the actually developed values. The discrepancy is generally greater for pure compression. The theoretical values should therefore be regarded as an upper limit.

The critical load or stress of a plate depends on the shapes and boundary conditions as well as on loading combinations of the plate. Solutions by reduction method are presented by Shuleshko in *Solution of Buckling Problems by Reduction Method* [6]. He provides stability coefficients for orthotropic rectangular plates with several combinations of boundary conditions. Some basic cases of buckling of thin plates are provided in this section. More detailed solutions can be found in Chapter 9 of *Theory of Elastic Stability* [1]. Note that the critical factor K is dimensionless and Poisson's ratio ($\nu = 0.3$) has been absorbed in the numerical factor.

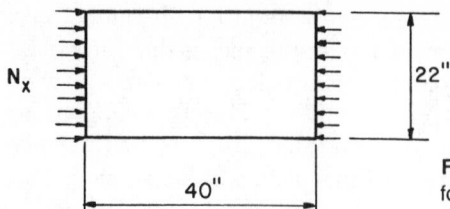

Figure 11-8. Uniformly compressed plate for Example 11-2.

Example 11-2

Calculate the critical stress of the simply supported rectangular steel plate shown in Figure 11-8, given a plate thickness of 0.25 in.

a/b = 40/22 = 1.818

The ratio a/b is greater than 1.414 but smaller than 2.449; the plate buckles in two half-waves.

$$a' = a/m = 40/2 = 20$$

$$k = \left(\frac{a'}{b} + \frac{b}{a'}\right)^2 = \left(\frac{20}{22} + \frac{22}{20}\right)^2 = 4.036$$

$$\sigma_{cr} = \frac{\pi^2 E k}{12(1 - \nu^2)} \left(\frac{t}{b}\right)^2 = \frac{\pi^2 \times 29 \times 10^3 \times 4.036}{12 \times 0.91} \left(\frac{0.25}{22}\right)^2$$

$$\sigma_{cr} = 13.66 \text{ kips/in.}^2$$

$$(N_x)_{cr} = \sigma_{cr} t = 13.66 \times 0.25 = 3.415 \text{ kips/in.}$$

REFERENCES

1. Timoshenko, S. and Gere, J., *Theory of Elastic Stability*, McGraw-Hill Book Company, 1964.
2. Roark, R. J., *Formulas for Stress and Strain*, Fourth Edition, McGraw-Hill Book Company, 1965.
3. Roark, R. and Young, W., *Formulas for Stress and Strain*, Fifth Edition, McGraw-Hill Book Company, 1975.
4. Brockenbrough, R. and Johnston, B., *Steel Design Manual*, U.S. Steel Corporation, 1968.
5. Johnston, B., *Guide to Stability Design Criteria for Metal Structures*, John Wiley & Sons, 1976.
6. Shuleshko, P., "Solution of Buckling Problems by Reduction Method," Proceedings of American Society of Civil Engineers, Vol. 90, June 1964.

Cases for Chapter 11

Case 1 Simply supported rectangular plate

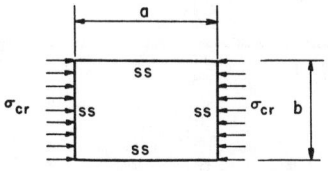

$$\sigma_{cr} = \frac{(N_{x})_{cr}}{t} = KE\left(\frac{t}{b}\right)^2$$

a/b	0.4	0.6	0.8	1.0	1.2	1.4	1.6	1.8	2.0	3.0	∞
K	7.601	4.646	3.796	3.616	3.733	4.040	3.798	3.656	3.615	3.616	3.616

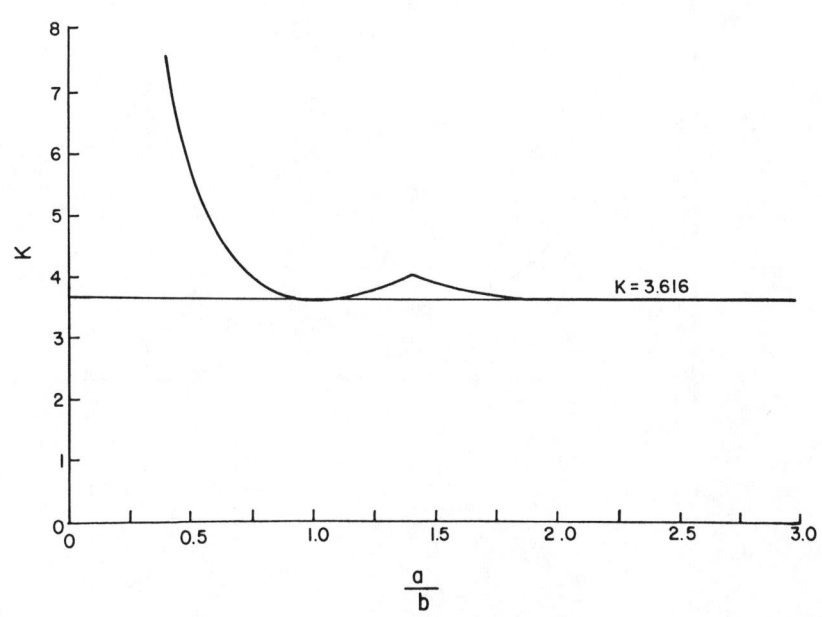

Case 2   Rectangular plate free along one of the unloaded sides and simply supported along the other three sides

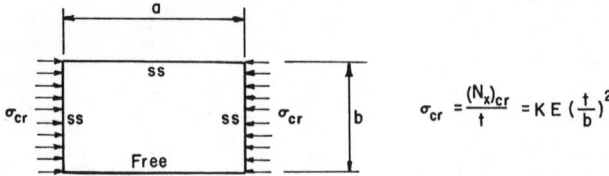

$$\sigma_{cr} = \frac{(N_x)_{cr}}{t} = K E \left(\frac{t}{b}\right)^2$$

a/b	0.5	1.0	1.2	1.4	1.6	1.8	2.0	3.0	4.0
K	3.977	1.301	1.026	0.860	0.755	0.682	0.631	0.510	0.466

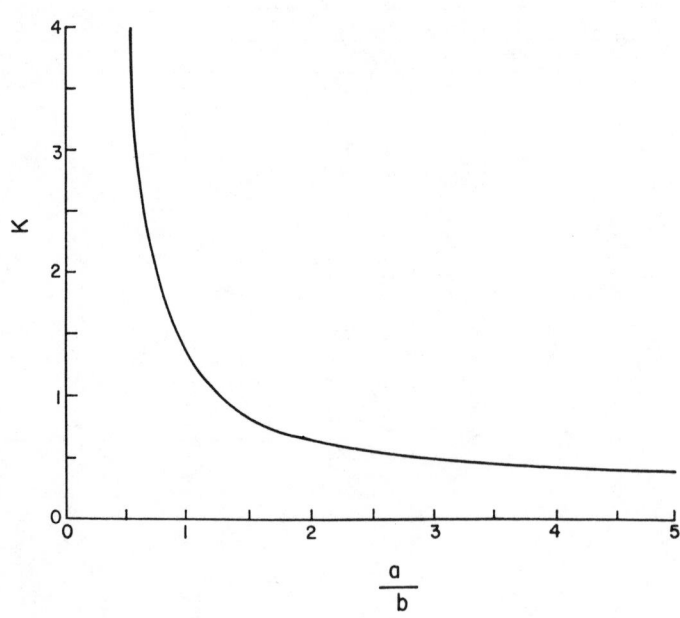

Case 3   Rectangular plate simply supported along
loaded sides fixed and free along the other sides

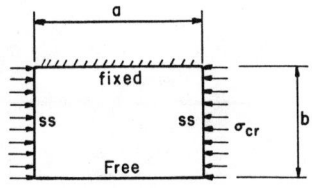

$$\sigma_{cr} = \frac{(N_x)_{cr}}{t} = KE\left(\frac{t}{b}\right)^2$$

a/b	1.0	1.1	1.2	1.3	1.4	1.5	1.6	1.8	2.0	2.2	2.4
K	1.536	1.410	1.329	1.274	1.229	1.211	1.202	1.211	1.247	1.311	1.329

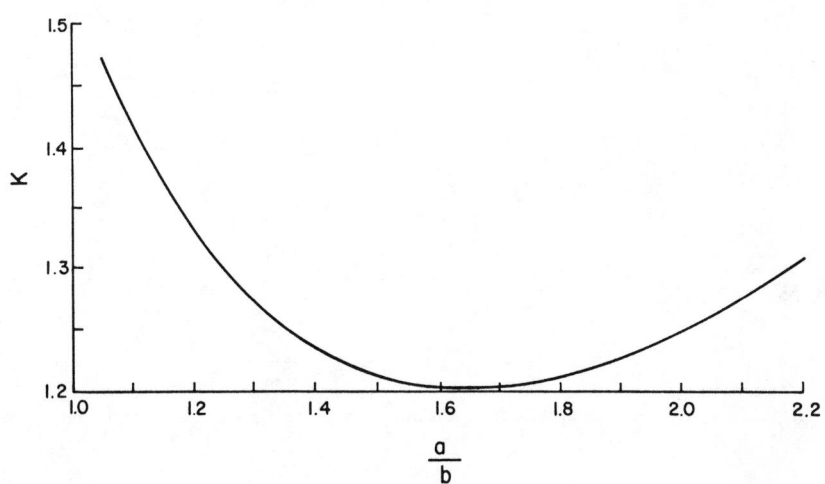

Case 4   Rectangular plate simply supported along the loaded sides and fixed along the other two sides

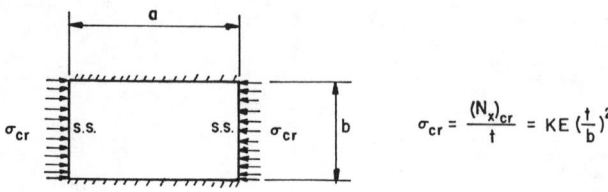

$$\sigma_{cr} = \frac{(N_x)_{cr}}{t} = KE\left(\frac{t}{b}\right)^2$$

a/b	0.4	0.5	0.6	0.7	0.8	0.9	1.0	1.4	1.8	2.1
K	8.532	6.950	6.372	6.327	6.589	7.077	6.950	6.330	6.374	6.330

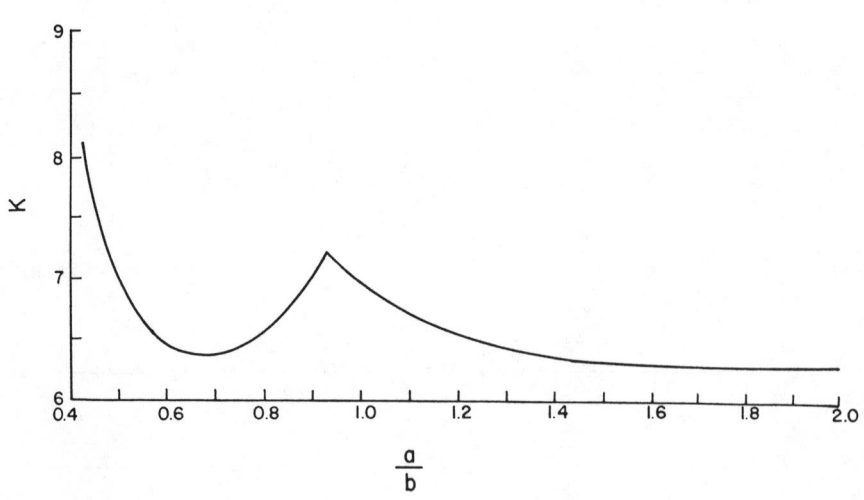

Case 5   Rectangular plate with clamped sides

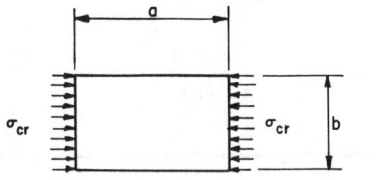

$$\sigma_{cr} = \frac{(N_x)_{cr}}{t} = KE\left(\frac{t}{b}\right)^2$$

a/b	0.75	1.00	1.25	1.5	1.75	2.00	2.25	2.50	3.00	3.50	4.00
K	10.57	9.10	8.36	7.53	7.33	7.12	6.90	6.84	6.66	6.57	6.54

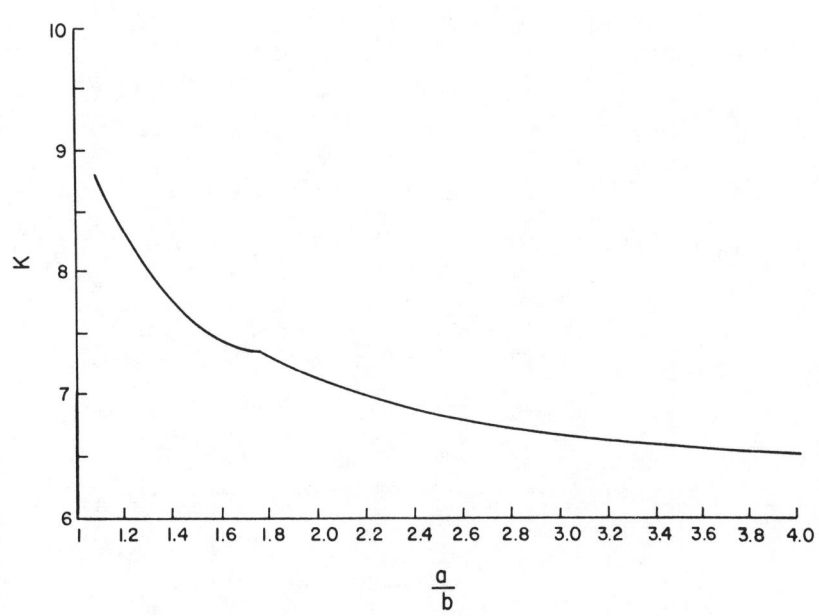

Case 6   Simply supported rectangular plate under pure bending

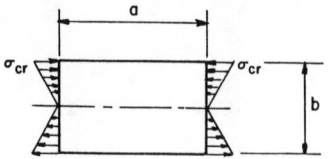

$$\sigma_{cr} = K E \left(\frac{t}{b}\right)^2$$

a/b	0.4	0.5	0.6	0.75	0.8	0.9	1.0	1.2	1.4	1.6
K	26.30	23.14	21.78	21.78	22.05	23.14	23.14	21.69	21.51	22.06

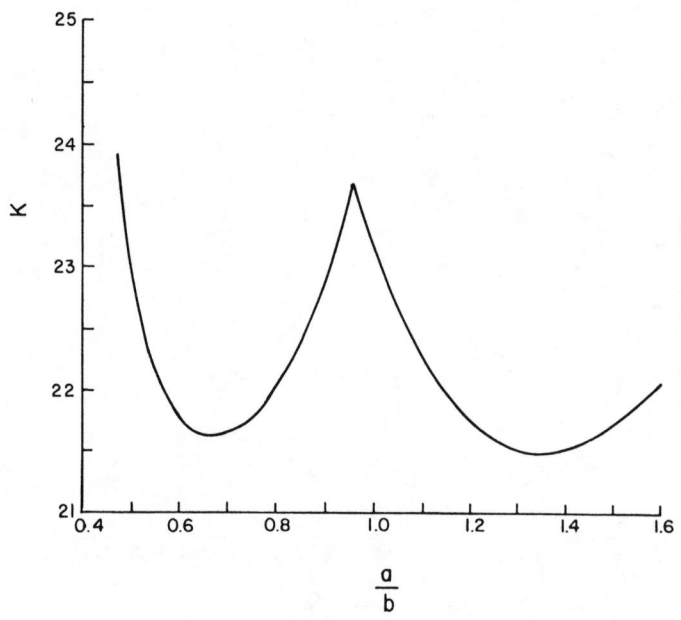

Case 7 Simply supported rectangular plate under
shearing stress

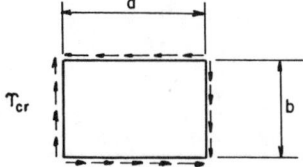

$$T_{cr} = K E \left(\frac{t}{b}\right)^2$$

a/b	1.0	1.2	1.4	1.6	1.8	2.0	2.5	3.0	4.0
K	8.44	7.23	6.60	6.33	6.15	5.97	5.51	5.33	5.15

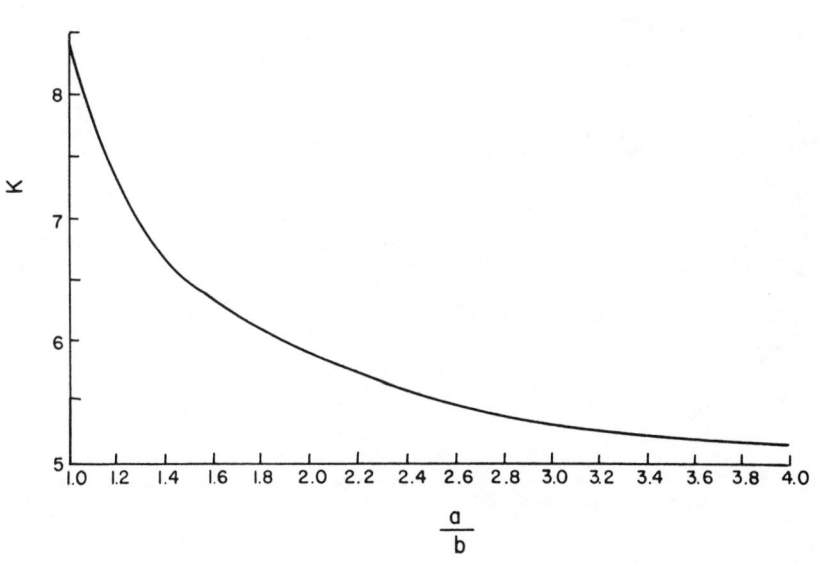

Case 8   Fixed rectangular plate under shearing stress

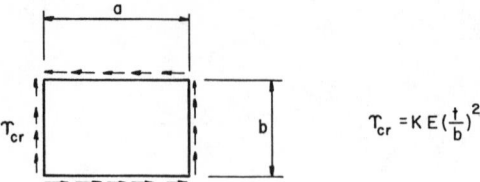

$$\tau_{cr} = K E \left(\frac{t}{b}\right)^2$$

a/b	1.0	1.5	2.0	2.5	∞
K	13.295	10.394	9.345	9.806	8.125

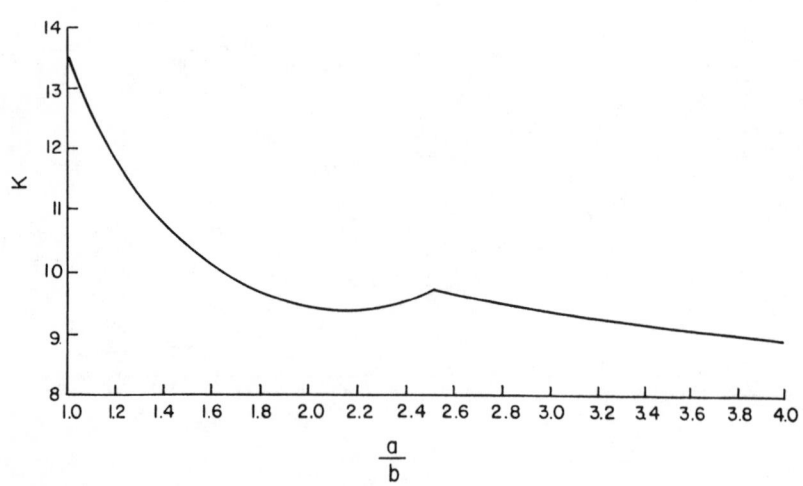

Case 9   Simply supported rectangular plate compressed
by two equal and opposite forces

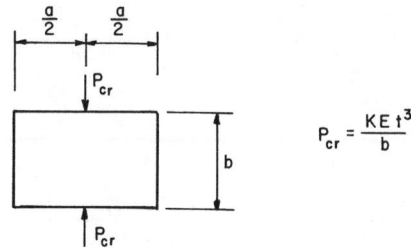

$$P_{cr} = \frac{KEt^3}{b}$$

a/b	0.50	0.75	1.00	1.25	1.50	1.75	2.00	3.00	∞
K	5.542	2.528	1.725	1.416	1.282	1.215	1.183	1.153	1.151

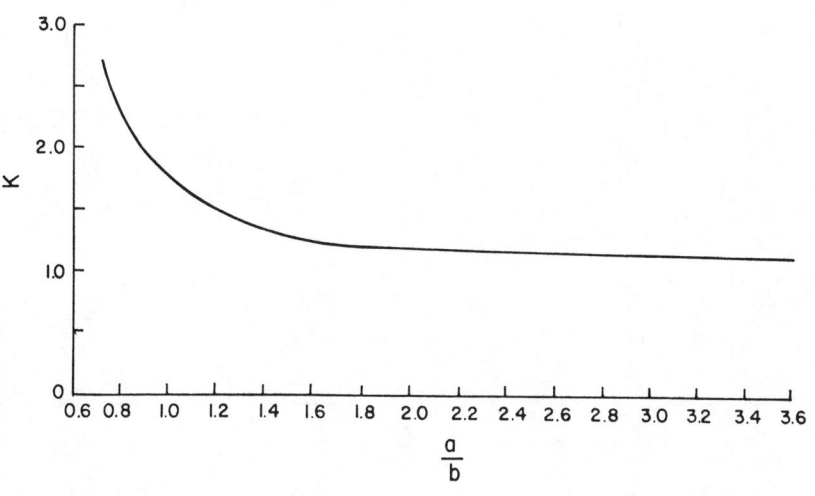

Case 10   Skew plate with clamped edges

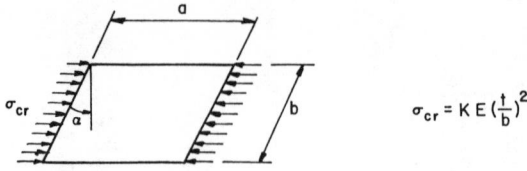

$$\sigma_{cr} = K E \left(\frac{t}{b}\right)^2$$

a/b	0.6	0.8	1.0	1.2	1.4	1.6	1.8	2.0
$\alpha = 0$	13.83	10.39	9.10	8.41	7.94	7.43	7.30	7.12
$\alpha = 30$	11.71	7.97	6.73	6.48	5.86	5.61	5.50	5.36
$\alpha = 45$	9.03	5.73	4.73	4.83	4.36	4.24	4.16	4.06

K

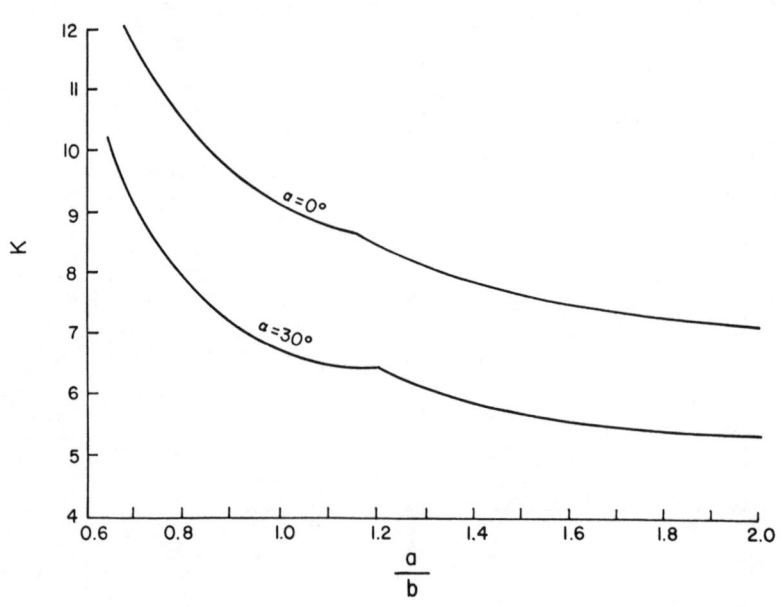

Case 11   Circular plate with clamped edge

$$\sigma_{cr} = \frac{(N_r)_{cr}}{t} = 1.344 \, E \left(\frac{t}{R}\right)^2$$

Case 12   Circular plate with simply supported edge

$$\sigma_{cr} = \frac{(N_r)_{cr}}{t} = 0.385 \, E \left(\frac{t}{R}\right)^2$$

Chapter 12

Thermal Stress

THERMAL STRESS IN BARS AND PLATES [1,6]

When the temperature of a body is uniformly increased, the body expands. Whenever the expansion or contraction resulting from heating or cooling of a body is prevented, stresses develop. When an unstrained body expands, the normal strain is

$$e_x = e_y = e_z = \alpha \, (\Delta T) \tag{12-1}$$

where ΔT is the temperature change in degrees and α is the coefficient of thermal expansion, which is the change in length per unit length for a change of one degree in temperature. The coefficient of surface expansion is approximately two times the linear coefficient, and the coefficient of volume expansion for a solid is approximately three times the linear coefficient. An unstrained bar will increase in length when temperature is increased, and the change in length is

$$\Delta L = \alpha(\Delta T) \, L \tag{12-2}$$

where L is the length of the bar. If the ends of a bar are fixed, a change in temperature ΔT will cause a change in unit stress of

$$\sigma = \alpha(\Delta T) \, E \tag{12-3}$$

In a similar manner, when the temperature of a restrained flat plate is increased, the compressive stress developed is

$$\sigma = \frac{\alpha(\Delta T)E}{1 - \nu} \tag{12-4}$$

If the temperature difference between two faces of a restrained flat plate is ΔT, the plate will be held flat by uniform edge moment; the bending moment is solved in Chapter 2 of *Theory of Plates and Shells* [3]. If the edges of the plate are entirely free, the plate will bend to a spherical sur-

364

face. The difference between the maximum thermal expansion and the middle surface expansion is $\alpha(\Delta T)/2$, and the curvature resulting from the heat can be expressed as

$$\rho = \frac{\alpha(\Delta T)}{t} \qquad (12\text{-}5)$$

In Chapter 7 it was mentioned that when a plate is under pure bending along the edges, the plate is bent to a spherical surface with a curvature of

$$\rho = \frac{M}{D(1 + \nu)} \qquad (12\text{-}6)$$

From Equations 12-5 and 12-6, we obtain the bending moment

$$M = \frac{\alpha(\Delta T)(1 + \nu)D}{t} = \frac{\alpha(\Delta T)Et^2}{12(1 - \nu)} \qquad (12\text{-}7)$$

The corresponding bending stress is

$$\sigma = \frac{6M}{t^2} = \frac{\alpha(\Delta T)E}{2(1 - \nu)} \qquad (12\text{-}8)$$

It should be noted that the bending stress is proportional to α, ΔT, and E. The thickness t is not covered, but the temperature difference ΔT is proportional to the plate thickness. Thicker plates have greater ΔT and thermal stress.

If the plate is rectangular and simply supported, the thermal stress at the boundary is

$$\sigma = \frac{(\Delta T)\alpha E}{2} \qquad (12\text{-}9)$$

If the plate is an equilateral triangle of altitude h and is rigidly restrained along the edges, the maximum resulting bending stress at the corner is

$$\sigma = \frac{3(\Delta T)\alpha E}{4} \qquad (12\text{-}10)$$

and it is a compression on the hot face, tension on the cold face. There are also high shear stresses near the corners.

THERMAL STRESS IN SHELLS [1,2]

Assume that T_1 and T_2 are the uniform temperatures of the inner and outer surfaces of a long cylindrical shell, respectively; and that the temperature gradient is in the radial direction, and the variation of the temperature through the thickness is linear. At points remote from the ends, the maximum circumferential stress is

$$\sigma_\phi = \frac{E\alpha(\Delta T)}{2(1 - \nu)} \tag{12-11}$$

where $\Delta T = T_1 - T_2 \ (T_1 > T_2)$. The stress σ_ϕ is compression at the inner surface and tension at the outer surface. The longitudinal stress $\sigma_x = \sigma_\phi$ and is compression inside, tension outside.

At the free ends, the maximum thermal stress acts in the circumferential direction and is given by

$$\sigma_\phi = \frac{E\alpha(\Delta T)}{2(1 - \nu)} (1 - \nu + 0.58 \sqrt{1 - \nu^2}) \tag{12-12}$$

If the wall is thick, the temperature gradient is not linear. Let r and R be the inner and outer radii of the hollow cylinder. The maximum circumferential stress at the inner surface is

$$\sigma_\phi = \frac{(\Delta T)\alpha E}{2(1 - \nu)} K_i \quad \text{(compression)} \tag{12-13}$$

The maximum circumferential stress at the outer surface is

$$\sigma_\phi = \frac{(\Delta T)\alpha E}{2(1 - \nu)} K_o \quad \text{(tension)} \tag{12-14}$$

where K_i and K_o are dimensionless parameters, which are functions of R/r. K_i and K_o are provided in References 2 and 7.

$$K_i = \frac{1 - \dfrac{2 R^2}{R^2 - r^2} \ln \dfrac{R}{r}}{\ln(R/r)}$$

$$K_o = \frac{1 - \dfrac{2 r^2}{R^2 - r^2} \ln \dfrac{R}{r}}{\ln(R/r)}$$

COEFFICIENTS OF THERMAL EXPANSION [4,5]

The average coefficient of thermal expansion for structural steel between room temperature and 100°F is 6.5×10^{-6} in./in./°F. For temperatures of 100°F to 1,200°F the coefficient of thermal expansion can be calculated by

$$\alpha = (6.1 + 0.0019 \text{ T}) \times 10^{-6}$$

where T = temperature in degrees Fahrenheit. The following are approximate values of the coefficient for various engineering materials for the temperature range of 32° to 212°F:

Material	Coefficient (in./in./°F)
Aluminum	13×10^{-6}
Brass	10.4×10^{-6}
Carbon steel	6.5×10^{-6}
Cast Iron	5.9×10^{-6}
Nickel steel	7.0×10^{-6}
Stainless steel	9.9×10^{-6}
Wire (iron)	6.9×10^{-6}
Zinc (rolled)	17.3×10^{-6}

The tensile modulus of elasticity of the alloy ferritic steels is 30,000 kips/in.^{2} at room temperature. It decreases linearly to about 25,000 kips/in.^{2} at 900°F, and then begins to drop at an increasing rate at higher temperatures.

REFERENCES

1. Shigley, J., *Mechanical Engineering Design*, Second Edition, McGraw-Hill Book Company, 1972.
2. Roark, R., *Formulas for Stress and Strain*, Fourth Edition, McGraw-Hill Book Company, 1965.
3. Timoshenko, S. and Woinowsky-Krieger, S., *Theory of Plates and Shells*, Second Edition, McGraw-Hill Book Company, 1959.
4. *Steel for Elevated Temperature Service*, U.S. Steel Corporation, 1966.
5. *Manual of Steel Construction*, Eighth Edition, American Institute of Steel Construction, Inc., 1980.
6. Maulbetsch, J., *Thermal Stresses in Plates*, American Society of Mechanical Engineers Journal of Applied Mechanics, Vol. 2, No. 4, December 1935.
7. Goodier, J., "Thermal Stress," American Society of Mechanical Engineers Journal of Applied Mechanics, Vol. 4, No. 1, March 1937.

Index